T0073175

Springer Series in the Data Sciences

Series Editors

David Banks, Duke University, Durham, NC, USA

Jianqing Fan, Department of Financial Engineering, Princeton University, Princeton, NJ, USA

Michael Jordan, University of California, Berkeley, CA, USA

Ravi Kannan, Microsoft Research Labs, Bangalore, India

Yurii Nesterov, CORE, Universite Catholique de Louvain, Louvain-la-Neuve, Belgium

Christopher Ré, Department of Computer Science, Stanford University, Stanford, USA

Ryan J. Tibshirani, Department of Statistics, Carnegie Melon University, Pittsburgh, PA, USA

Larry Wasserman, Department of Statistics, Carnegie Mellon University, Pittsburgh, PA, USA

Springer Series in the Data Sciences focuses primarily on monographs and graduate level textbooks. The target audience includes students and researchers working in and across the fields of mathematics, theoretical computer science, and statistics. Data Analysis and Interpretation is a broad field encompassing some of the fastest-growing subjects in interdisciplinary statistics, mathematics and computer science. It encompasses a process of inspecting, cleaning, transforming, and modeling data with the goal of discovering useful information, suggesting conclusions, and supporting decision making. Data analysis has multiple facets and approaches, including diverse techniques under a variety of names, in different business, science, and social science domains. Springer Series in the Data Sciences addresses the needs of a broad spectrum of scientists and students who are utilizing quantitative methods in their daily research. The series is broad but structured, including topics within all core areas of the data sciences. The breadth of the series reflects the variation of scholarly projects currently underway in the field of machine learning.

More information about this series at http://www.springer.com/series/13852

Jeff M. Phillips

Mathematical Foundations for Data Analysis

 Springer

Jeff M. Phillips
School of Computing
University of Utah
Salt Lake City, UT, USA

ISSN 2365-5674   ISSN 2365-5682   (electronic)
Springer Series in the Data Sciences
ISBN 978-3-030-62340-1   ISBN 978-3-030-62341-8   (eBook)
https://doi.org/10.1007/978-3-030-62341-8

Mathematics Subject Classification: 62-07

This Springer imprint is published by the registered company Springer Nature Switzerland AG
The registered company address is: Gewerbestrasse 11, 6330 Cham, Switzerland

to Bei, Stanley, and Max

Preface

This book is meant for data science students, and in particular preparing students mathematically for advanced concepts in data analysis. It can be used for a self-contained course that introduces many of the basic mathematical principles and techniques needed for modern data analysis, and can go deeper in a variety of topics; the shorthand **math for data** may be appropriate. In particular, it was constructed from material taught mainly in two courses. The first is an early undergraduate course which is designed to prepare students to succeed in rigorous Machine Learning and Data Mining courses. The second course is the advanced Data Mining course. It should be useful for any combination of such courses. The book introduces key conceptual tools which are often absent or brief in the undergraduate curriculum, and for most students, helpful to see multiple times. On top of these, it introduces the generic versions of the most basic techniques that comprise the backbone of modern data analysis. And then it delves deeper into a few more advanced topics and techniques—still focusing on clear, intuitive, and lasting ideas, instead of specific details in the ever-evolving state of the art.

Notation

Consistent, clear, and crisp mathematical notation is essential for intuitive learning. The domains which comprise modern data analysis (e.g., statistics, machine learning, and algorithms) until recently had matured somewhat separately with their own conventions for ways to write the same or similar concepts. Moreover, it is commonplace for researchers to adapt notation to best highlight ideas within specific papers. As such, much of the existing literature on the topics covered in this book has varied, sometimes inconsistent notation, and as a whole can be confusing. This text attempts to establish a common, simple, and consistent notation for these ideas, yet not veer too far from how concepts are consistently represented in the research literature, and as they will be in more advanced courses.

Indeed, the most prevalent sources of confusion in earlier uses of this text in class have arisen around overloaded notation.

Interaction with Other Courses

This book is written for students who have already taken *calculus*, for several topics it relies on integration (continuous probability) and differentiation (gradient descent). However, this book does not lean heavily on calculus, and as data science introductory courses are being developed which are not calculus-forward, students following this curriculum may still find many parts of this book useful.

For some advanced material in Sampling, Nearest Neighbors, Regression, Clustering, Classification, and especially Big Data, some basic familiarity with *programming and algorithms* will be useful to fully understand these concepts. These topics are deeply integrated with computational techniques beyond numerical linear algebra. When the implementation is short and sweet, several implementations are provided in Python. This is not meant to provide a full introduction to programming, but rather to help break down any barriers among students worrying that the programming aspects need to be difficult—many times they do not!

Probability and *Linear Algebra* are essential foundations for much of data analysis, and hence also in this book. This text includes reviews of these topics. This is partially to keep the book more self-contained, and partially to ensure that there is a consistent notation for all critical concepts. This material should be suitable for a review on these topics but is not meant to replace full courses. It is recommended that students take courses on these topics before, or potentially concurrently with, a more introductory course from this book.

If appropriately planned for, it is the hope that a first course taught from this book could be taken as early as the undergraduate sophomore level, so that more rigorous and advanced data analysis classes can be taken during the junior year.

Although we touch on Bayesian inference, we do not cover most of *classical statistics*; neither frequentist hypothesis testing nor similar Bayesian perspectives. Most universities have well-developed courses on these topics that are also very useful, and provide a complementary view of data analysis.

Scope and Topics

Vital concepts introduced include the concentration of measure and PAC bounds, cross-validation, gradient descent, a variety of distances, principal component analysis, and graph-structured data. These ideas are essential for modern data analysis, but are not often taught in other introductory mathematics classes in a computer science or math department, or if these concepts are taught, they are presented in a very different context.

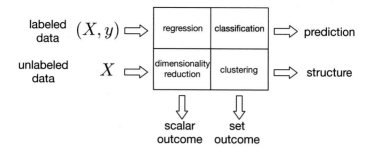

We also survey basic techniques in supervised (regression and classification) and unsupervised (dimensionality reduction and clustering) learning. We make an effort to keep simple the presentation and concepts on these topics. The book initially describes those methods which attempt to minimize the sum of squared errors. It leads with classic but magical algorithms like Lloyd's algorithm for k-means, the power method for eigenvectors, and perceptron for linear classification. For many students (even those in a computer science program), these are the first iterative, non-discrete algorithms they will have encountered. After the basic concepts, the book ventures into more advanced concepts like regularization and lasso, locality-sensitive hashing, multidimensional scaling, spectral clustering, kernel methods, and data sketching. These surveys of advanced techniques are chosen mostly among the ones which have solid mathematical footing, are built on initial concepts in this book, and are actually used. These can be sprinkled in, to allow courses to go deeper and more advanced as is suitable for the level of students.

On Data

While this text is mainly focused on mathematical preparation, what would data analysis be without data? As such we provide a discussion on how to use these tools and techniques on actual data, with simple examples given in Python. We choose Python since it has increasingly many powerful libraries often with efficient backends in low-level languages like C or Fortran. So for most data sets, this provides the proper interface for working with these tools. Data sets can be found here: https://mathfordata.github.io/data/.

But arguably, more important than writing the code itself is a discussion on when and *when not* to use techniques from the immense toolbox available. This is one of the main ongoing questions a data scientist must ask. And so, the text attempts to introduce the readers to this ongoing discussion—but resists diving into an explicit discussion of the actual tools available.

Examples, Geometry, and Ethics

Three themes that this text highlights to aid in a broader understanding of these fundamentals are examples, geometry, and ethical connotations. These are each offset in colored boxes.

Example: with focus on Simplicity

This book provides numerous simple and sometimes fun examples to demonstrate key concepts. It aims to be as simple as possible (but not simpler), and make data examples small, so they can be fully digested. These are illustrated with figures and plots, and often the supporting Python code is integrated when it is illustrative. For brevity, the standard import commands from Python are only written once per chapter, and state is assumed carried forward throughout the examples within a chapter. Although most such Python parts are fully self-contained.

Geometry of Data and Proofs

Many of the ideas in this text are inherently geometric, and hence we attempt to provide many geometric illustrations and descriptions to use visual intuition to shed light on otherwise abstract concepts. These boxes often go more in depth into what is going on, and include the most technical proofs. Occasionally the proofs are not really geometric, yet for consistency the book retains this format in those cases.

Ethical Questions with Data Analysis

As data analysis glides into an abstract, nebulous, and ubiquitous place, with a role in automatic decision making, the surrounding ethical questions are becoming more important. As such, we highlight various ethical questions which may arise in the course of using the analysis described in this text. We intentionally do not offer solutions, since there may be no single good answer to some of the dilemmas presented. Moreover, we believe the most important part of instilling positive ethics is to make sure analysts at least *think* about the consequences, which we hope is partially achieved via these highlighting boxes and ensuing discussions.

Salt Lake City, USA   Jeff M. Phillips
August 2020

Acknowledgements

I would like to thank the gracious support from NSF in the form of grants CCF-1350888, IIS-1251019, ACI-1443046, CNS-1514520, CNS-1564287, and IIS-1816149, which have funded my cumulative research efforts during the writing of this text. I would also like to thank the University of Utah, as well as the Simons Institute for Theory of Computing, for providing excellent work environments while this text was written. And thanks to Natalie Cottrill, Yang Gao, Koriann South, and many other students for a careful reading and feedback.

Contents

Chapter 1
Probability Review

Abstract Probability is a critical tool for modern data analysis. It arises in dealing with uncertainty, in prediction, in randomized algorithms, and Bayesian analysis. To understand any of these concepts correctly, it is paramount to have a solid and rigorous statistical foundation. This chapter reviews the basic definitions necessary for data analysis foundations.

1.1 Sample Spaces

We define probability through set theory, starting with a *sample space* Ω. This represents the set of all things that might happen in the setting we consider. One such potential outcome $\omega \in \Omega$ is a *sample outcome*; it is an element of the set Ω. We are usually interested in an *event* that is a subset $A \subseteq \Omega$ of the sample space.

Example: Discrete Sample Space for a 6-sided Die

Consider rolling a single fair, 6-sided die. Then $\Omega = \{1, 2, 3, 4, 5, 6\}$. One roll may produce an outcome $\omega = 3$, rolling a 3. An event might be $A = \{1, 3, 5\}$, any odd number.

The probability of rolling an odd number is then the ratio of the size of the event to the size of the sample space

$$\mathbf{Pr}(A) = |\{1, 3, 5\}|/|\{1, 2, 3, 4, 5, 6\}| = 1/2.$$

The *probability* of an event $\mathbf{Pr}(A)$ satisfies the following properties:

- $0 \leq \mathbf{Pr}(A) \leq 1$ for any A,
- $\mathbf{Pr}(\Omega) = 1$, and

© Springer Nature Switzerland AG 2021
J. M. Phillips, *Mathematical Foundations for Data Analysis*,
Springer Series in the Data Sciences,
https://doi.org/10.1007/978-3-030-62341-8_1

- The probability of the union of disjoint events is equivalent to the sum of their individual probabilities. Formally, for any sequence $A_1, A_2, \ldots$ where for all $i \neq j$ that $A_i \cap A_j = \emptyset$, then

$$\mathbf{Pr}\left(\bigcup_{i=1} A_i\right) = \sum_{i=1} \mathbf{Pr}(A_i).$$

Example: Probability for a Biased Coin

Now consider flipping a biased coin with two possible events $\Omega = \{H, T\}$ (i.e., heads $H = A_1$ and tails $T = A_2$). The coin is biased so the probabilities $\mathbf{Pr}(H) = 0.6$ and $\mathbf{Pr}(T) = 0.4$ of these events are not equal. However we notice still $0 \leq \mathbf{Pr}(T)$, $\mathbf{Pr}(H) \leq 1$, and that $\mathbf{Pr}(\Omega) = \mathbf{Pr}(H \cup T) = \mathbf{Pr}(H) + \mathbf{Pr}(T) = 0.6 + 0.4 = 1$. That is, the sample space Ω is the union of these two events, which cannot both occur (i.e., $H \cap T = \emptyset$), so they are disjoint. Thus Ω's probability can be written as the sum of the probability of those two events.

Sample spaces Ω can also be continuous, representing some quantity like water, time, and landmass which does not have discrete quantities. All of the above definitions hold for this setting.

Example: Continuous Sample Space

Assume you are riding a Swiss train that is always on time, but its departure is only specified to the minute (specifically, 1:37 pm). The true departure is then in the state space $\Omega = [1\!:\!37\!:\!00, 1\!:\!38\!:\!00]$. A continuous event may be $A = [1\!:\!37\!:\!00 - 1\!:\!37\!:\!40)$, the first 40 seconds of that minute.

Perhaps the train operators are risk averse, so $\mathbf{Pr}(A) = 0.80$. That indicates that 0.8 fraction of trains depart in the first $2/3$ of that minute (less than the 0.666 expected from a uniform distribution).

A *random variable* X is an important concept in probability. It can intuitively be thought of as a variable as in an algebra equation or a computer program, but where its value has not yet been set and will be determined by a random process. The random process determining its value could be a yet-to-be-performed experiment or a planned observation or the outcome of a randomized algorithm.

More formally, a random variable X is a measurable **function** $X : \Lambda \to \Omega$ which maps from one type of sample space Λ to another Ω. It endows the outcomes (of a random process) in Λ with a value in Ω, commonly the mapped-to sample space $\Omega \subseteq \mathbb{R}$, where $\mathbb{R}$ is the space of real numbers. The *measurable* term in "measurable function" is formalized through measure theory; intuitively it means that sets within both spaces Λ and Ω have well-defined notions of size (e.g., area and volume—you can measure them), and the function provides correspondences between these sets.

Example: Random Variables for a Fair Coin

Consider flipping a fair coin with $\Lambda = \{H, T\}$. If it lands as heads H, then I get 1 point, and if it lands as T, then I get 4 points. Thus the mapped-to sample space is $\Omega = \{1, 4\}$. This describes the random variable X, defined as $X(H) = 1$ and $X(T) = 4$.

Geometry of Sample Spaces

It may be useful to generically imagine a sample space Ω as a square. Then, as shown in (a), an event A may be an arbitrary subset of this space.

When the sample space describes the joint sample space over two random variables X and Y, then it may be convenient to parameterize Ω by its x- and y-coordinates, so that the X value is represented along one side of the square, and the Y value along the other, as in (b). Then for an event $A \subset \Omega$ which only pertains to the x value (e.g., $A = \{(x, y) \in \Omega \mid 0.2 \leq x \leq 0.5\}$), event A is represented as a rectangular strip defined by A's intersection with x domain.

If there is also an event B that only pertains to the y-coordinate (e.g., $B = \{(x, y) \in \Omega \mid 0.1 \leq 0.6\}$), and hence only the random variable Y, then this gives rise to another rectangular strip in the other direction, as drawn in (c). When these events are independent, then these strips intersect only in another rectangle $A \cap B$. When random variables X and Y are independent, then *all such strips*, defined by events $A \subset X$ and $B \subset Y$, intersect in a rectangle. If the events are not independent, then the associated picture of events will not look as organized, as in (a).

Given such independent events $A \subset X$ and $B \subset Y$, observe that $A \mid B$ can be realized, as in (d), with the rectangle $A \cap B$ restricted to the strip defined by B. Furthermore, imagining the area as being proportional to probability, we can see that $\mathbf{Pr}(A \mid B) = \mathbf{Pr}(A \cap B) / \mathbf{Pr}(B)$ since the strip B induces a new restricted sample space Ω_B, and an event only occurs in the strip-induced rectangle defined and further restricted by A which is precisely $A \cap B$.

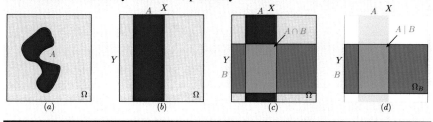

1.2 Conditional Probability and Independence

Now consider two events A and B. The *conditional probability* of A given B is written as $\mathbf{Pr}(A \mid B)$, and can be interpreted as the probability of A, restricted to the setting where we know B is true. It is defined in simpler terms as $\mathbf{Pr}(A \mid B) = \frac{\mathbf{Pr}(A \cap B)}{\mathbf{Pr}(B)}$, that is, the probability that A and B are both true divided by (normalized by) the probability that B is true. Be careful, this is only defined when $\mathbf{Pr}(B) \neq 0$.

Two **events** A and B are *independent* of each other if and only if

$$\mathbf{Pr}(A \mid B) = \mathbf{Pr}(A).$$

Equivalently, by algebraic manipulation, they are independent if and only if $\mathbf{Pr}(B \mid A) = \mathbf{Pr}(B)$ and also if and only if $\mathbf{Pr}(A \cap B) = \mathbf{Pr}(A) \mathbf{Pr}(B)$. This implies that knowledge about B has no effect on the probability of A (and vice versa from A to B).

Example: Conditional Probability

Consider two random variables T and C. Variable T is 1 if a test for cancer is positive, and 0 otherwise. Variable C is 1 if a patient has cancer, and 0 otherwise. The joint probability of the events is captured in the following table:

	cancer $C = 1$	no cancer $C = 0$
tests positive for cancer $T = 1$	0.1	0.02
tests negative for cancer $T = 0$	0.05	0.83

Note that the sum of all cells (the joint sample space Ω) is 1. The conditional probability of having cancer, given a positive test, is $\mathbf{Pr}(C = 1 \mid T = 1) = \frac{0.1}{0.1 + 0.02} = 0.8333$. The probability of cancer (ignoring the test) is $\mathbf{Pr}(C = 1) = 0.1 + 0.05 = 0.15$. Since $\mathbf{Pr}(C = 1 \mid T = 1) \neq \mathbf{Pr}(C = 1)$, then events $T = 1$ and $C = 1$ are not independent.

Two **random variables** X and Y are *independent* if and only if, for *all* possible events $A \subseteq \Omega_X$ and $B \subseteq \Omega_Y$, these events A and B are independent: $\mathbf{Pr}(A \cap B) = \mathbf{Pr}(A) \mathbf{Pr}(B)$.

1.3 Density Functions

Discrete random variables can often be defined through tables (as in the above cancer example), or we can define a function $f_X(k)$ as the probability that random variable X is equal to k. For continuous random variables, we need to be more careful: we will use calculus. We will next develop probability density functions (pdfs) and cumulative density functions (cdfs) for continuous random variables;

the same constructions are sometimes useful for discrete random variables as well, which basically just replace an integral with a sum.

We consider a continuous sample space Ω, and a random variable X with outcomes in Ω. The probability density function of a random variable X is written as f_X. It is defined with respect to any event $A \subset \Omega$ so the probability X is an element of A is $\mathbf{Pr}(X \in A) = \int_{\omega \in A} f_X(\omega) d\omega$. The value $f_X(\omega)$ *is not equal to* $\mathbf{Pr}(X = \omega)$ in general, since for continuous functions $\mathbf{Pr}(X = \omega) = 0$ for any single value $\omega \in \Omega$. Yet, we can interpret f_X as a *likelihood* function; its value has no units, but they can be compared and outcomes ω with larger $f_X(\omega)$ values are more likely.

Next we will define the *cumulative density function* $F_X(t)$; it is the probability that X takes on a value of t or smaller. Here it is typical to have $\Omega = \mathbb{R}$, the set of real numbers. Now define $F_X(t) = \int_{\omega = -\infty}^{t} f_X(\omega) d\omega$.

When the cdf is differentiable, we can then define a pdf in terms of a cdf as $f_X(\omega) = \frac{dF_X(\omega)}{d\omega}$.

Example: Normal Random Variable

A *normal* random variable X is a very common distribution to model noise. It has domain $\Omega = \mathbb{R}$. The pdf of the standard normal distribution is defined as $f_X(\omega) = \frac{1}{\sqrt{2\pi}} \exp(-\omega^2/2) = \frac{1}{\sqrt{2\pi}} e^{-\omega^2/2}$, and its cdf has no closed-form solution. We have plotted the cdf and pdf in the range $[-3, 3]$ where most of the mass lies.

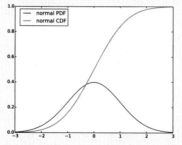

Plotting functions is an extremely important way to understand them. We provide the Python code for implementing the above plot.

```python
import matplotlib.pyplot as plt
from SciPy.stats import norm
import numpy as np
import math

mean = 0
variance = 1
sigma = math.sqrt(variance)
x = np.linspace(-3, 3, 201)

plt.plot(x, norm.pdf((x-mean)/sigma),label='normal_PDF')
plt.plot(x, norm.cdf((x-mean)/sigma),label='normal_CDF')
plt.legend(bbox_to_anchor=(.35,1))
plt.show()
```

1.4 Expected Value

The expected value of a random variable $X : \Lambda \to \Omega$ is a very important concept, basically a weighted average of Ω, weighted by the likelihood of X. Recall, usually $\Omega = \mathbb{R}$ are just real values. For a discrete random variable X, it is defined as the sum over all outcomes ω in the mapped-to sample space, or their value $\omega \in \Omega$ times their probability

$$\mathsf{E}[X] = \sum_{\omega \in \Omega} (\omega \cdot \mathsf{Pr}[X = \omega]).$$

For a continuous random variable X, it is defined as

$$\mathsf{E}[X] = \int_{\omega \in \Omega} \omega \cdot f_X(\omega) d\omega.$$

▷ Linearity of Expectation ◁

A key property of expectation is that it is a linear operation. That means for two random variables X and Y, we have $\mathsf{E}[X + Y] = \mathsf{E}[X] + \mathsf{E}[Y]$. For a scalar value α, we also have $\mathsf{E}[\alpha X] = \alpha \mathsf{E}[X]$.

Example: Expectation

A fair die has a sample space of $\Omega = \{\omega_1 = 1, \omega_2 = 2, \omega_3 = 3, \omega_4 = 4, \omega_5 = 5, \omega_6 = 6\}$, and the probability of each outcome w_i is $\mathsf{Pr}[D = w_i] = 1/6$. The expected value of a random variable D, the result of a roll of such a die, is

$$\mathsf{E}[D] = \sum_{\omega_i \in \Omega} \omega_i \cdot \mathsf{Pr}[D = \omega_i] = 1 \cdot \frac{1}{6} + 2 \cdot \frac{1}{6} + 3 \cdot \frac{1}{6} + 4 \cdot \frac{1}{6} + 5 \cdot \frac{1}{6} + 6 \cdot \frac{1}{6} = \frac{21}{6} = 3.5.$$

Example: Linearity of Expectation

Let H be the random variable of the height of a man in meters without shoes. Let the pdf f_H of H be a normal distribution with expected value $\mu = 1.755$m and with standard deviation 0.1m. Let S be the random variable of the height added by wearing a pair of shoes in centimeters (1 meter is 100 centimeters); its pdf is given by the following table:

$S = 1$	$S = 2$	$S = 3$	$S = 4$
0.1	0.1	0.5	0.3

Then the expected height of someone wearing shoes in centimeters is

$$\begin{aligned}
\mathbf{E}[100 \cdot H + S] &= 100 \cdot \mathbf{E}[H] + \mathbf{E}[S] \\
&= 100 \cdot 1.755 + (0.1 \cdot 1 + 0.1 \cdot 2 + 0.5 \cdot 3 + 0.3 \cdot 4) \\
&= 175.5 + 3 = 178.5
\end{aligned}$$

Note how the linearity of expectation allowed us to decompose the expression $100 \cdot H + S$ into its components, and take the expectation of each one individually. This trick is immensely powerful when analyzing complex scenarios with many factors.

1.5 Variance

The *variance* of a random variable X describes how spread out it is, with respect to its mean $\mathbf{E}[X]$. It is defined as

$$\begin{aligned}
\mathbf{Var}[X] &= \mathbf{E}[(X - \mathbf{E}[X])^2] \\
&= \mathbf{E}[X^2] - \mathbf{E}[X]^2.
\end{aligned}$$

The equivalence of those two above common forms above uses that $\mathbf{E}[X]$ is a fixed scalar:

$$\begin{aligned}
\mathbf{E}[(X - \mathbf{E}[X])^2] &= \mathbf{E}[X^2 - 2X\mathbf{E}[X] + \mathbf{E}[X]^2] = \mathbf{E}[X^2] - 2\mathbf{E}[X]\mathbf{E}[X] + \mathbf{E}[X]^2 \\
&= \mathbf{E}[X^2] - \mathbf{E}[X]^2.
\end{aligned}$$

For any scalar $\alpha \in \mathbb{R}$, then $\mathbf{Var}[\alpha X] = \alpha^2 \mathbf{Var}[X]$.

Note that the variance does not have the same units as the random variable or the expectation; it is that unit squared. As such, we also often discuss the *standard deviation* $\sigma_X = \sqrt{\mathbf{Var}[X]}$. A low value of $\mathbf{Var}[X]$ or σ_X indicates that most values are close to the mean, while a large value indicates that the data has a higher probability of being further from the mean.

Example: Variance

Consider again the random variable S for height added by a shoe:

$S = 1$	$S = 2$	$S = 3$	$S = 4$
0.1	0.1	0.5	0.3

Its expected value is $\mathbf{E}[S] = 3$ (a fixed scalar), and its variance is

$$\begin{aligned}
\mathbf{Var}[S] &= 0.1 \cdot (1-3)^2 + 0.1 \cdot (2-3)^2 + 0.5 \cdot (3-3)^2 + 0.3 \cdot (4-3)^2 \\
&= 0.1 \cdot (-2)^2 + 0.1 \cdot (-1)^2 + 0 + 0.3(1)^2 = 0.4 + 0.1 + 0.3 = 0.8.
\end{aligned}$$

Then the standard deviation is $\sigma_S = \sqrt{0.8} \approx 0.894$.

The *covariance* of two random variables X and Y is defined as $\textbf{Cov}[X, Y] = \textbf{E}[(X - \textbf{E}[X])(Y - \textbf{E}[Y])]$. It measures how much these random variables vary in accordance with each other; that is, if both are consistently away from the mean at the same time (in the same direction), then the covariance is high.

1.6 Joint, Marginal, and Conditional Distributions

Now through the idea of joint distributions, we generalize these concepts to more than one random variable, beyond the simple comparisons using the conditional probability of events and covariance. Consider two random variables X and Y. Their *joint pdf* is of the form $f_{X,Y} : \Omega_X \times \Omega_Y \rightarrow [0, \infty]$ and for discrete random variables, this is defined at values x, y by the probability that $X = x$ and $Y = y$, written as

$$f_{X,Y}(x, y) = \textbf{Pr}(X = x, Y = y).$$

In this discrete case, the domain of $f_{X,Y}$ is restricted so $f_{X,Y} \in [0, 1]$ and so $\sum_{x,y \in X \times Y} f_{X,Y}(x, y) = 1$, e.g., the sum of probabilities over the joint sample space is 1. Sometimes this discrete function is referred to as a *probability mass function* (pmf) instead of a probability density function; we will use pdf for both.

The *marginal pdf* in the discrete case is defined by summing over one variable

$$f_X(x) = \sum_{y \in \Omega_Y} f_{X,Y}(x, y) = \sum_{y \in \Omega_Y} \textbf{Pr}(X = x, Y = y).$$

Marginalizing removes the effect of a random variable (Y in the above definitions).

In this continuous case, it is more natural to first define the joint and marginal cdfs. When $\Omega_X = \Omega_Y = \mathbb{R}$, the *joint cdf* is defined as $F_{X,Y}(x, y) = \textbf{Pr}(X \le x, Y \le y)$. The *marginal cdf*s of $F_{X,Y}$ are defined as $F_X(x) = \lim_{y \to \infty} F_{X,Y}(x, y)$ and $F_Y(y) = \lim_{x \to \infty} F_{X,Y}(x, y)$. Then for continuous random variables, we can now define $f_{X,Y}(x, y) = \frac{d^2 F_{X,Y}(x,y)}{dxdy}$. And then the marginal pdf of X (when $\Omega_Y = \mathbb{R}$) is defined as $f_X(x) = \int_{y=-\infty}^{\infty} f_{X,Y}(x, y) dy$.

Now we can say that the random variables X and Y are independent if and only if $f_{X,Y}(x, y) = f_X(x) \cdot f_Y(y)$ for all x and y.

Then a *conditional distribution* of X given $Y = y$ is defined as $f_{X|Y}(x \mid y) = f_{X,Y}(x, y)/f_Y(y)$ (given that $f_Y(y) \ne 0$).

Example: Marginal and Conditional Distributions

Consider a student who randomly chooses his pants and shirt every day. Let P be a random variable for the color of pants, and S a random variable for the color of the shirt. Their joint probability is described by this table:

	S=green	S=red	S=blue
P=blue	0.3	0.1	0.2
P=white	0.05	0.2	0.15

Adding up along columns, the marginal distribution f_S for the color of the shirt is described by the following table:

S=green	S=red	S=blue
0.35	0.3	0.35

Isolating and renormalizing the middle "S=red" column, the conditional distribution $f_{P|S}(\cdot \mid S=\text{red})$ is described by the following table:

P=blue	$\frac{0.1}{0.3} = 0.3333$
P=white	$\frac{0.2}{0.3} = 0.6666$

Example: Gaussian Distribution

The *Gaussian distribution* is a d-variate distribution $\mathcal{G}_d : \mathbb{R}^d \to \mathbb{R}$ that generalizes the 1-dimensional normal distribution. The definition of the symmetric version (we will generalize to non-trivial covariance later on) depends on a mean $\mu \in \mathbb{R}^d$ and a variance σ^2. For any vector $v \in \mathbb{R}^d$, it is defined as

$$\mathcal{G}_d(v) = \frac{1}{\sigma^d \sqrt{(2\pi)^d}} \exp\left(-\frac{\|v - \mu\|^2}{2\sigma^2}\right),$$

where $\|v - \mu\|$ is the Euclidian distance between v and μ (see Section 4.2). For the 2-dimensional case where $v = (v_x, v_y)$ and $\mu = (\mu_x, \mu_y)$, this is defined as

$$\mathcal{G}_2(v) = \frac{1}{2\pi\sigma^2} \exp\left(\frac{-(v_x - \mu_x)^2 - (v_y - \mu_y)^2}{2\sigma^2}\right).$$

The example below shows a 2-variate Gaussian with $\mu = (2, 3)$ and $\sigma = 1$.

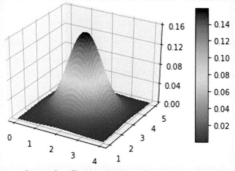

A *magical* property about the Gaussian distribution is that all conditional versions of it are also Gaussian, of a lower dimension. For instance, in the 2-dimensional case, $\mathcal{G}_2(v_x \mid v_y = 1)$ is a 1-dimensional Gaussian, which is precisely a normal distribution $\mathcal{N} = \mathcal{G}_1$. There are many other essential properties of the Gaussian that we will see throughout this text, including that it is invariant under all basis transformations and that it is the limiting distribution for central limit theorem bounds.

When the dimension d is implicit, but useful to write the mean μ and variance σ^2 (or standard deviation σ) in the definition of the pdf explicitly, we can write $\mathcal{G}_{\mu,\sigma}$ or $\mathcal{N}_{\mu,\sigma}$.

1.7 Bayes' Rule

Bayes' rule is the key component in how to build likelihood functions, which when optimized are essential to evaluating "models" based on data. Bayesian reasoning is a much broader area that can go well beyond just finding the single "most optimal" model. This line of work, which this section will only introduce, reasons about the many possible models and can make predictions with this uncertainty taken into account.

Given two events M and D, *Bayes' rule* states that

$$\mathbf{Pr}(M \mid D) = \frac{\mathbf{Pr}(D \mid M) \cdot \mathbf{Pr}(M)}{\mathbf{Pr}(D)}.$$

Mechanically, this provides an algebraic way to invert the direction of the conditioning of random variables, from (D given M) to (M given D). It assumes nothing about the independence of M and D (otherwise, it is pretty uninteresting). To derive this, we use

$$\mathbf{Pr}(M \cap D) = \mathbf{Pr}(M \mid D)\,\mathbf{Pr}(D)$$

and also

$$\mathbf{Pr}(M \cap D) = \mathbf{Pr}(D \cap M) = \mathbf{Pr}(D \mid M)\,\mathbf{Pr}(M).$$

Combining these, we obtain $\mathbf{Pr}(M \mid D)\,\mathbf{Pr}(D) = \mathbf{Pr}(D \mid M)\,\mathbf{Pr}(M)$, from which we can divide by $\mathbf{Pr}(D)$ to solve for $\mathbf{Pr}(M \mid D)$. So Bayes' rule is uncontroversially true; any "frequentist versus Bayesian" debate is about how to model data and perform analysis, not the specifics or correctness of this rule.

Example: Checking Bayes' Rule

Consider two events M and D with the following joint probability table:

	$M = 1$	$M = 0$
$D = 1$	0.25	0.5
$D = 0$	0.2	0.05

We can observe that indeed $\mathbf{Pr}(M \mid D) = \mathbf{Pr}(M \cap D)/\mathbf{Pr}(D) = \frac{0.25}{0.75} = \frac{1}{3}$, which is equal to

$$\frac{\mathbf{Pr}(D \mid M)\,\mathbf{Pr}(M)}{\mathbf{Pr}(D)} = \frac{\frac{.25}{.2+.25}(.2 + .25)}{.25 + .5} = \frac{.25}{.75} = \frac{1}{3}.$$

But Bayes' rule is not very interesting in the above example. In that example, it is actually *more* complicated to calculate the right side of Bayes' rule than the left side.

Example: Cracked Windshield and Bayes' Rule

Consider you bought a new car and its windshield was cracked, the event W. If the car was assembled at one of the three factories A, B, or C, you would like to know which factory was the most likely point of origin.

Assume that near you 50% of cars are from factory A (that is, $\mathbf{Pr}(A) = 0.5$) and 30% are from factory B ($\mathbf{Pr}(B) = 0.3$), and 20% are from factory C ($\mathbf{Pr}(C) = 0.2$).

Then you look up statistics online and find the following rates of cracked windshields for each factory—apparently this is a problem! In factory A, only 1% are cracked, in factory B 10% are cracked, and in factory C 2% are cracked. That is, $\mathbf{Pr}(W \mid A) = 0.01$, $\mathbf{Pr}(W \mid B) = 0.1$, and $\mathbf{Pr}(W \mid C) = 0.02$.

We can now calculate the probability that the car came from each factory:

- $\mathbf{Pr}(A \mid W) = \mathbf{Pr}(W \mid A) \cdot \mathbf{Pr}(A)/\mathbf{Pr}(W) = 0.01 \cdot 0.5/\mathbf{Pr}(W) = 0.005/\mathbf{Pr}(W)$.
- $\mathbf{Pr}(B \mid W) = \mathbf{Pr}(W \mid B) \cdot \mathbf{Pr}(B)/\mathbf{Pr}(W) = 0.1 \cdot 0.3/\mathbf{Pr}(W) = 0.03/\mathbf{Pr}(W)$.
- $\mathbf{Pr}(C \mid W) = \mathbf{Pr}(W \mid C) \cdot \mathbf{Pr}(C)/\mathbf{Pr}(W) = 0.02 \cdot 0.2/\mathbf{Pr}(W) = 0.004/\mathbf{Pr}(W)$.

We did not calculate $\mathbf{Pr}(W)$, but it must be the same for all factory events, so to find the highest probability factory we can ignore it. The probability $\mathbf{Pr}(B \mid W) = 0.03/\mathbf{Pr}(W)$ is the largest, and B is the most likely factory.

1.7.1 Model Given Data

In data analysis, M represents a "model" and D "data." Then $\mathbf{Pr}(M \mid D)$ is interpreted as the probability of model M given that we have observed D. A *maximum a posteriori* (or MAP) estimate is the model $M \in \Omega_M$ that maximizes $\mathbf{Pr}(M \mid D)$. That is, [1]

$$M^* = \underset{M \in \Omega_M}{\mathrm{argmax}}\ \mathbf{Pr}(M \mid D) = \underset{M \in \Omega_M}{\mathrm{argmax}}\ \frac{\mathbf{Pr}(D \mid M)\,\mathbf{Pr}(M)}{\mathbf{Pr}(D)} = \underset{M \in \Omega_M}{\mathrm{argmax}}\ \mathbf{Pr}(D \mid M)\,\mathbf{Pr}(M).$$

Thus, by using Bayes' rule, we can maximize $\mathbf{Pr}(M \mid D)$ using $\mathbf{Pr}(M)$ and $\mathbf{Pr}(D \mid M)$. We do not need $\mathbf{Pr}(D)$ since our data is given to us and fixed for all models.

In some settings, we may also ignore $\mathbf{Pr}(M)$, as we may assume all possible models are equally likely. This is not always the case, and we will come back to this. In this setting, we just need to calculate $\mathbf{Pr}(D \mid M)$. This function $L(M) = \mathbf{Pr}(D \mid M)$ is called the *likelihood* of model M.

---

[1] Consider a set S and a function $f : S \rightarrow \mathbb{R}$. The $\max_{s \in S} f(s)$ returns the *value* $f(s^*)$ for some element $s^* \in S$ which results in the largest valued $f(s)$. The $\mathrm{argmax}_{s \in S} f(s)$ returns the *element* $s^* \in S$ which results in the largest valued $f(s)$; if this is not unique, it may return any such $s^* \in S$.

So what is a "model" and what is "data?"

A *model* is usually a simple pattern from which we think data is generated, but then observed with some noise. Classic examples, which we will explore in this text, include the following:

- The model M is a single point in $\mathbb{R}^d$; the data is a set of points in $\mathbb{R}^d$ near M.
- **linear regression:** The model M is a line in $\mathbb{R}^2$; the data is a set of points such that for each x-coordinate, the y-coordinate is the value of the line at that x-coordinate with some added noise in the y value.
- **clustering:** The model M is a small set of points in $\mathbb{R}^d$; the data is a large set of points in $\mathbb{R}^d$, where each point is near one of the points in M.
- **PCA:** The model M is a k-dimensional subspace in $\mathbb{R}^d$ (for $k \ll d$); the data is a set of points in $\mathbb{R}^d$, where each point is near M.
- **linear classification:** The model M is a halfspace in $\mathbb{R}^d$; the data is a set of labeled points (with labels $+$ or $-$), so the $+$ points are mostly in M, and the $-$ points are mainly not in M.

Log-likelihoods

An important trick used in understanding the likelihood, and in finding the MAP model M^*, is to take the logarithm of the posterior. Since the logarithm operator $\log(\cdot)$ is monotonically increasing on positive values, and all probabilities (and more generally pdf values) are non-negative (treat $\log(0)$ as $-\infty$), then $\text{argmax}_{M \in \Omega_M} \ \mathbf{Pr}(M \mid D) = \text{argmax}_{M \in \Omega_M} \log(\mathbf{Pr}(M \mid D))$. It is commonly applied on only the likelihood function $L(M)$, and $\log(L(M))$ is called the *log-likelihood*. Since $\log(a \cdot b) = \log(a) + \log(b)$, this is useful in transforming definitions of probabilities, which are often written as products $\Pi_{i=1}^k P_i$ into sums $\log(\Pi_{i=1}^k P_i) = \sum_{i=1}^k \log(P_i)$, which are easier to manipulate algebraically.

Moreover, the base of the log is unimportant in model selection using the MAP estimate because $\log_{b_1}(x) = \log_{b_2}(x)/\log_{b_2}(b_1)$, and so $1/\log_{b_2}(b_1)$ is a coefficient that does not affect the choice of M^*. The same is true for the *maximum likelihood estimate* (MLE): $M^* = \text{argmax}_{M \in \Omega_M} L(M)$.

Example: Gaussian MLE

Let the data D be a set of points in $\mathbb{R}^1$: $\{1, 3, 12, 5, 9\}$. Let Ω_M be $\mathbb{R}$ so that the model is parametrized by a point $M \in \mathbb{R}$. If we assume that each data point is observed with independent Gaussian noise (with $\sigma = 2$, its pdf is described as $\mathcal{N}_{M,2}(x) = g(x) = \frac{1}{\sqrt{8\pi}} \exp(-\frac{1}{8}(M - x)^2)$, then

$$\mathbf{Pr}(D \mid M) = \prod_{x \in D} g(x) = \prod_{x \in D} \left(\frac{1}{\sqrt{8\pi}} \exp(-\frac{1}{8}(M - x)^2) \right).$$

Recall that we can take the product $\Pi_{x \in D} g(x)$ since we assume independence of $x \in D$! To find $M^* = \text{argmax}_M \ \mathbf{Pr}(D \mid M)$ is equivalent to $\text{argmax}_M \ \ln(\mathbf{Pr}(D \mid M))$, the *log-likelihood* which is

$$\ln(\mathbf{Pr}(D \mid M)) = \ln\left(\prod_{x \in D}\left(\frac{1}{\sqrt{8\pi}}\exp(-\frac{1}{8}(M - x)^2)\right)\right)$$

$$= \sum_{x \in D}\left(-\frac{1}{8}(M - x)^2\right) + |D|\ln(\frac{1}{\sqrt{8\pi}}).$$

We can ignore the last term in the sum since it does not depend on M. The first term is maximized when $\sum_{x \in D}(M - x)^2$ is minimized, which occurs precisely as $\mathbf{E}[D] = \frac{1}{|D|}\sum_{x \in D} x$, the mean of the data set D. That is, the maximum likelihood model is exactly the mean of the data D, and is quite easy to calculate.

Geometry of Average Minimizing Sum of Squared Errors

Here we will show that sum of squared errors from n values $X = \{x_1, x_2, \ldots, x_n\}$ is minimized at their average. We do so by showing that the first derivative is 0 at the average. Consider the sum of squared errors cost $s_X(z)$ at some value z

$$s_X(z) = \sum_{i=1}^{n}(x_i - z)^2 = \sum_{i=1}^{n} z^2 - 2x_i z + x_i^2 = nz^2 - (2\sum_{i=1}^{n} x_i)z + \sum_{i=1}^{n} x_i^2$$

and its derivative, set to 0, as

$$\frac{ds_X(z)}{dz} = 2nz - 2\sum_{i=1}^{n} x_i = 0.$$

Solving for z yields $z = \frac{1}{n}\sum_{i=1}^{n} x_i$. And this is exactly the average X. The second derivative ($\frac{d^2}{dz^2} s_X(z) = 2n$) is positive, so this is a minimum point, not a maximum point.

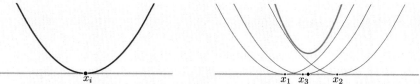

A single squared error function $s_i(z) = (x_i - z)^2$ is a parabola. And the sum of squared error functions is also a parabola, shown in green in the figure. This is because we can expand the sum of squared errors as a single quadratic equation in z, with coefficients determined by the x_i values. This green parabola is exactly minimized at the average of X, the ♦ mark.

1.8 Bayesian Inference

Bayesian inference focuses on a simplified version of Bayes's rule:

$$\mathbf{Pr}(M \mid D) \propto \mathbf{Pr}(D \mid M) \cdot \mathbf{Pr}(M).$$

The symbol $\propto$ means *proportional to*; that is, there is a fixed (but possibly unknown) constant factor c multiplied on the right (in this case, $c = 1/\mathbf{Pr}(D)$) to make them equal: $\mathbf{Pr}(M \mid D) = c \cdot \mathbf{Pr}(D \mid M) \cdot \mathbf{Pr}(M)$.

However, we may want to use continuous random variables, so strictly using probability $\mathbf{Pr}$ at a single point is not always correct. So we can replace each of these with pdfs

$$p(M \mid D) \propto f(D \mid M) \cdot \pi(M).$$
$$\text{posterior} \quad \text{likelihood} \quad \text{prior}$$

Each of these terms has a common name. As above, the conditional probability or pdf $\mathbf{Pr}(D \mid M) \propto f(D \mid M)$ is called the *likelihood*; it evaluates the effectiveness of the model M using the observed data D. The probability or pdf of the model $\mathbf{Pr}(M) \propto \pi(M)$ is called the *prior*; it is the assumption about the relative propensity of a model M, before or independent of the observed data. And the left-hand side $\mathbf{Pr}(M \mid D) \propto p(M \mid D)$ is called the *posterior*; it is the combined evaluation of the model that incorporates how well the observed data fits the model and the independent assumptions about the model.

Again it is common to be in a situation where, given a fixed model M, it is possible to calculate the likelihood $f(D \mid M)$. And again, the goal is to be able to compute $p(M \mid D)$, as this allows us to evaluate potential models M, given the data, we have seen, D.

The main difference is a careful analysis of $\pi(M)$, the prior—which is not necessarily assumed uniform or "flat." The prior allows us to encode our assumptions.

Example: Average Height

Let us estimate the height H of a typical student from our university. We can construct a data set $D = \{x_1, \ldots, x_n\}$ by measuring the height of a set of random students in inches. There may be an error in the measurement, and this is an incomplete set, so we do not entirely trust the data.

So we introduce a prior $\pi(M)$. Consider we read that the average height of a fully grown person is 66 inches, with a standard deviation of $\sigma = 6$ inches. So we assume that

$$\pi(M) = \mathcal{N}_{66,6}(M) = \frac{1}{\sqrt{\pi 72}} \exp(-(\mu_M - 66)^2/(2 \cdot 6^2))$$

is normally distributed around 66 inches.

Now, given this knowledge, we adjust the MLE example from the last subsection using this prior.

- *What if our MLE estimate without the prior (e.g., $\frac{1}{|D|}\sum_{x \in D} x$) provides a value of 5.5?*
  The data is very far from the prior! Usually this means something is wrong. We could find $\text{argmax}_M \, p(M \mid D)$ using this information, but that may give us an estimate of say 20 (that does not seem correct). A more likely explanation is a mistake somewhere: probably we measured in feet instead of inches!

Weight for Prior

So how important is the prior? In the average height example, it will turn out to be worth only $(1/9)$th of one student's measurement. But we can give it more weight.

Example: Weighted Prior for Height

Let us continue the example about the height of an average student from our university, and assume (as in the MLE example) the data is generated independently from a model M with Gaussian noise with $\sigma = 2$. Thus the likelihood of the model, given the data, is

$$f(D \mid M) = \prod_{x \in D} g(x) = \prod_{x \in D} \left(\frac{1}{\sqrt{8\pi}} \exp(-\frac{1}{8}(\mu_M - x)^2) \right).$$

Now using that the prior of the model is $\pi(M) = \frac{1}{\sqrt{\pi 72}} \exp(-(\mu_M - 66)^2/72)$, the posterior is given by

$$p(M \mid D) \propto f(D \mid M) \cdot \frac{1}{\sqrt{\pi 72}} \exp(-(\mu_M - 66)^2/72).$$

It is again easier to work with the log-posterior which is monotonic with the posterior, using some unspecified constant C (which can be effectively ignored):

$$\ln(p(M \mid D)) \propto \ln(f(D \mid M)) + \ln(\pi(M)) + C$$

$$\propto \sum_{x \in D} \left(-\frac{1}{8}(\mu_M - x)^2) \right) - \frac{1}{72}(\mu_M - 66)^2 + C$$

$$\propto -\sum_{x \in D} 9(\mu_M - x)^2 - (\mu_M - 66)^2 + C$$

So the maximum likelihood estimator occurs at the average of 66 and 9 copies of the student data; that is, the value 66 counts as much as 1/9th of one student data point x_i in the average. In general, the solution for μ_M takes the form of a weighted average.

Why is student measurement data worth so much ? We assume the standard deviation of the measurement error is 2, whereas we assumed that the standard deviation of the full population was 6. In other words, our measurements had variance $2^2 = 4$, and the population had variance $6^2 = 36$ (technically, this is best to interpret as the variance when adapted to various subpopulations, e.g., students at our university): that is 9 times as much.

If instead we assumed that the standard deviation of our prior is 0.1, with variance 0.01, then this is 400 times smaller than our sample set's measurement error variance. If we were to redo the above calculations with this smaller variance, we would find that this assumption weights the prior 400 times the effect of each student measurement in the MLE.

In fact, a much smaller variance on the prior is probably more realistic since national estimates on height are probably drawn from a very large sample. And it is important to keep in mind that we are estimating the *average height* of a population, not the *height of a single person* randomly drawn from the population. In the next chapter, we will see how averages of random variables have much smaller variance— are much more concentrated—than individual random variables.

So what happens with more data? Let us consider increasing our sample set to 1,000 or 10,000 students. Then again the student data is overall worth more than the prior data. So with any prior, if we get enough data, it no longer becomes important. But with a small amount of data, it can have a large influence on our model.

Weighted Average

When combining the effects of iid data and a prior, we may need to use a weighted average. Given a set of n values $x_1, x_2, \ldots, x_n$ as well as n corresponding weights $w_1, w_2, \ldots, w_n$, the *weighted average* is defined as

$$\frac{\sum_{i=1}^{n} w_i x_i}{\sum_{i=1}^{n} w_i}.$$

Note that the denominator $W = \sum_{i=1}^{n} w_i$ is the total weight. Thus, for the special case when all $w_i = 1$, this is the same as the regular average since $W = n$ and each term in the numerator is simply $1 x_i = x_i$; hence the entire quantity is $\frac{1}{n} \sum_{i=1}^{n} x_i$.

Example: Weighted Average

My favorite barbecue spice mix is made from 2 cups salt, 1.5 cups black pepper, and 1 cup cayenne pepper. At the store, salt costs 16¢ per cup, black pepper costs 50¢ per cup, and cayenne pepper costs 175¢ per cup. How much does my barbecue spice mix cost per cup?

This would be a weighted average with weights $w_1 = 2$, $w_2 = 1.5$, and $w_3 = 1$ for salt, black pepper, and cayenne pepper, respectively. The values being averaged are

$x_1 = 16$¢, $x_2 = 50$¢, and $x_3 = 175$¢ again for salt, black pepper, and cayenne pepper, respectively. The units for the weights cancel out, so they do not matter (as long as they are the same). But the units for the values do matter. The resulting cost per cup is

$$\frac{2 \cdot 16¢ + 1.5 \cdot 50¢ + 1 \cdot 175¢}{2 + 1.5 + 1} = \frac{32¢ + 75¢ + 175¢}{4.5} = \frac{282¢}{4.5} \approx 63¢.$$

Power of the Posterior

Another aspect of Bayesian inference is that we not only can calculate the maximum likelihood model M^*, but can also provide a posterior value for any model! This value is not an absolute probability (it is not normalized, and regardless it may be of measure 0), but it is powerful in other ways:

- We can say (under our model assumptions, which are now clearly stated) that one model M_1 is twice as likely as another M_2, if $p(M_1 \mid D)/p(M_2 \mid D) = 2$.
- We can now use more than one model for the prediction of a value. Given a new data point x', we may want to map it onto our model as $M(x')$ or assign it a score of fit. Instead of doing this for just one "best" model M^*, we can take a weighted average of all models, weighted by their posterior, that is, marginalizing over models.
- We can define a range of parameter values (with more model assumptions) that likely contains the true model.

Example: Using the Posterior

Consider three possible models M_1, M_2, and M_3 that assume data D is observed with normal noise with standard deviation 2. The difference in the models is the mean of the distribution which is assumed to be 1.0, 2.0, and 3.0, respectively in these models.

We also are given a prior for each model, in the following table:

$\pi(M_1)$	$\pi(M_2)$	$\pi(M_3)$
0.4	0.5	0.1

Now given a single data point $x = 1.6$ in D, the posterior for each model is

$$p(M_1 \mid D) = C \frac{1}{4\sqrt{2\pi}} \exp(-(x - 1.0)^2/8) \cdot \pi(M_1) = C'(0.96)(0.4) = C' \cdot 0.384$$

$$p(M_2 \mid D) = C \frac{1}{4\sqrt{2\pi}} \exp(-(x - 2.0)^2/8) \cdot \pi(M_2) = C'(0.98)(0.5) = C' \cdot 0.49$$

$$p(M_3 \mid D) = C \frac{1}{4\sqrt{2\pi}} \exp(-(x - 3.0)^2/8) \cdot \pi(M_3) = C'(0.78)(0.1) = C' \cdot 0.078$$

where C accounts for the unknown probability of the data, and C' folds in other fixed normalizing constants.

As before, it follows easily that M_2 is the maximum posterior estimate—and our single best choice given the prior and our data. We also know it is about 6 times more likely than model M_3, but only about 1.3 times more likely than model M_1.

Say we want to predict the square of the mean of the true model. Instead of just returning the maximum posterior (i.e., 4 for M_2), we can take the weighted average:

$$\frac{1.0^2 \cdot 0.384C' + 2.0^2 \cdot 0.49C' + 3.0^2 \cdot 0.078C'}{0.384C' + 0.49C' + 0.078C'} = \frac{9.364}{1.654} = 5.66$$

Note how the unknown constant C' cancels out in the calculation.

If instead we put a prior on all possible mean values, we can provide a range of possible models which achieve some degree of likelihood. For instance, if it is a "flat" prior and so $\pi(M_v)$ is the same for all possible mean values v, then the posterior and likelihood are proportional. The models with likelihood at least 0.05 are those with mean values v satisfying $-3.3 \le v \le 6.5$.

Exercises

1.1 Consider the probability table below for the random variables X and Y. One entry is missing, but you should be able to derive it. Then calculate the following values.

1. $\mathbf{Pr}(X = 3 \cap Y = 2)$
2. $\mathbf{Pr}(Y = 1)$
3. $\mathbf{Pr}(X = 2 \cap Y = 1)$
4. $\mathbf{Pr}(X = 2 \mid Y = 1)$

	$X = 1$	$X = 2$	$X = 3$
$Y = 1$	0.25	0.1	0.15
$Y = 2$	0.1	0.2	??

1.2 An "adventurous" athlete has the following running routine every morning: He takes a bus to a random stop, then hitches a ride, and then runs all the way home. The bus, described by a random variable B, has four stops where the stops are at a distance of 1, 3, 4, and 7 miles from his house—he chooses each with probability 1/4. Then the random hitchhiking takes him further from his house with a uniform distribution between -1 and 4 miles; that is, it is represented as a random variable H with pdf described as

$$f(H = x) = \begin{cases} 1/5 & \text{if } x \in [-1, 4] \\ 0 & \text{if } x \notin [-1, 4]. \end{cases}$$

What is the expected distance he runs each morning (all the way home)?

1.3 Consider rolling two fair die D_1 and D_2; each has a probability space of $\Omega = \{1, 2, 3, 4, 5, 6\}$ which each value equally likely. What is the probability that D_1 has a larger value than D_2? What is the expected value of the sum of the two die?

1.4 Let X be a random variable with a uniform distribution over $[0, 2]$; its pdf is described as

$$f(X = x) = \begin{cases} 1/2 & \text{if } x \in [0, 2] \\ 0 & \text{if } x \notin [0, 2]. \end{cases}$$

What is the probability $f(X = 1)$?

1.5 Consider rolling two fair die D_1 and D_2; each has a probability space of $\Omega = \{1, 2, 3, 4, 5, 6\}$ which each value equally likely. What is the probability that D_1 has a larger value than D_2? What is the expected value of the sum of the two die?

1.6 Let L be a random variable describing the outcome of a lottery ticket. The lottery ticket costs 1, and with probability 0.75 returns 0 dollars. With probability 0.15, it returns 1 dollar. With probability 0.05, it returns 2 dollars. With probability 0.04, it returns 5 dollars. And with probability 0.01, it returns 100 dollars.

1. Given a lottery ticket (purchased for you), what is the expected money returned by playing it?

2. Factoring in the cost of buying the ticket, what is the expected money won by buying and playing the lottery ticket?
3. If you were running the lottery, and wanted to have the player expect to lose 0.01 dollars (1 cent) per ticket, how much should you charge?

1.7 Use Python to plot the pdf and cdf of the Laplace distribution ($f(x) = \frac{1}{2}\exp(-|x|)$) for values of x in the range $[-3, 3]$. The function $\texttt{SciPy.stats.laplace}$ may be useful.

1.8 Consider the random variables X and Y described by the joint probability table.

	$X = 1$	$X = 2$	$X = 3$
$Y = 1$	0.10	0.05	0.10
$Y = 2$	0.30	0.25	0.20

Derive the following values:

1. $\mathbf{Pr}(X = 1)$
2. $\mathbf{Pr}(X = 2 \cap Y = 1)$
3. $\mathbf{Pr}(X = 3 \mid Y = 2)$

Compute the following probability distributions:

4. What is the marginal distribution for X?
5. What is the conditional probability for Y, given that $X = 2$?

Answer the following questions about the joint distribution:

6. Are random variables X and Y independent?
7. Is $\mathbf{Pr}(X = 1)$ independent of $\mathbf{Pr}(Y = 1)$?

1.9 Consider two models M_1 and M_2, where from prior knowledge we believe that $\mathbf{Pr}(M_1) = 0.25$ and $\mathbf{Pr}(M_2) = 0.75$. We then observe a data set D. Given each model, we assess the likelihood of seeing that data, given the model, as $\mathbf{Pr}(D \mid M_1) = 0.5$ and $\mathbf{Pr}(D \mid M_2) = 0.01$. Now that we have the data, which model has a higher probability of being correct?

1.10 Assume I observe 3 data points x_1, x_2, and x_3 drawn independently from an unknown distribution. Given a model M, I can calculate the likelihood for each data point as $\mathbf{Pr}(x_1 \mid M) = 0.5$, $\mathbf{Pr}(x_2 \mid M) = 0.1$, and $\mathbf{Pr}(x_3 \mid M) = 0.2$. What is the likelihood of seeing all of these data points, given the model M: $\mathbf{Pr}(x_1, x_2, x_3 \mid M)$?

1.11 Consider a data set D with 10 data points $\{-1, 6, 0, 2, -1, 7, 7, 8, 4, -2\}$. We want to find a model for M from a restricted sample space $\Omega = \{0, 2, 4\}$. Assume the data has Laplace noise defined, so from a model M a data point's probability distribution is described as $f(x) = \frac{1}{4}\exp(-|M - x|/2)$. Also assume we have a prior assumption on the models so that $\mathbf{Pr}(M = 0) = 0.25$, $\mathbf{Pr}(M = 2) = 0.35$, and $\mathbf{Pr}(M = 4) = 0.4$. Assuming all data points in D are independent, which model is most likely?

1.12 A student realizes she needs to file some registration paperwork 5 minutes from the deadline. She will be fined \$1 for each minute that it is late. So if it takes her 7 minutes, she will be 2 minutes late, and be fined \$2. The time is rounded down to the nearest minute late (so 2.6 minutes late has a \$2 fine).

Her friend tells her that the usual time to finish is a uniform distribution between 4 minutes and 8 minutes. That is, the time to finish represents a random variable T with pdf

$$f(T = x) = \begin{cases} 1/4 & \text{if } x \in [4, 8] \\ 0 & \text{if } x \notin [4, 8]. \end{cases}$$

1. What is the expected amount of money she will be fined if she starts immediately?
2. What is the expected amount of money she will be fined if she starts in 2 minutes?

1.13 A glassblower creates an intricate artistic glass bottle and attempts to measure its volume in liters; the 10 measures are:

$$\{1.82, 1.71, 2.34, 2.21, 2.01, 1.95, 1.76, 1.94, 2.02, 1.89\}.$$

To sell the bottle, by regulation, she must label its volume up to 0.1 Liters (it could be 1.7L or 1.8L or 1.9L, and so on). Her prior estimate is that it is 2.0L, but with a normal distribution with a standard deviation of 0.1; that is, her prior for the volume V being x is described by the pdf

$$f_V(x) = C \cdot \exp(-(2.0 - x)^2 / (2 \cdot 0.1^2)),$$

for some unknown constant C (since it is only valid at increments of 0.1). Assuming the 10 empirical estimates of the volume are unbiased, but have a normal error with a standard deviation of 0.2, what is the most likely model for the volume V?

Chapter 2
Convergence and Sampling

Abstract This topic will overview a variety of extremely powerful analysis results that span statistics, estimation theory, and big data. It provides a framework to think about how to aggregate more and more data to get better and better estimates. It will cover the *Central Limit Theorem* (CLT), Chernoff-Hoeffding bounds, Probably Approximately Correct (PAC) algorithms, as well as analysis of importance sampling techniques which improve the concentration of random samples.

2.1 Sampling and Estimation

Most data analysis start with some data set; we will call this data set P. It will be composed of a set of n data points $P = \{p_1, p_2, \ldots, p_n\}$.

But underlying this data is almost always a very powerful assumption that this data comes iid from a fixed, but usually, unknown pdf, called f. Let's unpack this: What does "*iid*" mean: Identically and Independently Distributed. "Identically" means each data point was drawn from the same f. "Independently" means that the first points have no bearing on the value of the next point.

Example: Polling

Consider a poll of $n = 1000$ likely voters in an upcoming election. If we assume each polled person is chosen iid, then we can use this to understand something about the underlying distribution f, for instance, the distribution of all likely voters.

More generally, f could represent the outcome of a process, whether that is a randomized algorithm, a noisy sensing methodology, or the common behavior of a species of animals. In each of these cases, we essentially "poll" the process (algorithm, measurement, and thorough observation) having it provide a single sample, and repeat many times over.

© Springer Nature Switzerland AG 2021

J. M. Phillips, *Mathematical Foundations for Data Analysis*,
Springer Series in the Data Sciences,
https://doi.org/10.1007/978-3-030-62341-8_2

Here we will talk about estimating the mean of f. To discuss this, we will now introduce a random variable $X \sim f$: a hypothetical new data point. The *mean* of f is the expected value of X: $\mathsf{E}[X]$.

We will estimate the mean of f using the *sample mean*, defined as $\bar{P} = \frac{1}{n} \sum_{i=1}^{n} p_i$. The following diagram represents this common process: from an unknown process f, we consider n iid random variables $\{X_i\}$ corresponding to a set of n independent observations $\{p_i\}$, and take their average $\bar{P} = \frac{1}{n} \sum_{i=1}^{n} p_i$ to estimate the mean of f.

$$\bar{P} \underset{\frac{1}{n} \Sigma}{=} \{p_i\} \underset{\text{realize}}{\leftarrow} \{X_i\} \underset{\text{iid}}{\sim} f$$

Central Limit Theorem

The *central limit theorem* is about how well the sample mean approximates the true mean. But to discuss the sample mean $\bar{P}$ (which is a fixed value), we need to discuss random variables $\{X_1, X_2, \ldots, X_n\}$ and their mean $\bar{X} = \frac{1}{n} \sum_{i=1}^{n} X_i$. Note that again $\bar{X}$ is a random variable. If we are to draw a new iid data set P' and calculate a new sample mean $\bar{P}'$, it will likely not be exactly the same as $\bar{P}$; however, the distribution of where this $\bar{P}'$ is likely to be is precisely $\bar{X}$. Arguably, this distribution is more important than $\bar{P}$ itself.

There are many formal variants of the central limit theorem, but the basic form is as follows:

> ▷ **Central Limit Theorem** ◁

Consider n iid random variables $X_1, X_2, \ldots, X_n$, where each $X_i \sim f$ for any fixed distribution f with mean μ and bounded variance σ^2. Then $\bar{X} = \frac{1}{n} \sum_{i=1}^{n} X_i$ converges to the normal distribution with mean $\mu = \mathsf{E}[X_i]$ and variance σ^2/n.

Let's highlight some important consequences:

- For any f (that is not too wild, since σ^2 is not infinite), $\bar{X}$ looks like a normal distribution.
- The mean of the normal distribution, which is the expected value of $\bar{X}$ satisfies $\mathsf{E}[\bar{X}] = \mu$, the mean of f (that is, for each i, $\mathsf{E}[\bar{X}] = \mathsf{E}[X_i]$). This implies that we can use $\bar{X}$ (and then also $\bar{P}$) as a guess for μ.
- As n gets larger (we have more data points), the variance of $\bar{X}$ (our estimator) decreases. So keeping in mind that although $\bar{X}$ has the right expected value, it also has some error; this error is decreases as n increases.

Example: Central Limit Theorem

To visualize the central limit theorem, consider f as a uniform distribution over $[0, 100]$. Create n samples $\{p_1, \ldots, p_n\}$ and calculate their sample mean $\bar{P}$; then we repeat this 30,000 times and plot the output in histograms for different choices of n.

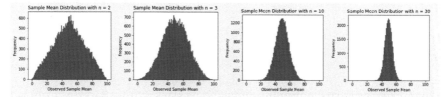

The leftmost chart shows this process for $n = 2$, and the next ones are for $n = 3$, $n = 10$, and $n = 30$. Each histogram shows sample distributions with the same approximate mean of 50. And as n increases, the variance decreases: for $n = 2$, there is some support close to 0 and 100, but for $n = 30$ the support is between 30 and 70, and mostly between 40 and 60. Moreover, all distributions appear vaguely like the normal distribution, with of course some error due to the histogram representation, because it is based on 30,000 samples of $\bar{P}$, and because n is finite.

This process is made more concrete in the following Python code.

```python
import matplotlib as mpl
import matplotlib.pyplot as plt
%matplotlib inline
import random


numBins = 100
numTrials = 30000
n = 10   #change for n = 2, 3, 10, and 30

sampleMeans = []
for j in range(numTrials):
  sampleSum = 0;
  for i in range(n):
    sampleSum += random.choice(range(numBins))
                 #uniform random in [0,numBins]
  sampleMeans.append(float(sampleSum)/ float(n))

plt.hist(sampleMeans, color="green", bins = range(numBins+1))
plt.title("Sample Mean Distribution with n = %s" % n)
plt.ylabel("Frequency")
plt.xlabel("Observed Sample Mean")
plt.show()
```

Remaining Mysteries

There should still be at least a few aspects of this not clear yet: (1) What does "convergence" mean? (2) How do we formalize or talk about this notion of error? (3) What does this say about our data $\bar{P}$?

First, *convergence* refers to what happens as some parameter increases, in this case n. As the number of data points increases and as n "goes to infinity," the above statement ($\bar{X}$ looks like a normal distribution) becomes more and more true. For small n, the distribution may not quite look normal, it may be more bumpy or maybe even multi-modal. The statistical definitions of convergence are varied, and we will not go into them here; we will instead replace it with more useful phrasing in explaining aspects (2) and (3).

Second, the error now has two components. We cannot simply say that $\bar{P}$ is at most some distance ε from μ. Something unusual may have happened (the sample is random after all). And it is not useful to try to write the probability that $\bar{P} = \mu$; for equality in continuous distributions, this probability is indeed 0. But we can combine these notions. We can say the distance between $\bar{P}$ and μ is more than ε, with probability at most δ. This is called "probably approximately correct" or PAC.

Third, we want to generate some sort of PAC bound (which is far more useful than "$\bar{X}$ looks kind of like a normal distribution"). While a frequentist may be happy with a confidence interval and a Bayesian may want a full normal posterior, these two options are not directly available since again, $\bar{X}$ is not exactly normal. So we will discuss some very common *concentration of measure* tools. These do not exactly capture the shape of the normal distribution, but provide upper bounds for its tails, and will allow us to state PAC bounds.

2.2 Probably Approximately Correct (PAC)

We will introduce shortly the three most common concentration of measure bounds, which provide increasingly strong bounds on the tails of distributions but require more and more information about the underlying distribution f. Each provides a *probability approximately correct* (PAC) bound of the following form:

$$\mathbf{Pr}[|\bar{X} - \mathbf{E}[\bar{X}]| \geq \varepsilon] \leq \delta.$$

That is, the probability that most δ that $\bar{X}$ (which is some random variable, often a sum of iid random variables) is further than ε to its expected value (which is μ, the expected value of f where $X_i \sim f$). Note that we do not try to say this probability is *exactly* δ; this is often too hard. In practice, there are a variety of tools, and a user may try each one and see which one gives the best bound.

It is useful to think of ε as the *error tolerance* and δ as the *probability of failure*, i.e., failure meaning that we exceed the error tolerance. However, often these bounds will allow us to write the required sample size n in terms of ε and δ. This allows us to trade these two terms off for any fixed known n; we can guarantee a smaller error tolerance if we are willing to allow more probability of failure and vice-versa.

2.3 Concentration of Measure

We will formally describe these bounds and give some intuition of why they are true (but not full proofs). But what will be the most important is what they imply. If you just know the distance of the expectation from the minimal value, you can get a very weak bound. If you know the variance of the data, you can get a stronger bound. If you know that the distribution f has a small and bounded range, then you can make the probability of failure (the δ in PAC bounds) very very small.

2.3.1 Markov Inequality

Let X be a random variable such that $X \geq 0$, that is, it cannot take on negative values. Then the *Markov inequality* states that for any parameter $\alpha > 0$,

$$\Pr[X > \alpha] \leq \frac{\mathsf{E}[X]}{\alpha}.$$

Note this is a PAC bound with $\varepsilon = \alpha - \mathsf{E}[X]$ and $\delta = \mathsf{E}[X]/\alpha$, or we can rephrase this bound as follows: $\Pr[X - \mathsf{E}[X] > \varepsilon] \leq \delta = \mathsf{E}[X]/(\varepsilon + \mathsf{E}[X])$.

Geometry of the Markov Inequality

Consider balancing the pdf of some random variable X on your finger at $\mathsf{E}[X]$, like a waiter balances a tray. If your finger is not under a value μ, $\mathsf{E}[X] = \mu$, then the pdf (and the waiter's tray) will tip and fall in the direction of μ—the "center of mass."

Now for some amount of probability α, how large can we increase its location so that we retain $\mathsf{E}[X] = \mu$? For each part of the pdf we increase, we must decrease some in proportion. However, by the assumption $X \geq 0$, the pdf must not have a mass below 0. In the limit of this, we can set $\Pr[X = 0] = 1 - \alpha$, and then move the remaining α probability as large as possible, to a location δ so $\mathsf{E}[X] = \mu$. That is,

$$\mathsf{E}[X] = 0 \cdot \Pr[X = 0] + \delta \cdot \Pr[X = \delta] = 0 \cdot (1 - \alpha) + \delta \cdot \alpha = \delta \cdot \alpha.$$

Solving for δ, we find $\delta = \mathsf{E}[X]/\alpha$.

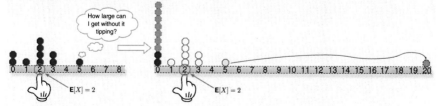

Imagine having 10 α-balls each representing $\alpha = 1/10$th of the probability mass. As in the figure, if these represent a distribution with $\mathsf{E}[X] = 2$ and this must stay fixed, how far can one ball increase if all other balls must take a value at least 0? One ball can move to 20.

If we instead know that $X \geq b$ for some constant b (instead of $X \geq 0$), then we state more generally $\Pr[X > \alpha] \leq (\mathsf{E}[X] - b)/(\alpha - b)$.

Example: Markov Inequality

Consider the pdf f drawn in blue in the following figures, with $\mathbf{E}[X]$ for $X \sim f$ marked as a red dot. The probability that X is greater than 5 (e.g., $\mathbf{Pr}[X \geq 5]$) is the shaded area.

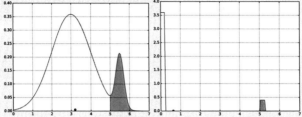

Notice that in both cases, $\mathbf{Pr}[X \geq 5]$ is about 0.1. This is the quantity we want to bound above by δ. But since $\mathbf{E}[X]$ is much larger in the first case (about 3.25), then the bound $\delta = \mathbf{E}[X]/\alpha$ is much larger, about 0.65. In the second case, $\mathbf{E}[X]$ is much smaller (about 0.6), so we get a much better bound of $\delta = 0.12$.

Example: Markov Inequality and Rainfall

Let R be a random variable describing the number of millimeters of rain that will fall in Salt Lake City the next June. Say we know $\mathbf{E}[R] = 20$ mm.

We use the Markov inequality to bound the probability that more than 50 mm will fall as

$$\mathbf{Pr}[R \geq 50] \leq \frac{\mathbf{E}[R]}{50} = \frac{20}{50} = 0.4$$

Hence, based on the expected value alone, we can bound with the probability less than 0.4 that it will rain more than 50 mm next June.

2.3.2 Chebyshev Inequality

Now let X be a random variable where we know $\mathbf{Var}[X]$ and $\mathbf{E}[X]$. Then the *Chebyshev inequality* states that for any parameter $\varepsilon > 0$,

$$\mathbf{Pr}[|X - \mathbf{E}[X]| \geq \varepsilon] \leq \frac{\mathbf{Var}[X]}{\varepsilon^2}.$$

Again, this clearly is a PAC bound with $\delta = \mathbf{Var}[X]/\varepsilon^2$. This bound is typically stronger than the Markov one since δ decreases quadratically in ε instead of linearly. Compared to the Markov inequality, it requires knowledge of $\mathbf{Var}[X]$, but it does not require $X \geq 0$.

Example: Chebyshev Inequality and Rainfall

Again let R be a random variable for the millimeters of rain in Salt Lake City next June with $E[R] = 20$. If we also know that the variance is not too large, specifically $Var[R] = 9$ (millimeters squared), then we can apply the Chebyshev inequality to get an improved bound:

$$Pr[R \geq 50] \leq Pr[|R - E[R]| \geq 30] \leq \frac{Var[R]}{30^2} = \frac{9}{900} = 0.01$$

That is, by using the expectation ($E[R] = 20$) and variance ($Var[R] = 9$), we can reduce the probability of exceeding 50 mm to at most probability 0.01.

Note that in the first inequality, we convert from a one-sided expression $R \geq 50$ to a two-sided expression $|R - E[R]| \geq 30$ (that is either $R - E[R] \geq 30$ or $E[R] - R \geq 30$). This is a bit wasteful, and stricter one-sided variants of Chebyshev inequality exist; we will not discuss these here in an effort for simplicity.

Recall that for an average of random variables $\bar{X} = (X_1 + X_2 + \ldots + X_n)/n$, where the X_is are iid, and have variance σ^2, then $Var[\bar{X}] = \sigma^2/n$. Hence

$$Pr[|\bar{X} - E[X_i]| \geq \varepsilon] \leq \frac{\sigma^2}{n\varepsilon^2}.$$

Consider now that we have input parameters ε and δ, our desired error tolerance and probability of failure. If we can draw $X_i \sim f$ (iid) for an unknown f (with known expected value and variance σ), then we can solve for how large n needs to be: $n = \sigma^2/(\varepsilon^2 \delta)$.

Since $E[\bar{X}] = E[X_i]$ for iid random variables $X_1, X_2, \ldots, X_n$, there is not a similar meaningfully different extension for the Markov inequality which demonstrates improved concentration with increase in data size n.

2.3.3 Chernoff-Hoeffding Inequality

Following the above extension of the Chebyshev inequality, we can consider a set of n iid random variables $X_1, X_2, \ldots, X_n$ where $\bar{X} = \frac{1}{n} \sum_{i=1}^{n} X_i$. Now assume we know that each X_i lies in a bounded domain $[b, t]$, and let $\Delta = t - b$. Then the *Chernoff-Hoeffding inequality* states for any parameter $\varepsilon > 0$,

$$Pr[|\bar{X} - E[\bar{X}]| > \varepsilon] \leq 2 \exp\left(\frac{-2\varepsilon^2 n}{\Delta^2}\right).$$

Again this is a PAC bound, now with $\delta = 2 \exp(-2\varepsilon^2 n/\Delta^2)$. For a desired error tolerance ε and failure probability δ, we can set $n = (\Delta^2/(2\varepsilon^2)) \ln(2/\delta)$. Note that this has a similar relationship with ε as the Chebyshev bound, but the dependence of n on δ is exponentially less for this bound.

Example: Chernoff-Hoeffding and Dice

Consider rolling a fair die 120 times and recording how many times a 3 is returned. Let T be the random variable for the total number of 3s rolled. Each roll is a 3 with probability $1/6$, so the expected number of 3s is $\mathbf{E}[T] = 20$. We would like to answer what the probability is that more than 40 rolls return a 3.

To do so, we analyze $n = 120$ iid random variables $T_1, T_2, \ldots, T_n$ associated with each roll. In particular, $T_i = 1$ if the ith roll is a 3 and is 0 otherwise. Thus $\mathbf{E}[T_i] = 1/6$ for each roll. Using $\bar{T} = T/n = \frac{1}{n} \sum_{i=1}^{n} T_i$ and noting that $\mathbf{Pr}[T \geq 40] = \mathbf{Pr}[\bar{T} \geq 1/3]$, we can now apply our Chernoff-Hoeffding bound as

$$\mathbf{Pr}[T \geq 40] \leq \mathbf{Pr}\left[\left|\bar{T} - \mathbf{E}[T_i]\right| \geq \frac{1}{6}\right] \leq 2\exp\left(\frac{-2(1/6)^2 \cdot 120}{1^2}\right)$$
$$= 2\exp(-20/3) \leq 0.0026.$$

So we can say that less than 3 out of 1000 times of running these 120 rolls, we should see at least 40 returned 3s.

In comparison, we could have also applied a Chebyshev inequality. The variance of a single random variable T_i is $\mathbf{Var}[T_i] \leq 5/36$, and hence $\mathbf{Var}[T] = n \cdot \mathbf{Var}[T_i] = 50/3$. Hence we can bound

$$\mathbf{Pr}[T \geq 40] \leq \mathbf{Pr}[|T - 20| \geq 20] \leq (50/3)/20^2 \leq 0.042.$$

That is, using the Chebyshev inequality, we were only able to claim that this event should occur at most 42 times out of 1000 trials.

Finally, we note that in both of these analyses, we only seek to bound the probability that the number of rolls of 3 exceeds some threshold (≥ 40), whereas the inequality we used bounded the absolute value of the deviation from the mean. That is, our goal was one-way, and the inequality was a stronger two-way bound. Indeed these results can be improved by roughly a factor of 2 by using similar one-way inequalities that we do not formally state here.

Relating this all back to the Gaussian distribution in the CLT, the Chebyshev bound only uses the variance information about the Gaussian, but the Chernoff-Hoeffding bound uses all of the "moments": this allows the probability of failure to decay exponentially.

These are the most basic and common PAC concentration of measure bounds, but are by no means exhaustive.

Example: Concentration on Samples from the Uniform Distribution

Consider a random variable $X \sim f$ where $f(x) = \{\frac{1}{2}$ if $x \in [0, 2]$ and 0 otherwise.$\}$, i.e., the uniform distribution on $[0, 2]$. We know $\mathbf{E}[X] = 1$ and $\mathbf{Var}[X] = \frac{1}{3}$.

- Using the Markov inequality, we can say that $\mathbf{Pr}[X > 1.5] \leq 1/(1.5) \approx 0.6666$
  and $\mathbf{Pr}[X > 3] \leq 1/3 \approx 0.33333$
  or $\mathbf{Pr}[X - \mu > 0.5] \leq \frac{2}{3}$ and $\mathbf{Pr}[X - \mu > 2] \leq \frac{1}{3}$.
- Using the Chebyshev inequality, we can say that $\mathbf{Pr}[|X - \mu| > 0.5] \leq (1/3)/0.5^2 = \frac{4}{3}$ (which is meaningless). But $\mathbf{Pr}[|X - \mu| > 2] \leq (1/3)/(2^2) = \frac{1}{12} \approx 0.08333$.

Now consider a set of $n = 100$ random variables $X_1, X_2, \ldots, X_n$ all drawn iid from the same pdf f as above. Now we can examine the random variable $\bar{X} = \frac{1}{n} \sum_{i=1}^{n} X_i$. We know that $\mu_n = \mathbf{E}[\bar{X}] = \mu$ and that $\sigma_n^2 = \mathbf{Var}[\bar{X}] = \sigma^2/n = 1/(3n) = 1/300$.

- Using the Chebyshev inequality, we can say that $\mathbf{Pr}[|\bar{X} - \mu| > 0.5] \leq \sigma_n^2/(0.5)^2 = \frac{1}{75} \approx 0.01333$ and $\mathbf{Pr}[|\bar{X} - \mu| > 2] \leq \sigma_n^2/2^2 = \frac{1}{1200} \approx 0.0008333$.
- Using the Chernoff-Hoeffding bound, we can say that $\mathbf{Pr}[|\bar{X} - \mu| > 0.5] \leq 2 \exp(-2(0.5)^2 n/\Delta^2) = 2 \exp(-100/8) \approx 0.0000074533$ and $\mathbf{Pr}[|\bar{X} - \mu| > 2] \leq 2 \exp(-2(2)^2 n/\Delta^2) = 2 \exp(-200) \approx 2.76 \cdot 10^{-87}$.

2.3.4 Union Bound and Examples

The *union bound* is the "Robin" to Chernoff-Hoeffding's "Batman." It is a simple helper bound that allows Chernoff-Hoeffding to be applied to much larger and more complex situations[1]. It may appear that is a crude and overly simplistic bound, but it can usually only be significantly improved if a fairly special structure is available for the specific class of problems considered.

Union Bound

Consider s possibly *dependent* random events $Z_1, \ldots, Z_s$. The probability that all events occur is at least

$$1 - \sum_{j=1}^{s}(1 - \mathbf{Pr}[Z_j]).$$

That is, all events are true if no event is not true.

Example: Union Bound and Dice

Returning to the example of rolling a fair die $n = 120$ times, and bounding the probability that a 3 was returned more than 40 times, let's now consider the probability that *no number* was returned more than 40 times. Each number corresponds with a

---

[1] I suppose these situations are the villains in this analogy, like "Riddler," "Joker," and "Poison Ivy." The union bound can also aid other concentration inequalities like Chebyshev, which I suppose is like "Catwoman."

random event $Z_1, Z_2, Z_3, Z_4, Z_5, and Z_6$, of that number occurring at most 40 times. These events are not independent, but nonetheless we can apply the union bound.

Using our Chebyshev inequality result that $\Pr[Z_3] \geq 1 - 0.042 = 0.958$, we can apply this symmetrically to all Z_j. Then by the union bound, we have the probability that all numbers occur less than 40 times on $n = 120$ independent rolls is at least

$$1 - \sum_{j=1}^{6}(1 - 0.958) = 0.748$$

Alternatively, we can use the result from the Chernoff-Hoeffding bound that $\Pr[Z_j] \geq 1 - 0.0026 = 0.9974$ inside the union bound to obtain that all numbers occurs no more than 40 times with probability at least

$$1 - \sum_{j=1}^{6}(1 - 0.9974) = 0.9844$$

So this joint event (of all numbers occurring at most 40 times) occurs more than 98% of the time, but using Chebyshev, we were unable to claim that it happened more than 75% of the time.

Quantile Estimation

An important use case of concentration inequalities and the union bound is to estimate distributions. For random variable X, let f_X describe its pdf and F_X its cdf. Suppose now we can draw iid samples $P = \{p_1, p_2, \ldots, p_n\}$ from f_X, then we can use these n data points to estimate the cdf F_X. To understand this approximation, recall that $F_X(t)$ is the probability that random variable X takes a value at most t. For any choice of t, we can estimate this from P as $\mathrm{nrank}_P(t) = |\{p_i \in P \mid p_i \leq t\}|/n$, i.e., as the fraction of samples with value at most t. The quantity $\mathrm{nrank}_P(t)$ is the *normalized rank* of P at value t, and the value of t for which $\mathrm{nrank}_P(t) \leq \phi < \mathrm{nrank}_P(t + \eta)$, for any $\eta > 0$, is known as the *ϕ-quantile* of P. For instance, when $\mathrm{nrank}_P(t) = 0.5$, it is the 0.5-quantile and thus the *median* of the data set. The *interquartile range* is the interval $[t_1, t_2]$ such that t_1 is the 0.25-quantile and t_2 is the 0.75-quantile. We can similarly define the ϕ-quantile (and hence the median and interquartile range) for a distribution f_X as the value t such that $F_X(t) = \phi$.

Example: CDF and Normalized Rank

The following illustration shows a cdf F_X (in blue) and its approximation via normalized rank nrank_P on a sampled point set P (in green). The median of the P and its interquartile range are marked (in red).

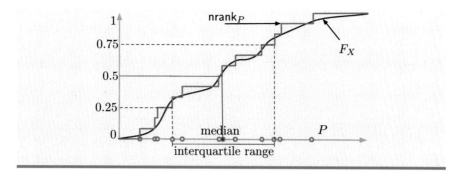

For a given value t, we can quantify how well $\text{nrank}_P(t)$ approximates $F_X(t)$ using a Chernoff-Hoeffding bound. For a given t, for each sample p_i, describe a random variable Y_i which is 1 if $p_i \leq t$ and 0 otherwise. Observe that $\mathbf{E}[Y_i] = F_X(t)$, since it is precisely the probability that a random variable $X_i \sim f_X$ (representing data point p_i, not yet realized) is less than t. Moreover, the random variable for $\text{nrank}_P(t)$ on a future iid sample P is precisely $\bar{Y} = \frac{1}{n} \sum_{i=1}^{n} Y_i$. Hence, we can provide a PAC bound on the probability (δ) of achieving more than ε error for any given t, as

$$\mathbf{Pr}[|\bar{Y} - F_X(t)| \geq \varepsilon] \leq 2 \exp\left(\frac{-2\varepsilon^2 n}{1^2}\right) = \delta.$$

If we have a desired error ε (e.g., $\varepsilon = 0.05$) and probability of failure δ (e.g., $\delta = 0.01$), we can solve for how many samples are required: these values are satisfied with

$$n \geq \frac{1}{2\varepsilon^2} \ln \frac{2}{\delta}.$$

Approximating *All* Quantiles

However, the above analysis only works for a single value of t. What if we wanted to show a similar analysis simultaneously for *all* values of t, that is, with how many samples n can we then ensure that with probability at least $1 - \delta$, for all values of t we will have $|\text{nrank}_P(t) - F_X(t)| \leq \varepsilon$?

We will apply the union bound, but, there is another challenge we face: there are an infinite number of possible values t for which we want this bound to hold! We address this by splitting the error component into two pieces ε_1 and ε_2 so $\varepsilon = \varepsilon_1 + \varepsilon_2$; we can set $\varepsilon_1 = \varepsilon_2 = \varepsilon/2$. Now we consider $1/\varepsilon_1$ different quantiles $\{\phi_1, \phi_2, \ldots, \phi_{1/\varepsilon_1}\}$ where $\phi_j = j \cdot \varepsilon_1 + \varepsilon_1/2$. This divides up the probability space (i.e., the interval $[0, 1]$) into $1/\varepsilon_1 + 1$ parts, so the gap between the boundaries of the parts is ε_1.

We will guarantee that each of these ϕ_j-quantiles are ε_2-approximated. Each ϕ_j corresponds with a t_j so t_j is the ϕ_j-quantile of f_X. We do not need to know what the precise values of t_j are; however, we do know that $t_j \leq t_{j+1}$, so they grow monotonically. In particular, this implies that for any t, $t_j \leq t \leq t_{j+1}$, then it must be that $F_X(t_j) \leq F_X(t) \leq F_X(t_{j+1})$; hence if both $F_X(t_j)$ and $F_X(t_{j+1})$ are within ε_2 of their estimated value, then $F_X(t)$ must be within $\varepsilon_2 + \varepsilon_1 = \varepsilon$ of its estimated value.

Example: Set of ϕ_j-quantiles to approximate

The illustration shows the set $\{\phi_1, \phi_2, \ldots, \phi_8\}$ of quantile points overlaid on the cdf F_X. With $\varepsilon_2 = 1/8$, these occur at values $\phi_1 = 1/16$, $\phi_2 = 3/16$, $\phi_3 = 5/16$, ..., and evenly divide the y-axis. The corresponding values $\{t_1, t_2, \ldots, t_8\}$ non-uniformly divide the x-axis. But as long as any consecutive pair t_j and t_{j+1} is approximated, because the cdf F_X and nrank are monotonic, then all intermediate values $t \in (t_j, t_{j+1}]$ are also approximated.

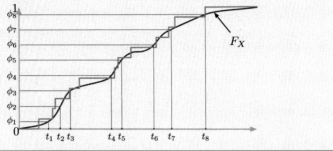

So what remains is to show that for all $t \in T = \{t_1, \ldots, t_{1/\varepsilon_1}\}$, a random variable $\bar{Y}(t_j)$ for $\text{nrank}_P(t_j)$ satisfies $\mathbf{Pr}[|\bar{Y}(t_j) - F_X(t_j)| \leq \varepsilon_2] \leq \delta$. By the above Chernoff-Hoeffding bound, this holds with probability $1 - \delta' = 1 - 2\exp(-2(\varepsilon_2)^2 n)$ for each t_j. Applying the union bound over these $s = 1/\varepsilon_1$ events, we find that they all hold with probability

$$1 - \sum_{j=1}^{s} 2\exp(-2(\varepsilon_2)^2 n) = 1 - \frac{1}{\varepsilon_1} 2\exp(-2(\varepsilon_2)^2 n).$$

Setting $\varepsilon_1 = \varepsilon_2 = \varepsilon/2$, a sample of size n will provide a nrank_P function so for any t we have $|\text{nrank}_P(t) - F_X(t)| \leq \varepsilon$, with probability at least $1 - \frac{4}{\varepsilon}\exp(-\frac{1}{2}\varepsilon^2 n)$. Setting the probability of failure $\frac{4}{\varepsilon}\exp(-\varepsilon^2 n/2) = \delta$, we can solve for n to see that we get at most ε error with probability at least $1 - \delta$ using $n = \frac{2}{\varepsilon^2}\ln(\frac{4}{\varepsilon\delta})$ samples. Using VC dimension-based arguments (see Section 9.4), we can reduce the number of samples required from proportional to $\frac{1}{\varepsilon^2}\log\frac{1}{\varepsilon}$ to $\frac{1}{\varepsilon^2}$.

2.4 Importance Sampling

Many important convergence bounds deal with approximating the mean of a distribution. When the samples are all uniformly drawn from an unknown distribution, the above bounds (and their generalizations) are the best way to understand convergence, up to small factors. In particular, this is true when the only access to the data is a new iid sample.

However, when more control can be made over how the sample is generated, then in some cases simple changes can dramatically improve the accuracy of the estimates. The key idea is called *importance sampling* and has the following principle: *sample larger weight data points more frequently, but in the estimate weigh them inverse to the sampling probability.*

Sample Average of Weights

Consider a discrete and very large set $A = \{a_1, a_2, \ldots, a_n\}$ where each element a_i has an associated weight $w(a_i)$. Our goal will be to estimate the expected (or average) weight

$$\bar{w} = \mathbf{E}[w(a_i)] = \frac{1}{n} \sum_{a_i \in A} w(a_i).$$

This set may be so large that we do not want to explicitly compute the sum, or perhaps soliciting the weight is expensive (e.g., like conducting a customer survey). So we want to avoid doing this calculation over all items. Rather, we sample k items iid $\{a'_1, a'_2, \ldots, a'_k\}$ (each a'_j uniformly and independently chosen from A, so some may be taken more than once), solicit the weight $w(a'_j)$ of each a'_j, and estimate the average weight as

$$\hat{w} = \frac{1}{k} \sum_{j=1}^{k} w(a'_j).$$

How accurately does $\hat{w}$ approximate $\bar{w}$? If all of the weights are roughly uniform or well-distributed in $[0, 1]$, then we can apply a Chernoff-Hoeffding bound so

$$\mathbf{Pr}[|\bar{w} - \hat{w}| \geq \varepsilon] \leq 2 \exp(-2\varepsilon^2 k).$$

So with probability at least $1 - \delta$, we have no more than ε errors using $k = \frac{1}{2\varepsilon^2} \ln \frac{2}{\delta}$ samples.

However, if we do not have a bound on the weights relative to the estimate, then this provides a poor approximation. For instance, if $\bar{w} = 0.0001$ since most of the weights are very small, then an $(\varepsilon = 0.01)$-approximation may not be very useful, or similarly, if most weights are near 1, and so $\bar{w} = 2$, but there are a few outlier weights with value $\Delta = 1,000$, then the Chernoff-Hoeffding bound only states

$$\mathbf{Pr}[|\bar{w} - \hat{w}| \geq \varepsilon] \leq 2 \exp\left(\frac{-2\varepsilon^2 k}{\Delta^2}\right) = \delta.$$

So this implies that instead we require $k = \frac{\Delta^2}{2\varepsilon^2} \ln(2/\delta)$, which is a factor $\Delta^2 = 1,000,000$ more samples than before!

Importance Sampling

We slightly recast the problem assuming a bit more information. There is a large set of items $A = \{a_1, a_2, \ldots, a_n\}$, and on sampling an item a'_j, its weight $w(a'_j)$ is revealed. Our goal is to estimate $\bar{w} = \frac{1}{n} \sum_{a_i \in A} w(a_i)$. We can treat W_i as a random variable for each sampled weight $w(a'_j)$, then $\bar{w} = \frac{1}{n} \sum_{i=1}^{n} W_i$ is also a random variable. In this setting, we also know for each a_i (before sampling) some information about the range of its weight. That is, we have a bounded range $[0, \psi_i]^2$, so $0 \leq w(a_i) \leq \psi_i$. This upper bound serves as an *importance* ψ_i for each a_i. Let $\Psi = \sum_{i=1}^{n} \psi_i$ be the sum of all importances.

As alluded to above, the solution is the following two-step procedure called *importance sampling*.

Algorithm 2.4.1 Importance Sampling

Sample k items $\{a'_1, a'_2, \ldots, a'_k\}$ independently from A, proportional to their importance ψ_i.

Estimate $w_I = \frac{1}{k} \sum_{j=1}^{k} \left(\frac{\Psi}{n\psi_j} \cdot w(a'_j) \right)$, where $\Psi = \sum_{i=1}^{n} \psi_i$.

We will first show that importance sampling provides an unbiased estimate, that is, $\mathbf{E}[w_I] = \bar{w}$. Define a random variable Z_j to be the value $\frac{\Psi}{n\psi_j} w(a'_j)$. By linearity of expectation and the independence of the samples, $\mathbf{E}[w_I] = \frac{1}{k} \sum_{j=1}^{k} \mathbf{E}[Z_j] = \mathbf{E}[Z_j]$. Sampling proportional to ψ_i means object a_i is chosen with probability ψ_i / Ψ. Summing over all elements, we get

$$\mathbf{E}[w_I] = \mathbf{E}[Z_j] = \sum_{i=1}^{n} \mathbf{Pr}[a'_j = a_i] \cdot \frac{\Psi}{n\psi_i} w(a_i) = \sum_{i=1}^{n} \frac{\psi_i}{\Psi} \cdot \frac{\Psi}{n\psi_i} w(a_i)$$

$$= \frac{1}{n} \sum_{i=1}^{n} w(a_i) = \bar{w}.$$

Note that this worked for any choice of ψ_i. Indeed, uniform sampling (which implicitly has $\psi_i = 1$ for each a_i) also is an unbiased estimator. The real power of importance sampling is that it improves the concentration of the estimates.

Improved Concentration

To improve the concentration, the critical step is to analyze the range of each estimator $\frac{\Psi}{n\psi_j} \cdot w(a'_j)$. Since we have that $w(a'_j) \in [0, \psi_j]$, then as a result

---

[2] We can more generally allow $w(a_i)$ to have any bounded range $[L_i, U_i]$. In this case, we set $\psi_i = U_i - L_i$, add $\frac{1}{n} \sum_{i=1}^{n} L_i$ to the final estimate, and if a_i is the jth sample let $w(a'_j) = w(a_i) - L_i$.

$$\frac{\Psi}{n\psi_j} \cdot w(a_j') \in \frac{\Psi}{n\psi_j} \cdot [0, \psi_j] = [0, \frac{\Psi}{n}].$$

Now applying a Chernoff-Hoeffding bound, we can upper-bound the probability that w_I has more than ε errors with respect to $\bar{w}$:

$$\mathbf{Pr}[|\bar{w} - \hat{w}| \geq \varepsilon] \leq 2 \exp\left(\frac{-2\varepsilon^2 k}{(\Psi/n)^2}\right) = \delta$$

Fixing the allowed error ε and probability of failure δ, we can solve for the number of samples required as

$$k = \frac{(\Psi/n)^2}{2\varepsilon^2} \ln(2/\delta).$$

Now instead of depending quadratically on the *largest* possible value Δ as in the uniform sampling, this now depends quadratically on the *average* upper bound on all values Ψ/n. In other words, with importance sampling, we reduce the sample complexity from depending on the maximum importance $\max_i \psi_i$ to on the *average importance* Ψ/n.

Example: Company Salary Estimation

Consider a company with $n = 10,000$ employees and we want to estimate the average salary. However the salaries are very imbalanced; the CEO makes way more than the typical employee does. Say we know the CEO makes at most 2 million a year, but the other 9,999 employees make at most 50 thousand a year.

Using just uniform sampling of $k = 100$ employees, we can apply a Chernoff-Hoeffding bound to estimate the average salary $\hat{w}$ from the true average salary $\bar{w}$ with error more than \$8,000 with probability

$$\mathbf{Pr}[|\hat{w} - \bar{w}| \geq 8,000] \leq 2 \exp\left(\frac{-2(8,000)^2 \cdot 100}{(2 \text{ million})^2}\right) = 2 \exp\left(\frac{-2}{625}\right) \approx 1.99$$

This is a useless bound, since the probability is greater than 1. If we increase the error tolerance to half a million, we still only get a good estimate with probability 0.42. The problem hinges on if we sample the CEO, and our estimate is too high; if we do not, then the estimate is too low.

Now using importance sampling, the CEO gets an importance of 2 million, and the other employees all get an importance of 50 thousand. The average importance is now $\Psi/n = 50,195$, and we can that bound the probability that the new estimate w_I is more than \$8,000 from $\bar{w}$ is at most

$$\mathbf{Pr}[|w_I - \bar{w}| \geq 8,000] \leq 2 \exp\left(\frac{-2(8,000)^2 \cdot 100}{(51,950)^2}\right) \leq 0.017.$$

So now for 98.3% of the time, we get an estimate within \$8,000. In fact, we get an estimate within \$4,000 for at least 38% of the time. Basically, this works because we expect to sample the CEO about twice, but then weigh that contribution slightly higher. On the other hand, when we sample a different employee, we increase the effect of their salary by about 4%.

Implementation

It is standard for most programming languages to have built-in functions to generate random numbers in a uniform distribution $u \sim \text{unif}(0, 1]$. This can easily be transformed to select an element from a large discrete set. If there are k elements in the set, then $i = \lceil uk \rceil$ (multiply by k and take the ceiling[3]) is a uniform selection of an integer from 1 to k, and represents selecting the ith element from the set.

This idea can easily be generalized to selecting an element proportional to its weight. Let $W = n\bar{w} = \sum_{i=1}^{n} w(a_i)$. Our goal is to sample element a_i with probability $w(a_i)/W$. We can also define a probability $t_j = \sum_{i=1}^{j} w(a_i)/W$, the probability that an object of index j or less should be selected. Once we calculate t_j for each a_j, and set $t_0 = 0$, a uniform random value $u \sim \text{unif}(0, 1)$ is all that is needed to select an object proportional to its weight. We just return the item a_j such that $t_{j-1} \leq u \leq t_j$ (the correct index j can be found in time proportional to $\log n$ if these t_j are stored in a balanced binary tree).

Geometry of Partition of Unity

In this illustration, 6 elements with normalized weights $w(a_i)/W$ are depicted in a bar chart on the left. These bars are then stacked end-to-end in a unit interval on the right, precisely stretched from 0.00 to 1.00. The t_i values mark the accumulation of probability that one of the first i values is chosen. Now when a random value $u \sim \text{unif}(0, 1]$ is chosen at random, it maps into this *"partition of unity"* and selects an item. In this case, it selects item a_4 since $u = 0.68$ and $t_3 = 0.58$ and $t_4 = 0.74$ for $t_3 < u \leq t_4$.

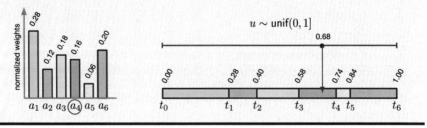

---

[3] The ceiling operation $\lceil x \rceil$ returns the smallest integer larger than x. For instance, $\lceil 7.342 \rceil = 8$.

2.4.1 Sampling Without Replacement with Priority Sampling

Many examples discussed in this book analyze data assumed to be k elements drawn iid from a distribution. However, when algorithmically generating samples, it can be advantageous to sample k elements without replacement from a known distribution. While this can make the analysis slightly more complicated, variants of Chernoff-Hoeffding bound exist for without-replacement random samples instead of independent random variables. Moreover, in practice the concentration and approximation quality is often improved using without-replacement samples and is especially true when drawing weighted samples.

When sampling data proportional to weights, if elements exist with sufficiently large weights, then it is best to always sample these high-weight elements. The low-weight ones need to be selected with some probability, and this should be proportional to their weights, and then re-weighted as in importance sampling. A technique called *priority sampling* elegantly combines these properties.

Algorithm 2.4.2 Priority Sampling

---

for item $a_i \in A$ **do**
  Generate $u_i \sim \text{unif}(0, 1]$.
  Set priority $\rho_i = w(a_i)/u_i$, using weight $w(a_i)$.
Let $\tau = \rho'_{k+1}$, the $(k + 1)$th largest priority.
Assign new weights $w'(a_i) = \begin{cases} \max(w(a_i), \tau) & \text{if } \rho_i > \tau \\ 0 & \text{if } \rho_i \leq \tau \end{cases}$.

The new weight function $w' : A \to [0, \infty)$ has only k items with non-zero values; only those need to be retained in the sample A'. This has many nice properties, most importantly $\mathbf{E}[w'(a_i)] = w(a_i)$. Thus for any subset $S \subset A$, we can estimate the sum of weights in that subset $\sum_{a_i \in S} w(a_i)$ using only $\sum_{a_i \in S \cap A'} w'(a_i)$, and this has the correct expected value. Thus for $w_P = \frac{1}{n} \sum_{i=1}^{n} w'(a_i)$ as an analog to importance sampling, we also have $\mathbf{E}[w_P] = \bar{w}$.

Additionally, the elements with very large weights (those with weight above τ) are always retained. This is because $\rho_i \geq w(a_i)$ for all i (since $1/u_i \geq 1$), so if $w(a_i) > \tau$ then $\rho_i > \tau$ and it is always chosen, and its new weight is $w'(a_i) = \max(w(a_i), \tau) = w(a_i)$ which is the same as before. Hence, for a fixed τ this item has no variance in its effect on the estimate. The remaining items have weights assigned as if in importance sampling, and so the overall estimate has small (and indeed near-optimal) variance.

Example: Priority Sampling

In this example, 10 items are shown with weights from $w(a_{10}) = 0.08$ to $w(a_1) = 1.80$. For a clearer picture, they are sorted in decreasing order. Each is then given a priority by dividing the weight by a different $u_i \sim \text{unif}(0, 1]$ for each element. To sample $k = 4$ items, the 5th largest priority value $\rho_4 = \tau = 1.10$ (belonging to a_4) is marked by a horizontal dashed line. Then all elements with priorities above τ are given non-zero weights. The largest weight element a_1 retains its original weight $w(a_1) = w'(a_1) = 1.80$ because it is above τ. The other retained elements have weight below τ and so are given new weights $w'(a_2) = w'(a_4) = w'(a_5) = \tau = 1.10$. The other elements are implicitly given new weights of 0.

Notice that $W' = \sum_{i=1}^{10} w'(a_i) = 5.10$ is very close to $W = \sum_{i=1}^{10} w(a_i) = 5.09$.

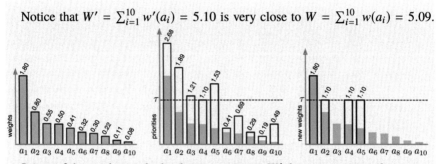

It is useful to understand why the new estimate W' does not necessarily increase if more elements are retained. In this case if $k = 5$ elements are retained instead of $k = 4$, then τ would become $\rho_7 = 0.69$, the 6th largest priority. So then the new weights for several of the elements would decrease from 1.10 to 0.69.

Exercises

2.1 Consider a pdf f so that a random variable $X \sim f$ has expected value $E[X] = 3$ and variance $Var[X] = 10$. Now consider $n = 10$ iid random variables $X_1, X_2, \ldots, X_{10}$ drawn from f. Let $\bar{X} = \frac{1}{10} \sum_{i=1}^{10} X_i$.

1. What is $E[\bar{X}]$?
2. What is $Var[\bar{X}]$?
3. What is the standard deviation of $\bar{X}$?
4. Which is larger, $Pr[X > 4]$ or $Pr[\bar{X} > 4]$?
5. Which is larger, $Pr[X > 2]$ or $Pr[\bar{X} > 2]$?

2.2 Let X be a random variable that you know is in the range $[-1, 2]$ and you know has an expected value of $E[X] = 0$. Use the Markov inequality to upper-bound $Pr[X > 1.5]$.
(Hint: you will need to use a change of variables.)

2.3 For a random variable X, let $Y = Var[X] = E[(X - E[X])^2]$ be another random variable representing its variance. Apply the Markov inequality to Y, and some algebra, to derive the Chebyshev inequality for X.

2.4 Consider a pdf f so that a random variable $X \sim f$ has expected value $E[X] = 5$ and variance $Var[X] = 100$. Now consider $n = 16$ iid random variables $X_1, X_2, \ldots, X_{16}$ drawn from f. Let $\bar{X} = \frac{1}{16} \sum_{i=1}^{16} X_i$.

1. What is $E[\bar{X}]$?
2. What is $Var[\bar{X}]$?
   Assume we know that X is never smaller than 0 and never larger than 20.
3. Use the Markov inequality to upper-bound $Pr[\bar{X} > 8]$.
4. Use the Chebyshev inequality to upper-bound $Pr[\bar{X} > 8]$.
5. Use the Chernoff-Hoeffding inequality to upper-bound $Pr[\bar{X} > 8]$.
6. If we increase n to 100, how will the above three bounds be affected.

2.5 Consider a (parked) self-driving car that returns n iid estimates to the distance of a tree. We will model these n estimates as a set of n scalar random variables $X_1, X_2, \ldots, X_n$ taken iid from an unknown pdf f, which we assume models the true distance plus unbiased noise (the sensor can take many iid estimates in rapid fire fashion.). The sensor is programmed to only return values between 0 and 20 feet, and that the variance of the sensing noise is 64 feet squared. Let $\bar{X} = \frac{1}{n} \sum_{i=1}^{n} X_i$. We want to understand as a function of n how close $\bar{X}$ is to μ, which is the true distance to the tree.

1. Use Chebyshev's inequality to determine a value n so that $Pr[|\bar{X} - \mu| \geq 1] \leq 0.5$.
2. Use Chebyshev's inequality to determine a value n so that $Pr[|\bar{X} - \mu| \geq 0.1] \leq 0.1$.
3. Use the Chernoff-Hoeffding bound to determine a value n so that $Pr[|\bar{X} - \mu| \geq 1] \leq 0.5$.
4. Use the Chernoff-Hoeffding bound to determine a value n so that $Pr[|\bar{X} - \mu| \geq 0.1] \leq 0.1$.

2.6 Consider two random variables C and T describing how many coffees and teas I will buy in the coming week; clearly neither can be smaller than 0. Based on personal experience, I know the following summary statistics about my coffee and tea buying habits: $\mathbf{E}[C] = 3$ and $\mathbf{Var}[C] = 1$ also $\mathbf{E}[T] = 2$ and $\mathbf{Var}[T] = 5$.

1. Use Markov's inequality to upper-bound the probability that I buy 4 or more coffees, and the same for teas: $\mathbf{Pr}[C \geq 4]$ and $\mathbf{Pr}[T \geq 4]$.
2. Use Chebyshev's inequality to upper-bound the probability that I buy 4 or more coffees, and the same for teas: $\mathbf{Pr}[C \geq 4]$ and $\mathbf{Pr}[T \geq 4]$.

2.7 The average score on a test is 82 with a standard deviation of 4 percentage points. All tests have scores between 0 and 100.

1. Using Chebyshev's inequality, what percentage of the tests have a grade of at least 70 and at most 94?
2. Using Markov's inequality, what is the highest percentage of tests which could have a score less than 60?

2.8 Consider a random variable X with expected values $\mathbf{E}[X] = 7$ and variance $\mathbf{Var}[X] = 2$. We would like to upper-bound the probability $\mathbf{Pr}[X < 5]$.

1. Which bound can and cannot be used with what we know about X (Markov, Chebyshev, or Chernoff-Hoeffding), and why?
2. Using that bound, calculate an upper bound for $\mathbf{Pr}[X < 5]$.
3. Describe a probability distribution for X where the other two bounds are definitely not applicable.

2.9 Consider n iid random variables $X_1, X_2, \ldots, X_n$ with expected value $\mathbf{E}[X_i] = 20$ and variance $\mathbf{Var}[X_i] = 2$. Assume we also know that each X_i must satisfy $15 \leq X_i \leq 22$. We now want to analyze the random variable of their average $\bar{X} = \frac{1}{n} \sum_{i=1}^{n} X_i$.
Assume first that $n = 20$ (the number of random variables).
1. Use the Chebyshev inequality to upper-bound $\mathbf{Pr}[\bar{X} > 21]$.
2. Use the Chernoff-Hoeffding inequality to upper-bound $\mathbf{Pr}[\bar{X} > 21]$.
   Now assume first that $n = 200$ (the number of random variables).
3. Use the Chebyshev inequality to upper-bound $\mathbf{Pr}[\bar{X} > 21]$.
4. Use the Chernoff-Hoeffding inequality to upper-bound $\mathbf{Pr}[\bar{X} > 21]$.

2.10 Consider 5 items a_1, a_2, a_3, a_4, a_5 with weights $w_1 = 2$, $w_2 = 3$, $w_3 = 10$, $w_4 = 8$, and $w_5 = 1$. But before sampling, we only know upper bounds on the weights with $w_1 \in [0, 4]$, $w_2 \in [0, 4]$, $w_3 \in [0, 10]$, $w_4 \in [0, 10]$, and $w_5 \in [0, 2]$.

1. In the context of importance sampling, list the importance ψ_i for each item.
2. Use the Partition of Unity approach to sample 2 items according to their importance. Use $u_1 = 0.377$ and $u_2 = 0.852$ as the "random" values in unif$(0, 1]$.
3. Report the estimate of the average weight using importance sampling, based on the two items sampled.

Chapter 3
Linear Algebra Review

Abstract We briefly review many key definitions and aspects of linear algebra that will be necessary for the remainder of the book. This review includes basic operations for vectors and matrices. A highlight is the dot product and its intuitive geometric properties.

3.1 Vectors and Matrices

For the context of data analysis, the critical part of linear algebra deals with vectors and matrices of real numbers.

In this context, a *vector* $v = (v_1, v_2, \ldots, v_d)$ is equivalent to a point in $\mathbb{R}^d$. By default, a vector will be a column of d numbers (where d is context specific)

$$v = \begin{bmatrix} v_1 \\ v_2 \\ \vdots \\ v_n \end{bmatrix}$$

but in some cases we will assume the vector is a row

$$v^T = [v_1 \; v_2 \; \ldots \; v_d].$$

From a data structure perspective, these representations can be stored equivalently as an array.

An $n \times d$ *matrix* A is then an ordered set of n row vectors $a_1, a_2, \ldots, a_n$

$$A = [a_1; \; a_2; \; \ldots; \; a_n] = \begin{bmatrix} - \; a_1 \; - \\ - \; a_2 \; - \\ \vdots \\ - \; a_n \; - \end{bmatrix} = \begin{bmatrix} A_{1,1} & A_{1,2} & \ldots & A_{1,d} \\ A_{2,1} & A_{2,2} & \ldots & A_{2,d} \\ \vdots & \vdots & \ddots & \vdots \\ A_{n,1} & A_{n,2} & \ldots & A_{n,d} \end{bmatrix},$$

© Springer Nature Switzerland AG 2021
J. M. Phillips, *Mathematical Foundations for Data Analysis*,
Springer Series in the Data Sciences,
https://doi.org/10.1007/978-3-030-62341-8_3

where vector $a_i = [A_{i,1}, A_{i,2}, \ldots, A_{i,d}]$, and $A_{i,j}$ is the element of the matrix in the ith row and jth column. We can write $A \in \mathbb{R}^{n \times d}$ when it is defined on the reals.

An example 4-dimensional vector v and 2×3 matrix A are

$$v = \begin{bmatrix} 1 \\ 2 \\ 7 \\ 5 \end{bmatrix} \quad \text{and} \quad A = \begin{bmatrix} 3 & -7 & 2 \\ -1 & 2 & -5 \end{bmatrix}.$$

The numpy package is the default way to represent them in Python:

```
import numpy as np
v = np.array([1,2,7,5])
A = np.array([[3,-7,2],[-1,2,-5]])
```

It is often useful to print their contents on the command line using

```
print(v)
print(A)
```

Note that a vector will always print as a row vector even if it is a column vector. The transpose option reverses the role of columns and rows, and the operation in Python can be observed using

```
print(v.T)
print(A.T)
```

However, again since Python always prints vectors as rows, for the vector v those results will display the same.

An element of a vector or matrix, or submatrix can be printed as

```
print(v[2])
print(A[1,2])
print(A[:, 1:3])
```

Note that the indexing starts at 0, so these commands result in

$$7, \quad -5, \quad \text{and} \quad \begin{bmatrix} -7 & 2 \\ 2 & -5 \end{bmatrix}.$$

Geometry of Vectors and Matrices

It will be convenient to imagine length d, vectors v as points in $\mathbb{R}^d$. And subsequently, it will be convenient to think of an $n \times d$ matrix A by each of its rows $a_1, \ldots, a_n$ as a point each in $\mathbb{R}^d$. The "vector" is then the "arrow" from the origin $\mathbf{0} = (0, 0, \ldots, 0)$ to that point.

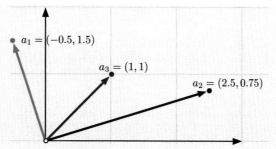

This picture with $n = 3$ points in $d = 2$ dimensions is equivalent to the 3×2 matrix representation

$$A = \begin{bmatrix} a_1 \\ a_2 \\ a_3 \end{bmatrix} = \begin{bmatrix} -0.5 & 1.5 \\ 2.5 & 0.75 \\ 1 & 1 \end{bmatrix}.$$

A *transpose* operation $(\cdot)^T$ reverses the roles of the rows and columns, as seen above with vector v. For a matrix, we can write

$$A^T = \begin{bmatrix} | & | & & | \\ a_1 & a_2 & \dots & a_n \\ | & | & & | \end{bmatrix} = \begin{bmatrix} A_{1,1} & A_{2,1} & \dots & A_{n,1} \\ A_{1,2} & A_{2,2} & \dots & A_{n,2} \\ \vdots & \vdots & \ddots & \vdots \\ A_{1,d} & A_{2,d} & \dots & A_{n,d} \end{bmatrix}.$$

Example: Linear Equations

A simple place these objects arise is in linear equations. For instance,

$$\begin{aligned} 3x_1 - 7x_2 + 2x_3 &= -2 \\ -1x_1 + 2x_2 - 5x_3 &= 6 \end{aligned}$$

is a system of $n = 2$ linear equations, each with $d = 3$ variables. We can represent this system in matrix-vector notation as

$$Ax = b$$

where

$$b = \begin{bmatrix} -2 \\ 6 \end{bmatrix} \quad x = \begin{bmatrix} x_1 \\ x_2 \\ x_3 \end{bmatrix} \quad \text{and} \quad A = \begin{bmatrix} 3 & -7 & 2 \\ -1 & 2 & -5 \end{bmatrix}.$$

3.2 Addition and Multiplication

We can add together two vectors or two matrices only if they have the same dimensions. For vectors $x = (x_1, x_2, \ldots, x_d) \in \mathbb{R}^d$ and $y = (y_1, y_2, \ldots, y_d) \in \mathbb{R}^d$, then vector

$$z = x + y = (x_1 + y_1, x_2 + y_2, \ldots, x_d + y_d) \in \mathbb{R}^d.$$

Geometry of Vector Addition

Vector addition can be geometrically realized as just chaining two vectors together. It is easy to see that this operation is commutative. That is $x + y = y + x$, since it does not matter which order we chain the vectors, both result in the same summed-to point.

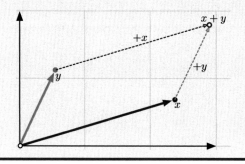

Similarly, for two matrices $A, B \in \mathbb{R}^{n \times d}$, then $C = A + B$ is defined where $C_{i,j} = A_{i,j} + B_{i,j}$ for all i, j.

Example: Adding Vectors

We can add two matrices

$$A = \begin{bmatrix} 3 & -7 & 2 \\ -1 & 2 & -5 \end{bmatrix} \quad \text{and} \quad B = \begin{bmatrix} 2 & 5 & 3 \\ 4 & 8 & 1 \end{bmatrix}$$

in Python as

```
B = np.array([[2,5,3],[4,8,1]])
A+B
```

and the result is

$$A + B = \begin{bmatrix} 5 & -2 & 5 \\ 3 & 10 & -4 \end{bmatrix}.$$

Matrix-Matrix Products

Multiplication only requires alignment along one dimension. For two matrices $A \in \mathbb{R}^{n \times d}$ and $B \in \mathbb{R}^{d \times m}$ we can obtain a new matrix $C = AB \in \mathbb{R}^{n \times m}$ where $C_{i,j}$, the element in the ith row and jth column of C, is defined as

$$C_{i,j} = \sum_{k=1}^{d} A_{i,k} B_{k,j}.$$

To multiply A times B (where A is to the left of B, the order matters!), we require the row dimension d of A to match the column dimension d of B. If $n \neq m$, then we *cannot* multiply BA. Keep in mind:

- Matrix multiplication is *associative* $(AB)C = A(BC)$.
- Matrix multiplication is *distributive* $A(B + C) = AB + AC$.
- Matrix multiplication is **not** commutative $AB \neq BA$.

Example: Matrix Multiplication

Consider two matrices $G \in \mathbb{R}^{2 \times 2}$ and $B \in \mathbb{R}^{2 \times 3}$

$$G = \begin{bmatrix} 1 & 3 \\ 4 & 7 \end{bmatrix} \quad \text{and} \quad B = \begin{bmatrix} 2 & 5 & 3 \\ 4 & 8 & 1 \end{bmatrix}.$$

Their product is $GB = C \in \mathbb{R}^{2 \times 3}$ which can be taken in Python as

```
G = np.array([[1,3],[4,7]])
C = G @ B        #or
C = G.dot(B)     #or
C = np.dot(G,B)
```

resulting in

$$C = \begin{bmatrix} 14 & 29 & 6 \\ 36 & 76 & 19 \end{bmatrix},$$

where the upper right corner $C_{1,1} = G_{1,1} \cdot B_{1,1} + G_{1,2} \cdot B_{2,1} = 1 \cdot 2 + 3 \cdot 4 = 14$.

We can also multiply a matrix A by a scalar α. In this setting, $\alpha A = A \alpha$ and is defined by a new matrix B where $B_{i,j} = \alpha A_{i,j}$.

Vector-Vector Products

There are two types of vector-vector products, and their definitions follow directly from that of matrix-matrix multiplication (since a vector is a matrix where one of the dimensions is 1). But it is worth highlighting these.

Given two <u>column</u> vectors $x, y \in \mathbb{R}^{d}$, the *inner product* or *dot product* is written as

$$x^T y = x \cdot y = \langle x, y \rangle = [x_1\ x_2\ \ldots\ x_d] \begin{bmatrix} y_1 \\ y_2 \\ \vdots \\ y_d \end{bmatrix} = \sum_{i=1}^{d} x_i y_i,$$

where x_i is the ith element of x and is similar for y_i. This text will prefer the notation $\langle x, y \rangle$ since the same can be used for row vectors, and there is no confusion with scalar multiplication in using $x \cdot y$. Whether a vector is a row or a column is often arbitrary; in a computer, they are typically stored the same way in memory.

Note that this dot product operation produces a single scalar value. And it is a linear operator. So this means for any scalar value α and three vectors $x, y, z \in \mathbb{R}^d$, we have

$$\langle \alpha x, y + z \rangle = \alpha \langle x, y + z \rangle = \alpha \left(\langle x, y \rangle + \langle x, z \rangle \right).$$

This operation is associative, distributive, and commutative.

Geometry of the Dot Product

The *dot product* is my favorite mathematical operation! It encodes a lot of geometry. Consider two vectors $u = (\frac{3}{5}, \frac{4}{5})$ and $v = (2, 1)$, with an angle θ between them. Then it holds

$$\langle u, v \rangle = \text{length}(u) \cdot \text{length}(v) \cdot \cos(\theta).$$

Here length($\cdot$) measures the distance from the origin. We will see how to measure length with a "norm" $\| \cdot \|$ soon.

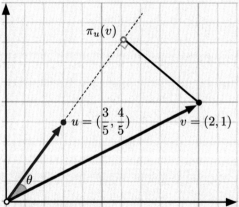

Moreover, since $\|u\| = \text{length}(u) = 1$, then we can also interpret $\langle u, v \rangle$ as the length of v projected onto the line through u. That is, let $\pi_u(v)$ be the closest point to v on the line through u (the line through u and the line segment from v to $\pi_u(v)$ make a right angle). Then

$$\langle u, v \rangle = \text{length}(\pi_u(v)) = \|\pi_u(v)\|.$$

For two column vectors $x \in \mathbb{R}^n$ and $y \in \mathbb{R}^d$, the *outer product* is written as

$$
xy^T =
\begin{bmatrix}
x_1 \\
x_2 \\
\vdots \\
x_n
\end{bmatrix}
\begin{bmatrix} y_1 & y_2 & \cdots & y_d \end{bmatrix}
=
\begin{bmatrix}
x_1 y_1 & x_1 y_2 & \cdots & x_1 y_d \\
x_2 y_1 & x_2 y_2 & \cdots & x_2 y_d \\
\vdots & \vdots & \ddots & \vdots \\
x_n y_1 & x_n y_2 & \cdots & x_n y_d
\end{bmatrix}
\in \mathbb{R}^{n \times d}.
$$

Note that the result here is a matrix, not a scalar. The dimensions are not required to match.

Matrix-Vector Products

Another important and common operation is a matrix-vector product. Given a matrix $A \in \mathbb{R}^{n \times d}$ and a vector $x \in \mathbb{R}^d$, their product $y = Ax \in \mathbb{R}^n$.

When A is composed of row vectors $[a_1; a_2; \ldots; a_n]$, then it is useful to imagine this as transposing x (which should be a column vector here, so a row vector after transposing), and taking the dot product with each row of A. I like to think of this as x^T sliding down the rows of A, and for each row a_i outputting a scalar value $\langle a_i, x \rangle$ into the corresponding output vector.

$$
y = Ax =
\begin{bmatrix}
- & a_1 & - \\
- & a_2 & - \\
 & \vdots & \\
- & a_n & -
\end{bmatrix}
x =
\begin{bmatrix}
\langle a_1, x \rangle \\
\langle a_2, x \rangle \\
\vdots \\
\langle a_n, x \rangle
\end{bmatrix}.
$$

3.3 Norms

The standard *Euclidean norm* (think "length") of a vector $v = (v_1, v_2, \ldots, v_d) \in \mathbb{R}^d$ is defined as

$$
\|v\| = \sqrt{\sum_{i=1}^{d} v_i^2} = \sqrt{v_1 v_1 + v_2 v_2 + \ldots + v_d v_d} = \sqrt{\langle v, v \rangle}.
$$

This measures the "straight-line" distance from the origin to the point at v. A vector v with norm $\|v\| = 1$ is said to be a *unit vector*; sometimes a vector x with $\|x\| = 1$ is said to be *normalized*.

However, a "norm" is a more general concept. A class called L_p *norms* is well-defined for any parameter $p \in [1, \infty)$ as

$$
\|v\|_p = \left(\sum_{i=1}^{d} |v_i|^p \right)^{1/p}.
$$

Thus, when no p is specified, it is assumed to be $p = 2$. It is also common to denote $\|v\|_\infty = \max_{i=1}^d |v_i|$, which is also a norm. Indeed this is the result of taking the limit of p to ∞.

Because subtraction is well-defined between vectors $v, u \in \mathbb{R}^d$ of the same dimension, we can also take the norm of $\|v - u\|_p$. While this is technically the norm of the vector resulting from the subtraction of u from v, it also provides a distance between u and v. Section 4.2 elaborates on the properties of these distances in far more detail.

We can also define norms for matrix A. These take on slightly different notational conventions. The two most common are the spectral norm $\|A\| = \|A\|_2$ and the Frobenius norm $\|A\|_F$. The *Frobenius norm* is the most natural extension of the $p = 2$ norm for vectors but uses a subscript F instead. It is defined for matrix $A \in \mathbb{R}^{n \times d}$ as

$$\|A\|_F = \sqrt{\sum_{i=1}^n \sum_{j=1}^d A_{i,j}^2} = \sqrt{\sum_{i=1}^n \|a_i\|^2},$$

where $A_{i,j}$ is the element in the ith row and jth column of A, and where a_i is the ith row vector of A. The *spectral norm* is defined for a matrix $A \in \mathbb{R}^{n \times d}$ as

$$\|A\| = \|A\|_2 = \max_{\substack{x \in \mathbb{R}^d \\ \|x\| \neq 0}} \|Ax\|/\|x\| = \max_{\substack{y \in \mathbb{R}^n \\ \|y\| \neq 0}} \|yA\|/\|y\|.$$

It is useful to think of these x and y vectors as being unit vectors, then the denominator can be ignored (as they are 1). Then we see that x and y only contain "directional" information, and the arg max vector (e.g., the x which maximizes $\|Ax\|/\|x\|$) points in the directions that maximize the norm.

Example: Vector and Matrix Norms

For vector $v = [-1, 2, 7, 5]$, its standard norms are

$$\|v\| = \|v\|_2 = 8.889, \qquad \|v\|_1 = 15 \qquad \text{and,} \quad \|v\|_\infty = 7.$$

In Python, these are computed as

```
from numpy import linalg as LA
LA.norm(v)
LA.norm(v,1)
LA.norm(v,np.inf)
```

For matrix $A = \begin{bmatrix} 3 & -7 & 2 \\ -1 & 2 & -5 \end{bmatrix}$, the standard norms are

$$\|A\|_2 = 10.70 \qquad \text{and} \qquad \|A\|_F = 10.95.$$

In Python, these are computed as

```
LA.norm(A,1)
LA.norm(A, 'fro')
```

3.4 Linear Independence

Consider a set of k vectors $x_1, x_2, \ldots, x_k \in \mathbb{R}^d$ and a set of k scalars $\alpha_1, \alpha_2, \ldots, \alpha_k \in \mathbb{R}$. Then a central property of linear algebra allows us to write a new vector in $\mathbb{R}^d$ as

$$z = \sum_{i=1}^{k} \alpha_i x_i.$$

For a set of vectors $X = \{x_1, x_2, \ldots, x_k\}$, we say a vector z is *linear dependent* on X if there exists a set of scalars $\alpha = \{a_1, a_2, \ldots, a_k\}$ so z can be written as the above summation. If z **cannot** be written with any choice of α_is, then we say z is *linearly independent* of X. All vectors $z \in \mathbb{R}^d$ which are linearly dependent on X are said to be in its *span*.

$$\text{span}(X) = \left\{ z \mid z = \sum_{i=1}^{k} \alpha_i x_i, \ \alpha_i \in \mathbb{R} \right\}.$$

If $\text{span}(X) = \mathbb{R}^d$ (that is, for vectors $X = x_1, x_2, \ldots, x_k \in \mathbb{R}^d$ *all* vectors are in the span), then we say X forms a *basis*.

Example: Linear Independence

Consider input vectors in a set X as

$$x_1 = \begin{bmatrix} 1 \\ 3 \\ 4 \end{bmatrix} \qquad x_2 = \begin{bmatrix} 2 \\ 4 \\ 1 \end{bmatrix}$$

And two other vectors

$$z_1 = \begin{bmatrix} -3 \\ -5 \\ 2 \end{bmatrix} \qquad z_2 = \begin{bmatrix} 3 \\ 7 \\ 1 \end{bmatrix}$$

Note that z_1 is linearly dependent on X since it can be written as $z_1 = x_1 - 2x_2$ (here $\alpha_1 = 1$ and $\alpha_2 = -2$). However z_2 is linearly independent of X since there are no scalars α_1 and α_2 so that $z_2 = \alpha_1 x_1 + \alpha_2 x_2$ (we need $\alpha_1 = \alpha_2 = 1$ so the first two coordinates align, but then the third coordinate cannot).

Also the set X is linearly independent, since there is no way to write $x_2 = \alpha_1 x_1$.

A set of vectors $X = \{x_1, x_2, \ldots, x_n\}$ is *linearly independent* if there is no way to write any vector $x_i \in X$ with scalars $\{\alpha_1, \ldots, \alpha_{i-1}, \alpha_{i+1}, \ldots, \alpha_n\}$ as the sum

$$x_i = \sum_{\substack{j=1 \\ j \neq i}}^{n} \alpha_j x_j$$

of the other vectors in the set.

Geometry of Linear Dependence

A geometric way to understand linear dependence is if there is a lower-dimensional subspace that passes through all of the points. Consider the example of the 3×2 matrix A which corresponds with 3 points in 2 dimensions.

$$A = \begin{bmatrix} a_1 \\ a_2 \\ a_3 \end{bmatrix} = \begin{bmatrix} 1.5 & 0 \\ 2 & 0.5 \\ 1 & 1 \end{bmatrix}.$$

Then these points are linearly dependent because there exists a line $\ell : y = -0.5x + 1.5$ which passes through all points.

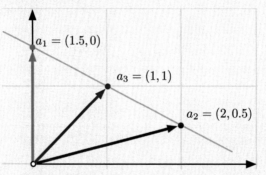

In general, a linear subspace of a data set also includes the origin $\mathbf{0} = (0, 0)$. The line (or in general the k-dimensional subspace) which passes through all data points, but not necessarily the origin (as shown here), is known as an *affine subspace*.

3.5 Rank

The *rank* of a set of vectors $X = \{x_1, \ldots, x_n\}$ is the size of the largest subset $X' \subset X$ which is linearly independent. Usually we report rank(A) as the rank of a matrix A. It is defined as the rank of the rows of the matrix, or the rank of its columns; it turns out these quantities are always the same.

If $A \in \mathbb{R}^{n \times d}$, then rank$(A) \leq \min\{n, d\}$. If rank$(A) = \min\{n, d\}$, then A is said to be *full rank*. For instance, if $d < n$, then using the rows of $A = [a_1; a_2; \ldots; a_n]$, we can describe *any* vector $z \in \mathbb{R}^d$ as the linear combination of these rows: $z = \sum_{i=1}^{n} \alpha_i a_i$ for some set $\{\alpha_1, \ldots, \alpha_n\}$. In fact, if A is full rank, we can do so and set all but d of these scalars to 0.

Example: Matrix Rank

Both of the matrices

$$A = \begin{bmatrix} 3 & -7 & 2 \\ -1 & 2 & -5 \end{bmatrix} \quad \text{and} \quad G = \begin{bmatrix} 1 & 3 \\ 4 & 7 \end{bmatrix}$$

have rank 2. Hence, both are full rank. We can compute this in Python as

```
np.matrix_rank(A)
np.matrix_rank(G)
```

3.6 Square Matrices and Properties

A matrix A is said to be *square* if it has the same number of columns as it has rows.

Inverse

A square matrix $A \in \mathbb{R}^{n \times n}$ may have an *inverse* denoted by A^{-1}. If it exists, it is a unique matrix which satisfies

$$A^{-1}A = I = AA^{-1}$$

where I is the $n \times n$ identity matrix

$$I = \begin{bmatrix} 1 & 0 & \dots & 0 & 0 \\ 0 & 1 & \dots & 0 & 0 \\ \vdots & \vdots & \ddots & \vdots & \vdots \\ 0 & 0 & \dots & 1 & 0 \\ 0 & 0 & \dots & 0 & 1 \end{bmatrix} = \mathrm{diag}(1, 1, \dots, 1).$$

Note that I serves the purpose of 1 in scalar algebra, so for any (non-zero) scalar α using $\alpha^{-1} = \frac{1}{\alpha}$, we have $\alpha\alpha^{-1} = 1 = \alpha^{-1}\alpha$.

A matrix is said to be *invertible* if it has an inverse. Only square, full-rank matrices are invertible; and a matrix is always invertible if it is square and full rank. If a matrix is not square, the inverse is not defined. If a matrix is not full rank, then it does not have an inverse.

Eigenvectors and Eigenvalues

An *eigenvector* of a square matrix M is a vector v such that there is some scalar λ that satisfies

$$Mv = \lambda v.$$

That is, multiplying M by v results in a scaled version of v. When v is normalized, $\|v\| = 1$ then the associated value λ is called the *eigenvalue*.

In general, a square matrix $M \in \mathbb{R}^{n \times n}$ may have up to n eigenvectors and eigenvalues. Some of the eigenvalues may be complex numbers (even when all of M's entries are real!).

Example: Eigenvectors and Eigenvalues

Consider the following 3×3 symmetric square matrix

$$M = \begin{bmatrix} 1 & 3 & 4 \\ 3 & 5 & 2 \\ 4 & 2 & 8 \end{bmatrix}.$$

Its *top* eigenvector is $v_1 = \begin{bmatrix} 0.43 \\ 0.44 \\ 0.78 \end{bmatrix}$ and the associated eigenvalue is $\lambda_1 = 11.36$. The top one refers to the eigenvector with the largest eigenvalue.

The other eigenvectors are $v_2 = \begin{bmatrix} -0.11 \\ -0.83 \\ 0.54 \end{bmatrix}$ and $v_3 = \begin{bmatrix} -0.90 \\ 0.31 \\ 0.31 \end{bmatrix}$ with associated eigenvectors $\lambda_2 = 4.10$ and $\lambda_3 = -0.46$.

Since λ_3 is negative, then M is not positive definite.

Positive Definite Matrices

A *positive definite matrix* $M \in \mathbb{R}^{n \times n}$ is a symmetric matrix with all real and positive eigenvalues. Another characterization is that for every vector $x \in \mathbb{R}^n$, then $x^T M x$ is positive. A *positive semidefinite matrix* $M \in \mathbb{R}^{n \times n}$ may have some eigenvalues at 0 but they are otherwise positive, and still all real; equivalently for any vector $x \in \mathbb{R}^n$, then $x^T M x$ may be zero or positive.

Determinant

A notion of matrix volume is the *determinant* of a matrix A and is denoted $|A|$. It is only defined for square matrices. It is most simply defined inductively. For a 2×2 matrix $B = \begin{bmatrix} a & b \\ c & d \end{bmatrix}$, the determinant is defined as $|B| = ad - bc$. In this case, it is the area of the parallelogram determined by vectors $b - a$ and $d - c$.

For an $n \times n$ matrix A define $n - 1 \times n - 1$ submatrix $\breve{A}_{i,j}$ as the same as A, but missing row i and column j. Now the determinant of A is

$$|A| = \sum_{i=1}^{n} (-1)^{i+1} A_{1,i} \cdot |\breve{A}_{1,i}| = A_{1,1} \cdot |\breve{A}_{1,1}| - A_{1,2} \cdot |\breve{A}_{1,2}| + A_{1,3} \cdot |\breve{A}_{1,3}| - A_{1,4} \cdot |\breve{A}_{1,4}| + \ldots$$

Example: Matrix Determinant

For a 3×3 matrix

$$M = \begin{bmatrix} a & b & c \\ d & e & f \\ g & h & i \end{bmatrix}$$

the determinant is defined as

$$|M| = a \left| \begin{bmatrix} e & f \\ h & i \end{bmatrix} \right| - b \left| \begin{bmatrix} d & f \\ g & i \end{bmatrix} \right| + c \left| \begin{bmatrix} d & e \\ g & h \end{bmatrix} \right|.$$

3.7 Orthogonality

Two vectors $x, y \in \mathbb{R}^d$ are *orthogonal* if $\langle x, y \rangle = 0$. This means those vectors are at a right angle to each other.

Example: Orthogonality

Consider two vectors $x = (2, -3, 4, -1, 6)$ and $y = (4, 5, 3, -7, -2)$. They are orthogonal since

$$\langle x, y \rangle = (2 \cdot 4) + (-3 \cdot 5) + (4 \cdot 3) + (-1 \cdot -7) + (6 \cdot -2)$$
$$= 8 - 15 + 12 + 7 - 12 = 0.$$

A set of columns which are normalized and all orthogonal to each other are said to be *orthonormal*. If $V \in \mathbb{R}^{n \times d}$ and has orthonormal columns, then $V^T V = I$ (here I is $d \times d$) but $VV^T \neq I$.

A square matrix $U \in \mathbb{R}^{n \times n}$ that has all orthonormal rows and all orthonormal columns is *orthogonal*. It follows that

$$U^T U = I = UU^T$$

since for any normalized vector u that $\langle u, u \rangle = \|u\| = 1$, and any two distinct orthonormal columns $u_i \neq u_j$ then $\langle u_i, u_j \rangle = 0$. Orthogonal matrices are norm preserving under multiplication. That means for an orthogonal matrix $U \in \mathbb{R}^{n \times n}$ and any vector $x \in \mathbb{R}^n$, $\|Ux\| = \|x\|$.

Moreover, the columns $[u_1, u_2, \ldots, u_n]$ of an orthogonal matrix $U \in \mathbb{R}^{n \times n}$ form a *basis* for $\mathbb{R}^n$. This means that for any vector $x \in \mathbb{R}^n$, there exists a set of scalars

$\alpha_1, \ldots, \alpha_n$ such that $x = \sum_{i=1}^{n} \alpha_i u_i$. More interestingly, since u_i are unit vectors, we also have $\|x\|^2 = \sum_{i=1}^{n} \alpha_i^2$.

This can be interpreted as U describing a *rotation* (with possible mirror flips) to a new set of coordinates. That is the old coordinates of x are $(x_1, x_2, \ldots, x_n)$ and the coordinates in the new orthogonal basis $[u_1, u_2, \ldots, u_n]$ are $(\alpha_1, \alpha_2, \ldots, \alpha_n)$.

Exercises

3.1 Consider a matrix

$$A = \begin{bmatrix} 2 & 2 & 3 \\ -2 & 7 & 4 \\ -3 & -3 & -4.5 \\ -8 & 2 & 3 \end{bmatrix}.$$

1. Add a column to A so that it is invertible.
2. Remove a row from A so that it is invertible (this can be done).
3. Is AA^T invertible?
4. Is $A^T A$ invertible?

3.2 Consider two vectors $u = (0.5, 0.4, 0.4, 0.5, 0.1, 0.4, 0.1)$ and $v = (-1, -2, 1, -2, 3, 1, -5)$.

1. Check if u or v is a unit vector.
2. Calculate the dot product $\langle u, v \rangle$.
3. Are u and v orthogonal?

3.3 Consider the following 3 vectors in $\mathbb{R}^9$:

$$v = (1, 2, 5, 2, -3, 1, 2, 6, 2)$$
$$u = (-4, 3, -2, 2, 1, -3, 4, 1, -2)$$
$$w = (3, 3, -3, -1, 6, -1, 2, -5, -7)$$

Report the following:

1. $\langle v, w \rangle$
2. Are any pair of vectors orthogonal, and if so which ones?
3. $\|u\|_2$
4. $\|w\|_\infty$

3.4 Consider the following 2 matrices:

$$A = \begin{bmatrix} 4 & -1 & 3 \\ 0 & -1 & 4 \\ 2 & -2 & 6 \end{bmatrix} \quad B = \begin{bmatrix} 2 & 3 & -1 \\ 5 & 3 & 5 \\ -2 & -2 & 1 \end{bmatrix}$$

Report the following:

1. $A^T B$
2. AB
3. BA
4. $B + A$
5. Which matrices are invertible? For any that are invertible, report the result.

3.5 Consider the following 2 vectors in $\mathbb{R}^4$:

$$p = (4, 2, -6, x)$$
$$q = (2, -4, 1, -2)$$

Report the following:

1. The value x so that p and q are orthogonal.
2. Calculate $\|q\|_1$
3. Calculate $\|q\|_2^2$

3.6 Consider the following 3 matrices:

$$A = \begin{bmatrix} 2 & -2 \\ -3 & 1 \\ 5 & -3 \end{bmatrix} \quad B = \begin{bmatrix} 4 & 4 & 4 \\ -2 & 3 & -7 \\ 2 & 5 & -7 \end{bmatrix} \quad C = \begin{bmatrix} 4 & -1 & 2 \\ -8 & 2 & -4 \\ 2 & 1 & -4 \end{bmatrix}$$

Report the following:

1. $A^T B$
2. $C + B$
3. Which matrices are full rank?
4. $\|C\|_F$
5. $\|A\|_2$
6. B^{-1}
7. $|C|$
8. Are matrices B or C positive definite?

Chapter 4
Distances and Nearest Neighbors

Abstract At the core of most data analysis tasks and their formulations is a distance. This **choice** anchors the meaning and the modeling inherent in the patterns found and the algorithms used. However, there are an enormous number of distances to choose from. We attempt to survey those most common within data analysis. This chapter also provides an overview of the most important properties of distances (e.g., is it a metric?) and how they are related to the dual notion of a similarity. We provide some common modeling dynamics which motivate some of the distances, and overview their direct uses in nearest neighbor approaches and how to algorithmically deal with the challenges that arise.

4.1 Metrics

So what makes a good distance? There are two aspects to answer this question. The first aspect is whether it captures the "right" properties of the data, but this is a sometimes ambiguous modeling problem. The second aspect is more well-defined; that is, does it satisfy the properties of a metric?

A *distance* $\mathbf{d} : \mathcal{X} \times \mathcal{X} \to \mathbb{R}^+$ is a bivariate operator (it takes in two arguments, say $a \in \mathcal{X}$ and $b \in \mathcal{X}$) that maps to $\mathbb{R}^+ = [0, \infty)$. It is a *metric* if

(M1)	$\mathbf{d}(a, b) \geq 0$	**(non-negativity)**
(M2)	$\mathbf{d}(a, b) = 0$ if and only if $a = b$	**(identity)**
(M3)	$\mathbf{d}(a, b) = \mathbf{d}(b, a)$	**(symmetry)**
(M4)	$\mathbf{d}(a, b) \leq \mathbf{d}(a, c) + \mathbf{d}(c, b)$	**(triangle inequality)**

A distance that satisfies (M1), (M3), and (M4) (but not necessarily (M2)) is called a *pseudometric*.

A distance that satisfies (M1), (M2), and (M4) (but not necessarily (M3)) is called a *quasimetric*.

In the next few sections we outline a variety of common distances used in data analysis, and provide examples of their use cases.

© Springer Nature Switzerland AG 2021

J. M. Phillips, *Mathematical Foundations for Data Analysis*,
Springer Series in the Data Sciences,
https://doi.org/10.1007/978-3-030-62341-8_4

4.2 L_p Distances and their Relatives

We next introduce a specific family of distances between vectors $a, b \in \mathbb{R}^d$. As they are defined between vectors, the most common ones are defined purely from notions of norms in linear algebra. But other variants will restrict vectors to model specific sorts of data like probability distribution, and then draw from more probabilistic elements.

4.2.1 L_p Distances

Consider two vectors $a = (a_1, a_2, \ldots, a_d)$ and $b = (b_1, b_2, \ldots, b_d)$ in $\mathbb{R}^d$. Now an L_p distance is defined as

$$\mathbf{d}_p(a, b) = \|a - b\|_p = \left(\sum_{i=1}^{d} (|a_i - b_i|)^p \right)^{1/p}.$$

- The most common is the L_2 distance, also know as the *Euclidean distance*,

$$\mathbf{d}_2(a, b) = \|a - b\| = \|a - b\|_2 = \sqrt{\sum_{i=1}^{d} (a_i - b_i)^2}.$$

It easy interpreted as the "straight-line" distance between two points or vectors, since if you draw a line segment between two points, its length measures the Euclidean distance.
It is also the only L_p distance that is invariant to the rotation of the coordinate system (which, as we will see, will often be useful but sometimes restrictive).
- Another common distance is the L_1 distance

$$\mathbf{d}_1(a, b) = \|a - b\|_1 = \sum_{i=1}^{d} |a_i - b_i|.$$

This is also known as the *Manhattan distance* since it is the sum of lengths on each coordinate axis; the distance you would need to walk in a city like Manhattan since you must stay on the streets and cannot cut through buildings.
- Another common modeling goal is the L_0 distance

$$\mathbf{d}_0(a, b) = \|a - b\|_0 = d - \sum_{i=1}^{d} \mathbb{1}(a_i = b_i),$$

where $\mathbb{1}(a_i = b_i) = \begin{cases} 1 & \textbf{if } a_i = b_i \\ 0 & \textbf{if } a_i \neq b_i. \end{cases}$ Unfortunately, $\mathbf{d}_0$ is not convex.

When each coordinate a_i or b_i is either 0 or 1, then this is known as the *Hamming distance*.

- Finally, another useful variation is the L_∞ distance

$$\mathbf{d}_\infty(a, b) = \|a - b\|_\infty = \max_{i=1\ldots d} |a_i - b_i|.$$

It is the maximum deviation along any one coordinate. Geometrically in $\mathbb{R}^2$, it is a rotation of the L_1 distance, so many algorithms designed for L_1 can be adapted to L_∞ in $\mathbb{R}^2$. However, in high dimensions they can act in surprisingly different ways. $\mathbf{d}_\infty$ is not technically an L_p distance, but is the limit of such distances as p goes to ∞.

Geometry of L_p Unit Balls

A useful way to imagine the geometry of these various L_p distances is by considering their unit balls. Much of the relevant information is conveyed even in $\mathbb{R}^2$. A *unit ball* is the set of points $a \in \mathbb{R}^2$ so that $\mathbf{d}(\mathbf{0}, a) = 1$ for our choice of distance $\mathbf{d}$.

The unit balls for L_1, L_2, and L_∞ distance are shown in $\mathbb{R}^2$. The most important properties to notice are that L_1 is never larger than L_2 which is never larger than L_∞. Moreover, they are always equal along the coordinate axis, and these are the only places they are the same. Indeed these principles hold for any L_p distance where the L_p ball is never greater than the $L_{p'}$ ball for $p < p'$, and they are only and always equal along the coordinate axis.

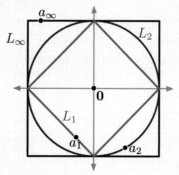

The L_2 ball is the only distance invariant to the choice of axis. This means, for instance, that if it is rotated it stays the same. This is not true for any other L_p balls.

It is also possible to draw L_p balls for $p < 1$. However, these balls are not convex; they "curve in" between the coordinate axis. Algorithmically, this makes them difficult to work with. It also, in effect, is the reason those L_p distances are not a metric; they violate the triangle inequality.

All of the above distances are metrics, and in general all L_p distances are metrics for $p \in [1, \infty)$.

Geometry of Metric Properties for L_p

Some of the metric requirements are easy to show hold for all L_p distances. (M1) and (M2) hold since the distances are, at the core, a sum of non-negative terms, and are only all 0 if all coordinates are identical. (M3) holds since $|a_i - b_i| = |b_i - a_i|$, the vector subtraction is symmetric.

Property (M4—triangle inequality) is a bit trickier to show, and the general proof is beyond the scope of this book. The proof for L_2 is more straight-forward, and naturally follows by the geometry of a triangle. Consider the line $\ell_{a,b}$ that goes through two points $a, b \in \mathbb{R}^d$. Let $\pi_{a,b}(c)$ be the orthogonal projection of any point $c \in \mathbb{R}^d$ onto this line $\ell_{a,b}$. Now we can decompose a squared distance $\mathbf{d}_2(a, c)^2$ (and symmetrically $\mathbf{d}_2(c, b)^2$) into $\mathbf{d}_2(a, c)^2 = \mathbf{d}_2(a, \pi_{a,b}(c))^2 + \mathbf{d}_2(\pi_{a,b}(c), c)^2$ by the Pythagorian theorem. Moreover, the distance $\mathbf{d}_2(a, b) \leq \mathbf{d}_2(a, \pi_{a,b}(c)) + \mathbf{d}_2(\pi_{a,b}(c), b)$, where there is an equality when $\pi_{a,b}(c)$ is on the line segment between a and b. Thus we can conclude

$$\mathbf{d}_2(a, b) \leq \mathbf{d}_2(a, \pi_{a,b}(c)) + \mathbf{d}_2(\pi_{a,b}(c), b) \leq \mathbf{d}_2(a, c) + \mathbf{d}_2(c, b).$$

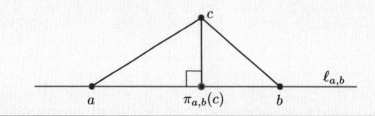

▷ Warning about L_p Distance ◁

These L_p distances should *not* be used to model data when the units and meaning on each coordinate are not the same. For instance, consider representing two people p_1 and p_2 as points in $\mathbb{R}^3$ where the x-coordinate represents height in inches, the y-coordinate represents weight in pounds, and the z-coordinate represents income in dollars per year. Then most likely this distance is dominated by the z-coordinate income which might vary on the order of 10,000 while the others vary on the order of 10.

However, for the same data we could change the units, so the x-coordinate represents height in meters, the y-coordinate represents weight in centigrams, and the z-coordinate represents income in dollars per hour. The information may be exactly the same, only the unit changed. It is now likely dominated by the y-coordinate representing weight.

These sorts of issues can hold for distances other than L_p as well. Some heuristics to overcome this are: set hand-tuned scaling of each coordinate, "standardize" the distance so they all have the same min and max value (e.g., all in the range [0, 1]), or "normalize" the distance so they all have the same mean and variance. Applying any of these mechanically without any inspection of their effects may have unintended consequences. For instance the [0, 1] standardization is at the mercy of outliers, and mean-variance normalization can have strange effects in multi-modal distributions. *Again, these are not solutions, they are modeling choices or heuristics!* This modeling challenge will be highlighted again with data matrices in Section 7.1.

With some additional information about which points are "close" or "far" one may be able to use the field of *distance metric learning* to address some of these problems. A simple solution can then be derived over all Mahalanobis distances (defined below), using some linear algebra and gradient descent. But without this information, there is no one right answer. If your axes are the numbers of apples (x-axis) and number of oranges (y-axis), then it is *literally comparing apples to oranges!*

4.2.2 Mahalanobis Distance

An extension to the L_2 distance is the *Mahalanobis distance* defined for two vectors $a, b \in \mathbb{R}^d$ and a $d \times d$ matrix M as

$$\mathbf{d}_M(a, b) = \sqrt{(a - b)^T M (a - b)}.$$

When $M = I$ (the identity matrix, so $I_{j,j} = 1$ and $I_{j,j'} = 0$ for $j \neq j'$), then $\mathbf{d}_M = \mathbf{d}_2$. When M is a diagonal matrix (so $M_{j,j'} = 0$ for $j \neq j'$) then $\mathbf{d}_M$ can be interpreted as skewing the Euclidean space (shrink some coordinates, and expanding others) based on the matrix M. When M is not diagonal, the skewing of Euclidean space still holds, but the skew is not aligned with the coordinate axis; instead it is defined through the eigenvectors and by the eigenvalues of M (see Chapter 7). As long as all eigenvalues are positive and real (implying M is positive definite) then $\mathbf{d}_M$ is a metric; since then the skew is well-defined and full-dimensional.

4.2.3 Cosine and Angular Distance

The *cosine distance* measures 1 minus the cosine of the "angle" between vectors $a = (a_1, a_2, \ldots, a_d)$ and $b = (b_1, b_2, \ldots, b_d)$ in $\mathbb{R}^d$

$$\mathbf{d}_{\cos}(a, b) = 1 - \frac{\langle a, b \rangle}{\|a\| \|b\|} = 1 - \frac{\sum_{i=1}^{d} a_i b_i}{\|a\| \|b\|}.$$

Recall that if $\theta_{a,b}$ is the angle between vectors a and b then $\cos(\theta_{a,b}) = \frac{\langle a, b \rangle}{\|a\| \|b\|}$. Hence $\mathbf{d}_{\cos}(a, b) = 1 - \cos(\theta_{a,b})$.

Note that $\mathbf{d}_{\cos}(A, B) \in [0, 2]$ and it does not depend on the magnitude $\|a\|$ of the vectors since this is normalized out. It only cares about their directions. This is useful when vectors represent data sets of different sizes and we want to compare how similar are those distributions, but not their size. This makes $\mathbf{d}_{\cos}$ at best a psuedo-metric since for two vectors a and $a' = (2a_1, 2a_2, \ldots, 2a_d)$ where $\|a'\| = 2\|a\|$ have $\mathbf{d}_{\cos}(a, a') = 0$, but they are not equal.

Sometimes $\mathbf{d}_{\cos}$ is defined only with respect to normalized vectors $a, b \in \mathbb{S}^{d-1}$, where

$$\mathbb{S}^{d-1} = \left\{ x \in \mathbb{R}^d \mid \|x\| = 1 \right\}.$$

In this case, then more simply

$$\mathbf{d}_{\cos}(a, b) = 1 - \langle a, b \rangle.$$

Restricted to vectors in $\mathbb{S}^{d-1}$, then $\mathbf{d}_{\cos}$ does not have the issue of two vectors $a \neq b$ such that $\mathbf{d}_{\cos}(a, b) = 0$. However, it is not yet a metric since it can violate the triangle inequality.

A simple isometric transformation of the cosine distance (this means the distance-induced ordering between any pairs of points are the same) is called the *angular distance* $\mathbf{d}_{\text{ang}}$; that is for $a, b, c, d \in \mathbb{S}^{d-1}$ if $\mathbf{d}_{\cos}(a, b) < \mathbf{d}_{\cos}(c, d)$ then $\mathbf{d}_{\text{ang}}(a, b) < \mathbf{d}_{\text{ang}}(c, d)$. Specifically we define for any $a, b \in \mathbb{S}^{d-1}$

$$\mathbf{d}_{\text{ang}}(a, b) = \cos^{-1}(\langle a, b \rangle) = \arccos(\langle a, b \rangle).$$

That is, this undoes the cosine-interpretation of the cosine distance, and only measures the angle. Since, the inner product for any $a, b \in \mathbb{S}^{-1}$ is in the range $[-1, 1]$, then the value of $\mathbf{d}_{\cos}$ is in the range $[0, 2]$, but the value of $\mathbf{d}_{\text{ang}}(a, b) \in [0, \pi]$. Moreover, $\mathbf{d}_{\text{ang}}$ is a metric over $\mathbb{S}^{d-1}$.

Geometry of Angular Distance

The angular interpretation of the cosine distance $\mathbf{d}_{\cos}(a, b) = 1 - \cos(a, b)$ and angular distance $\mathbf{d}_{\text{ang}}(a, b) = \theta_{a,b}$ is convenient to think of for points on the sphere. To understand the geometry of the angular distance between two points, it is sufficient to consider them as lying on the unit sphere $\mathbb{S}^1 \subset \mathbb{R}^2$. Now $\mathbf{d}_{\text{ang}}(a, b)$ is the radians of the angle $\theta_{a,b}$, or equivalently the arclength traveled between a and b if walking on the sphere.

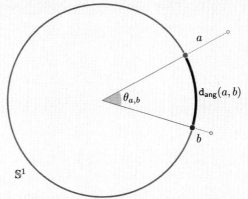

Now we can see why $\mathbf{d}_{\text{ang}}$ is a metric if restricted to vectors on $\mathbb{S}^{d-1}$. (M1) and (M3) hold by definition, and (M2) on $\mathbb{S}^{d-1}$ holds because no two distinct unit vectors have an angle of 0 radians between them. To show the triangle inequality (M4) (here we need to think in $\mathbb{S}^2 \subset \mathbb{R}^3$), observe that since $\mathbf{d}_{\text{ang}}$ measures the shortest distance restricted to the sphere, there is no way that $\mathbf{d}_{\text{ang}}(a, b)$ can be longer than $\mathbf{d}_{\text{ang}}(a, c) + \mathbf{d}_{\text{ang}}(c, b)$ since that would imply going through point $c \in \mathbb{S}^{d-1}$ makes the path from a to b shorter—which is not possible.

4.2.4 KL Divergence

The Kullback-Leibler Divergence (or KL Divergence) is a distance that is *not* a metric. Somewhat similar to the cosine distance, it considers as input discrete distributions X_a and X_b; they are d-variate random variables which can be instantiated at one of d possible values. Equivalently, it is useful to think of these objects as vectors $a, b \in \Delta_\circ^{d-1}$. Like the $(d - 1)$-dimensional sphere, the $(d - 1)$-dimensional open simplex $\Delta_\circ^{d-1}$ is a bounded subset of $\mathbb{R}^d$. Specifically

$$\Delta_\circ^{d-1} = \left\{ x = (x_1, x_2, \ldots x_d) \in \mathbb{R}^d \mid \|x\|_1 = 1 \text{ and } x_i > 0 \text{ for all } i \right\}.$$

That is $\Delta_\circ^{d-1}$ defines the set of d-dimensional discrete probability distributions where for $a \in \Delta_\circ^{d-1}$, the coordinate a_i is the probability that X_a takes the ith value. The $(d - 1)$-dimensional (closed) simplex Δ^{d-1} differs in that it also allows values a_i to be 0, i.e., to have 0 probability to take the ith value. But the KL Divergence is only defined over vectors on $\Delta_\circ^{d-1}$.

Then we can define the *Kullback-Leibler Divergence* (often written $\mathbf{d}_{KL}(X_a\|X_b)$) as

$$\mathbf{d}_{KL}(X_a, X_b) = \mathbf{d}_{KL}(a, b) = \sum_{i=1}^{d} a_i \ln(a_i/b_i).$$

It is the expectation (over the probability distribution a) of the natural log of the ratios between a and b, and can be interpreted as the "information gain" in using distribution b in place of a.

Note that $\mathbf{d}_{KL}$ is *not* a metric, violating (M3) since it is not symmetric. It also violates the triangle inequality (M4).

4.3 Distances for Sets and Strings

We now introduce some more general notions of distance, in particular ones that are heavily used to understand the relationship between text data: strings of words or characters. There are other techniques which draw more heavily on the semantic and fine-grained structural properties of text. We focus here on the ones which have simple mathematical connections and as a result are often more scalable and flexible.

Quick Review of Sets

A *set A* is a collection of elements $\{a_1, a_2, \ldots, a_k\}$. It is standard to use curly brackets $\{\ldots\}$ to denote a set. The ordering of elements does not matter, e.g., $\{a_1, a_2\} = \{a_2, a_1\}$. Multiplicity is not allowed or is ignored, e.g., $\{a_1, a_1, a_2\} = \{a_1, a_2\}$; if it is considered, then it is called a *multi-set* (these more naturally model counts, or after normalization probability distributions; see cosine distance and KL divergence).

Given a set A, the *cardinality* of A denoted as $|A|$ counts how many elements are in A. The *intersection* between two sets A and B is denoted as $A \cap B$ and reveals all items which are in *both* sets. The *union* between two sets A and B is denoted as $A \cup B$ and reveals all items which are in *either* set. The *set difference* of two sets A and B is denoted as $A \setminus B$ and is all elements in A which are not in B. The *symmetric distance* between two sets A and B is denoted as $A \triangle B$ and is the union of $A \setminus B$ and $B \setminus A$. Given a domain Ω (which contains all sets of interest, i.e., $A \subset \Omega$), the *complement of a set A* is defined as $\bar{A} = \Omega \setminus A$.

Example: Set operations

Observe the example with $A = \{1, 3, 5\}$ and $B = \{2, 3, 4\}$. These are represented as a Venn diagram with a blue region for A and a red one for B. Element 6 is in neither set. Then the cardinality of A is $|A| = 3$, and it is the same for B. The intersection

$A \cap B = \{3\}$ since it is the only object in both sets, and is visually represented as the purple region. The union $A \cup B = \{1, 2, 3, 4, 5\}$. The set difference $A \setminus B = \{1, 5\}$ and $B \setminus A = \{2, 4\}$. The complement $\bar{A} = \{2, 4, 6\}$ and the complement $\overline{A \cup B} = \{6\}$. Finally, the symmetric difference is $A \triangle B = \{1, 2, 4, 5\}$.

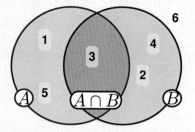

4.3.1 Jaccard Distance

The *Jaccard distance* between two sets A and B is defined as

$$\mathbf{d}_J(A, B) = 1 - \frac{|A \cap B|}{|A \cup B|}.$$

Since the union is always larger than the intersection, the fractional term is always between 0 and 1, and this distance always takes value in $[0, 1]$. It is only 1 if the sets have nothing in common, and is only 0 if the intersection equals the union and the sets are exactly the same.

Example: Jaccard Distance

Consider two sets $A = \{0, 1, 2, 5, 6\}$ and $B = \{0, 2, 3, 5, 7, 9\}$. The Jaccard distance between A and B is

$$\mathbf{d}_J(A, B) = 1 - \frac{|A \cap B|}{|A \cup B|}$$

$$= 1 - \frac{|\{0, 2, 5\}|}{|\{0, 1, 2, 3, 5, 6, 7, 9\}|} = 1 - \frac{3}{8} = 0.625.$$

Notice that if we add an element 7 to A (call this set A') that is already in B, then the numerator increases, but the denominator stays the same. So then $\mathbf{d}_J(A', B) = 1 - \frac{4}{8} = 0.5$ and the distance is smaller—they are closer to each other.

On the other hand, if we add an element 4 to A (call this set A'') which is in neither A or B, then the numerator stays the same, but the denominator increases. So then $\mathbf{d}_J(A'', B) = 1 - \frac{3}{9} \approx 0.666$ and then distance is larger—the sets are farther from each other.

The Jaccard distance is a popular distance between sets since it is a metric, and it is invariant to the size of the sets. It only depends on the fraction of the items among both sets which are the same. It also does not require knowledge of some larger universe Ω of elements that the sets may be from. For instance, as in the example, we can implicitly require that the sets contain only positive integers, but do not need to know an upper-bound on the largest positive integer allowed.

Proof for Metric Property of Jaccard Distance

To show that $\mathbf{d}_J$ is a metric, we need to show that the 4 properties each hold. The first three are direct. In particular (M1) and (M2) follow from $\mathbf{d}_J(A, B) \in [0, 1]$ and only being 0 if $A = B$. Property (M3) holds by the symmetry of set operations $\cap$ and $\cup$.

The triangle inequality (M4) requires a bit more effort to show, namely for any sets A, B, C that $\mathbf{d}_J(A, C) + \mathbf{d}_J(C, B) \geq \mathbf{d}_J(A, B)$. We will use the notation that

$$\mathbf{d}_J(A, B) = 1 - \frac{|A \cap B|}{|A \cup B|} = \frac{|A \triangle B|}{|A \cup B|}.$$

We first rule out that there are elements $c \in C$ which are not in A or not in B. Removing these elements from C will only decrease the left-hand side of the triangle inequality while not affecting the right-hand side. So if C violates this inequality, we can assume there is no such $c \in C$ and it will still violate it. So now we assume $C \subseteq A$ and $C \subseteq B$.

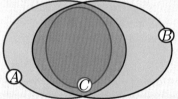

Now we have

$$
\begin{aligned}
\mathbf{d}_J(A, C) + \mathbf{d}_J(C, B) &= \frac{|A \setminus C|}{|A|} + \frac{|B \setminus C|}{|B|} \\
&\geq \frac{|A \setminus C| + |B \setminus C|}{|A \cup B|} \\
&\geq \frac{|A \triangle B|}{|A \cup B|} = \mathbf{d}_J(A, B).
\end{aligned}
$$

The first inequality follows since $|A|, |B| \leq |A \cup B|$. The second inequality holds since we assume $C \subseteq A$ and $C \subseteq B$, hence $C \subseteq A \cap B$, and thus $|A \setminus C| + |B \setminus C| \geq |A \setminus B| + |B \setminus A| = |A \triangle B|$.

4.3.2 Edit Distance

Let Σ be a set, in this case an alphabet of possible characters (e.g., all ASCII characters, or all lowercase letters so $\Sigma = \{a, b, \ldots, z\}$). Then we can say a string a of length d is an element in Σ^d; that is an ordered sequence of characters from the alphabet Σ. The *edit distance* considers two strings $a, b \in \Sigma^d$, and

$$\mathbf{d}_{\text{ed}}(a, b) = \# \text{ operations to make } a = b,$$

where an operation can delete a letter or insert a letter. In fact, the strings are not required to have the same length, since we can insert items in the shorter one to make up the difference.

Example: Edit Distance

Consider two strings $a = $ mines and $b = $ smiles. Here $\mathbf{d}_{\text{ed}}(a, b) = 3$.

mines	
1 : minles	insert l
2 : miles	delete n
3 : smiles	insert s

There are many alternative variations of operations. The insert operation may cost more than the delete operation. Or we could allow a replace operation at the same unit cost as either insert or delete; in this case the edit distance of mines and smiles is only 2.

Edit distance is a metric. (M1) holds since the number of edits is always non-negative. (M2) is true because there are no edits only if they are the same. (M3) holds because the operations can be reversed. (M4) follows because if c is an intermediate "word" then the $\mathbf{d}_{\text{ed}}(a, c) + \mathbf{d}_{\text{ed}}(c, b) = \mathbf{d}_{\text{ed}}(a, b)$, otherwise it requires more edits.

Is this good for large text documents? Not really. It is slow to compute—basically requiring quadratic time dynamic programing in the worst case to find the smallest set of edits. And removing one sentence can cause a large edit distance without changing meaning. But this *is* good for small strings. Some version is used in most spelling recommendation systems (e.g., a search engine's auto-correct). It is a good guide that usually $\mathbf{d}_{\text{ed}}(a, b) > 3$ is pretty large since, e.g., $\mathbf{d}_{\text{ed}}($cart, score$) = 4$.

4.4 Modeling Text with Distances

There are many many choices of distances. Which one to *choose* (it is definitely a choice) is based on (a) computational efficiency and (b) modeling effectiveness. The Euclidean distance d_2 and sometimes the Jaccard distance d_J are often chosen because various algorithmic and computational benefits are available for these— as we will see, this efficiency comes not just from time to compute the distance once, but how it can be used within more complex operations. However, each has other benefits due to modeling factors. Sometimes this modeling is just based on mathematical properties (is it a metric? are my vectors normalized?), sometimes it is intuitive, and sometimes it can be empirically validated by measuring performance on downstream applications. In this section we show how to arrive at various distances as the logical choice in an example case of modeling text.

As mentioned, edit distance d_{ed} is useful for shorter strings. The other variants will all be more useful when dealing with much larger texts.

Example: Running text example

In this section we will use as a running example the text from the following 4 short documents. In practice, these approaches are typically applied to much longer documents (e.g., text on a webpage, a newspaper article, and a person's bio).

D_1 : I am Pam.

D_2 : Pam I am.

D_3 : I do not like jelly and ham.

D_4 : I do not, do not, like them, Pam I am.

How can we measure the distance among these 4 documents?

4.4.1 Bag-of-Words Vectors

The simplest model for converting text into an abstract representation for applying a distance is the bag-of-words approach. Intuitively, each document creates a "bag" and throws each word in that "bag" (a multi-set data structure), and maintains only the count of each word. This transforms each document into a multi-set. However it is convenient to think of it as a (very sparse, meaning mostly 0s) vector.

That is, consider a vector $v \in \mathbb{R}^D$ for a very large D, where D is the number of all possible words. Each coordinate of such a vector corresponds to one word and records the count of that word.

These vector representations naturally suggest that one could use an L_p distance, most commonly d_2, to measure their distance. However, it is more common to use the cosine distance d_{cos}. This has the advantage of not penalizing certain documents for their length, in principle focusing more on their content. For instance, a document with a simple phrase would be identical under $\mathbf{d}_{\cos}$ to another document that repeated that phrase multiple times. Or two documents about, say, baseball would typically draw from a similar set of words (e.g., {bat, ball, hit, run, batter, inning}) and likely be close even if their lengths differ.

Example: Bag-of-Words

For the running example, consider D-dimensional space with $D = 11$; it could be much higher. For each coordinate, we list the corresponding word as

(am, and, do, ham, I, jelly, like, not, Pam, them, zebra).

Now each of the documents have the following representative vectors

$$v_1 = (1, 0, 0, 0, 1, 0, 0, 0, 1, 0, 0)$$
$$v_2 = (1, 0, 0, 0, 1, 0, 0, 0, 1, 0, 0)$$
$$v_3 = (0, 1, 1, 1, 1, 1, 1, 1, 0, 0, 0)$$
$$v_4 = (1, 0, 2, 0, 2, 0, 1, 2, 1, 1, 0).$$

We can now use the Euclidean distance $\mathbf{d}_2$ between these vectors (or any other L_p distance). We notice that $\mathbf{d}_2(v_1, v_2) = 0$, even though the text is different. However, since the bag-of-words only measures which words are present, and not how they are used, it cannot distinguish between these cases.

Also notice that the 11th coordinate for the word zebra is never used, and is 0 in all coordinates of the vectors. If this coordinate was omitted and only a 10-dimensional representation was used it would not change these distances. On the other hand, these vectors can be much larger and represent many other words, and this will not affect the distance.

	(v_1, v_2)	(v_1, v_3)	(v_1, v_4)	(v_2, v_3)	(v_2, v_4)	(v_3, v_4)
$\mathbf{d}_2$	0	2.83	3.32	2.83	3.32	3
$\mathbf{d}_{\cos}$	0	0.781	0.423	0.781	0.423	0.339

Alternatively, we can use the cosine distance $\mathbf{d}_{\cos}$. This models the text differently; the main difference is that it normalizes the vectors. So for instance D_3 and D_4 are now likely to be further apart simply because they contain more words, as is the case with $\mathbf{d}_2$. This metric still treats D_1 and D_2 as identical.

These distance calculations are simple to perform and print in Python. First the Euclidean distance is computed as

```python
import numpy as np
from numpy import linalg as LA

v1 = np.array([1,0,0,0,1,0,0,0,1,0,0])
v2 = np.array([1,0,0,0,1,0,0,0,1,0,0])
v3 = np.array([0,1,1,1,1,1,1,1,0,0,0])
v4 = np.array([1,0,2,0,2,0,1,2,1,1,0])
```

```
print([LA.norm(v1-v2), LA.norm(v1-v3), LA.norm(v1-v4),
    LA.norm(v2-v3), LA.norm(v2-v4), LA.norm(v3-v4)])
```

For the cosine distance, we first normalize the vectors as

```
v1n = v1/LA.norm(v1)
v2n = v2/LA.norm(v2)
v3n = v3/LA.norm(v3)
v4n = v4/LA.norm(v4)

print([1-(v1n @ v2n), 1-(v1n @ v3n), 1-(v1n @ v4n),
    1-(v2n @ v3n), 1-(v2n @ v4n), 1-(v3n @ v4n)])
```

Another potential use of the bag-of-words representation is as a probability distribution over the words in a document. That is, again consider a D-dimensional space where D is the number of possible words. But now for each word w_i, its representative coordinate (the ith coordinate) denotes the probability that a randomly drawn word from the document is that word. This is equivalent to normalizing a vector v by its L_1 norm so $\bar{v} = v/\|v\|_1$, and this ensures that $v \in \Delta^{D-1}$. This representation inclines a modeler to use a distance designed for probability distributions, such as the KL Divergence $\mathbf{d}_{KL}$.

Example: KL Divergence on Probability Measures

Using the same 11-dimensional representation of our running example documents, we can normalize by L_1 the vectors v_1, v_2, v_3, and v_4 so they lie on $\Delta_\circ^{10}$. However, this does not quite work, since the vectors have 0 values, which are not allowed in $\Delta_\circ^{10}$. The right way to address this (if we insist on using KL Divergence) is to add a so-called regularization term α to each vector value before L_1 normalization. Using $\alpha = 0.01$ we obtain

$$\bar{v}_1 = (0.325, 0.003, 0.003, 0.003, 0.325, 0.003, 0.003, 0.003, 0.325, 0.003, 0.003)$$
$$\bar{v}_2 = (0.325, 0.003, 0.003, 0.003, 0.325, 0.003, 0.003, 0.003, 0.325, 0.003, 0.003)$$
$$\bar{v}_3 = (0.001, 0.142, 0.142, 0.142, 0.142, 0.142, 0.142, 0.142, 0.001, 0.001, 0.001)$$
$$\bar{v}_4 = (0.100, 0.001, 0.199, 0.001, 0.199, 0.001, 0.100, 0.199, 0.100, 0.100, 0.001).$$

The KL divergence is not symmetric, so we report the distance in both directions for all pairs of vectors, where the row label is the first term and the column label is the second term (e.g., $\mathbf{d}_{KL}(\bar{v}_2\|\bar{v}_4) = 0.89$).

$\mathbf{d}_{KL}$	$\bar{v}_1$	$\bar{v}_2$	$\bar{v}_3$	$\bar{v}_4$
$\bar{v}_1$	0	0	3.74	0.89
$\bar{v}_2$	0	0	3.74	0.89
$\bar{v}_3$	3.09	3.09	0	2.01
$\bar{v}_4$	1.99	1.99	1.43	0

Again, this distance is simple in Python, using the SciPy stats package:

```
import SciPy as sp
from SciPy import stats

reg = 0.01
v1r = v1+reg
v2r = v2+reg
v3r = v3+reg
v4r = v4+reg

print([stats.entropy(v1r,v1r), stats.entropy(v1r,v2r),
    stats.entropy(v1r,v3r), stats.entropy(v1r,v4r)])
print([stats.entropy(v2r,v1r), stats.entropy(v2r,v2r),
    stats.entropy(v2r,v3r), stats.entropy(v2r,v4r)])
print([stats.entropy(v3r,v1r), stats.entropy(v3r,v2r),
    stats.entropy(v3r,v3r), stats.entropy(v3r,v4r)])
print([stats.entropy(v4r,v1r), stats.entropy(v4r,v2r),
    stats.entropy(v4r,v3r), stats.entropy(v4r,v4r)])
```

4.4.2 k-Grams

As an alternative to bag-of-words vectors, k-grams can provide a richer context for text. These convert a document into a set (not a multi-set), but does not use just the words. There are a few variants of how to apply k-grams, and we will focus on two common and simple versions over words and characters. The *word k-grams* over a document is the set of all witnessed consecutive sets of k words.

Example: Word 2-Grams

Using the running example, we show the 2-word grams of each of the documents. Each gram is shown as a set of two words in square brackets.

$D_1: \ G_1 = \{[\text{I am}], \ [\text{am Pam}]\}$

$D_2: \ G_2 = \{[\text{Pam I}], \ [\text{I am}]\}$

$D_3: \ G_3 = \{[\text{I do}], \ [\text{do not}], \ [\text{not like}], \ [\text{like jelly}],$
$\qquad\qquad [\text{jelly and}], \ [\text{and ham}]\}$

$D_4: \ G_4 = \{[\text{I do}], \ [\text{do not}], \ [\text{not do}], \ [\text{not like}], \ [\text{like them}],$
$\qquad\qquad [\text{them Pam}], \ [\text{Pam I}], \ [\text{I am}]\}$

In particular, note that in D_4 [do not] $\neq$ [not do] so both appear in G_4; however, even though the two-word sequence [do not] appears twice in D_4, it only appears once in the set G_4.

We can then compute the Jaccard distance between these sets as representations of the distances between documents.

	(D_1, D_2)	(D_1, D_3)	(D_1, D_4)	(D_2, D_3)	(D_2, D_4)	(D_3, D_4)
	$1 - \frac{1}{3}$	$1 - \frac{0}{8}$	$1 - \frac{1}{9}$	$1 - \frac{0}{8}$	$1 - \frac{2}{8}$	$1 - \frac{3}{11}$
d_J	≈ 0.667	$= 1$	≈ 0.889	$= 1$	$= 0.75$	≈ 0.727

Ethical Questions with Modeling Text

After a mechanism to transform a text document or a webpage into an abstract data type (e.g., a vector or a set) is implemented and a choice of distance is made, then many downstream algorithms and analysis can be applied. And these applications can be fairly automatically invoked with little or no modification based on the modeling decisions. Then these can be used to make recommendations about jobs, loans, and education opportunities.

Consider now you are processing essays to aid in college admission decisions, and you change a modeling choice from bag-of-words and cosine distance to k-grams and Jaccard distance. Then you realize that this change will sort admission applications differently, and a class of applicants will likely have their admission decisions changed because of this modeling. Do you have an obligation to those applicants to keep the decision the same? How could you alert decision makers to this effect, and what resolutions could you make knowing that this happens?

Variants and Modeling Choices

There are many variants of how to construct k-grams for words. The most prominent of which is to use k consecutive characters instead of k consecutive words; we call these *character k-grams*.

Many other modeling decisions go into constructing a k-gram. Should punctuation be included? Should whitespace be used as a character, or should sentence breaks be used as a word in word grams? Should differently capitalization characters or words represent distinct objects? And most notoriously, how large should k be?

Example: Character k-Grams

Characters k grams for sample document D_4, ignoring whitespace, punctuation, and capitalization.

Characters $k = 3$:
{[ido], [don], [ono], [not], [otd], [tdo], [otl], [tli], [lik], [ike], [ket], [eth], [the], [hem], [ems], [mpa], [pam], [ami], [mia], [iam]}

Characters $k = 4$:
{[idon], [dono], [onot], [notd], [otdo], [tdon], [notl],

[otli], [tlik], [like], [iket], [keth], [ethe], [them], [hemp],
[empa], [mpam], [pami], [amia], [miam]}

Through all of these variants, a few common rules apply. First, the more expressive the k-grams (e.g., keeping track of punctuation, capitalization, and whitespace), the larger the quantity of data that is required to get meaningful results—otherwise most documents will have a Jaccard distance 1 or very close to it unless large blocks of text are verbatim repeated (for instance, plagiarism). But for long and structured articles (e.g., newspaper articles with a rigid style guide), some of these more expressive choices can be effective.

Second, for longer articles it is better to use words and larger values of k, while for shorter articles (like tweets) it is better to use smaller k and possibly character k-grams. The values used for k are often perhaps-surprisingly short, such as $k = 3$ or $k = 4$ for both characters and words.

Finally, there are a few structured tricks which come up. It can be useful to particularly keep track of or emphasize starts of sentences, capitalized words, or word k-grams which start with "stop words" (that is, very common words like {a, for, the, to, and, that, it, ... } that often signal starts of expressions).

Continuous Bag-of-Words

As an aside, we mention a more recent trend. Instead of creating abstract representations of entire documents, we may use many such documents to create similar representations for each word, based on how it appears in the union of the documents. These methods typically aim for a Euclidean representation for words. For a set of words, these are called *word vector embeddings*. Then the distance between the abstract representations of these words can be shown to carry a variety of information.

There are numerous of these techniques, but at a high-level, they start by creating a bag-of-words representation for each instance of a word. That is within some sequential radius of r words in a text, for each instance of a word, it creates a D-dimensional vector (where we maintain counts over a size D vocabulary). This is like the $D = 11$ words in the running example, but typically D is tens or hundreds of thousands. This vector representation of each word is sparse, and counts the multiplicity of each other word in its neighborhood. Then these vectors are averaged over all instances of each word. This is called the *continuous bag-of-words* or CBOW model. Usually these representations are non-linearly compressed to a lower-dimensional representation, often using 300 or 1,000 dimensions.

Ethical Questions with using Abstract Language Models

These word vectors embedding have led to dramatic improvements for various natural language processing tasks such as translation between languages and comparing meaning at sentence level structures. As a result, they are now commonly incorporated into state-of-the-art text models. However, they have also been shown to implicitly encode bias into the embeddings, coming from the text corpuses on which they were built. For instance in many such embeddings, the word man is significantly closer to the word engineer than is the word woman. Such biases, even if unintentional, may affect automated hiring decisions based on gender. As a data analyst, should you use these models to help make predictions if they are known to include biases, even if they actually lead to better prediction and generalization?

4.5 Similarities

A notion that is dual to a distance **d** is a similarity **s**. It is still a bivariate function, but when the arguments are close, then the value should be large. Often, but not always, the definitions are normalized so the similarity of an object with itself is $\mathbf{s}(A, A) = 1$, and so the range of the function is in $[0, 1]$, indicating that pairs of values A, B which are totally different have value $\mathbf{s}(A, B) = 0$.

There are two standard ways to convert from a similarity to a distance. For set-based similarities, it is common to obtain a distance as $\mathbf{d}(A, B) = 1 - \mathbf{s}(A, B)$. In vectorized and norm-based similarities, it is common to obtain a distance as $\mathbf{d}(A, B) = \sqrt{\mathbf{s}(A, A) + \mathbf{s}(B, B) - 2\mathbf{s}(A, B)}$.

4.5.1 Set Similarities

Given two sets A and B, the *Jaccard similarity* is defined as

$$\mathbf{s}_J(A, B) = \frac{|A \cap B|}{|A \cup B|}.$$

Indeed the Jaccard distance is defined as $\mathbf{d}_J(A, B) = 1 - \mathbf{s}_J(A, B)$.

To generalize set similarities (at least those that are amenable to large-scale techniques) we consider a class of similarities which can be written in the following form based on parameters x, y, z, z':

$$\mathbf{s}_{x,y,z,z'}(A, B) = \frac{x|A \cap B| + y|\overline{A \cup B}| + z|A \triangle B|}{x|A \cap B| + y|\overline{A \cup B}| + z'|A \triangle B|}.$$

For instance $\mathbf{s}_J = \mathbf{s}_{1,0,0,1}$. Note that this family includes a complement operation of $\overline{A \cup B}$ and thus seems to require to know the size of the entire domain Ω from which A and B are subsets. However, in the case of Jaccard similarity and others when $y = 0$, this term is not required, and thus the domain Ω is not required to be defined.

Other common set similarities in this family include the following.

$$\text{Hamming: } \mathbf{s}_{\text{Ham}}(A, B) = \mathbf{s}_{1,1,0,1}(A, B) = \frac{|A \cap B| + |\overline{A \cup B}|}{|A \cap B| + |\overline{A \cup B}| + |A \triangle B|}$$

$$\text{Andberg: } \mathbf{s}_{\text{Andb}}(A, B) = \mathbf{s}_{1,0,0,2}(A, B) = \frac{|A \cap B|}{|A \cap B| + 2|A \triangle B|}$$

$$\text{Rogers-Tanimoto: } \mathbf{s}_{\text{RT}}(A, B) = \mathbf{s}_{1,1,0,2}(A, B) = \frac{|A \cap B| + |\overline{A \cup B}|}{|A \cap B| + |\overline{A \cup B}| + 2|A \triangle B|}$$

$$\text{Sørensen-Dice: } \mathbf{s}_{\text{Dice}}(A, B) = \mathbf{s}_{2,0,0,1}(A, B) = \frac{2|A \cap B|}{2|A \cap B| + |A \triangle B|}$$

For $\mathbf{s}_J$, $\mathbf{s}_{\text{Ham}}$, $\mathbf{s}_{\text{Andb}}$, and $\mathbf{s}_{\text{RT}}$, then $\mathbf{d}(A, B) = 1 - \mathbf{s}(A, B)$ is a metric. In particular, $\mathbf{d}_{\text{Ham}}(A, B) = |\Omega|(1 - \mathbf{s}_{\text{Ham}}(A, B))$ is known as the *Hamming distance*; it is typically applied between bit vectors. In the bit vector setting it counts the number of bits the vectors differ on. Indeed, if we represent each object $i \in \Omega$ by a coordinate b_i in $|\Omega|$-dimension vector $b = (b_1, b_2, \ldots, b_{|\Omega|})$ then these notions are equivalent.

Example: Set Similarities

Consider two sets $A = \{0, 1, 2, 5, 6\}$ and $B = \{0, 2, 3, 5, 7, 9\}$ in a domain $\Omega = \{0, 1, 2, 3, 4, 5, 6, 7, 8, 9\}$. We can compute each of our defined set similarities between these sets:

$$\begin{array}{rl}
\text{Jaccard:} & \mathbf{s}_J(A, B) = \frac{3}{8} = 0.375 \\
\text{Hamming:} & \mathbf{s}_{\text{Ham}}(A, B) = 1 - \frac{5}{10} = 0.5 \\
\text{Andberg:} & \mathbf{s}_{\text{Andb}}(A, B) = \frac{3}{13} \approx 0.231 \\
\text{Rogers-Tanimoto:} & \mathbf{s}_{\text{RT}}(A, B) = \frac{10-5}{10+5} \approx 0.333 \\
\text{Sørensen-Dice:} & \mathbf{s}_{\text{Dice}}(A, B) = \frac{2(3)}{5+6} \approx 0.545
\end{array}$$

4.5.2 Normed Similarities

The most basic normed similarity is the Euclidean dot product. For two vectors $p, q \in \mathbb{R}^d$, then we can define the *dot product similarity* as

$$\mathbf{s}_{\text{dot}}(p, q) = \langle p, q \rangle;$$

that is, just the dot product. And indeed this converts to the Euclidean distance as

$$\mathbf{d}_2(p, q) = \|p - q\|_2 = \sqrt{\mathbf{s}_{\text{dot}}(p, p) + \mathbf{s}_{\text{dot}}(q, q) - 2\mathbf{s}_{\text{dot}}(p, q)}$$
$$= \sqrt{\langle p, p \rangle + \langle q, q \rangle - 2\langle p, q \rangle}$$
$$= \sqrt{\|p\|_2^2 + \|q\|_2^2 - 2\langle p, q \rangle}.$$

However, in this case the similarity could be arbitrarily large, and indeed the similarity of a vector p with itself $\mathbf{s}_{\text{dot}}(p, p) = \langle p, p \rangle = \|p\|^2$ is the squared Euclidean norm. To enforce that the similarity is at most 1, we can normalize the vectors first, and we obtain the *cosine similarity* as

$$\mathbf{s}_{\cos}(p, q) = \langle \frac{p}{\|p\|}, \frac{q}{\|q\|} \rangle = \frac{\langle p, q \rangle}{\|p\|\|q\|}.$$

Converting to a distance via the normed transformation

$$\sqrt{\mathbf{s}_{\cos}(p, p) + \mathbf{s}_{\cos}(q, q) - 2\mathbf{s}_{\cos}(p, q)} = \mathbf{d}_2\left(\frac{p}{\|p\|}, \frac{q}{\|q\|}\right)$$

is still the Euclidean distance, but now between the normalized vectors $\frac{p}{\|p\|}$ and $\frac{q}{\|q\|}$. However, for this similarity, it is more common to instead use the set transformation to obtain cosine distance:

$$\mathbf{d}_{\cos}(p, q) = 1 - \frac{\langle p, q \rangle}{\|p\|\|q\|} = 1 - \mathbf{s}_{\cos}(p, q).$$

Another family of similarities in this class is from kernels. The most common of which is the Gaussian kernel $K(p, q) = \exp(-\|p - q\|^2/\sigma^2)$. The σ parameter is an important bandwidth modeling parameter which controls which distances $\|p - q\|$ are considered similar; however for most discussions for simplicity we will omit it. Then the *kernel distance* is defined as

$$\mathbf{d}_K(p, q) = \sqrt{K(p, p) + K(q, q) - 2K(p, q)}.$$

For Gaussian kernels, and a larger class called *characteristic kernels* (a substantial subset of positive definite kernels), this distance is a metric. A *positive definite kernel* K is one where for any point set $P = \{p_1, p_2, \ldots, p_n\}$ the $n \times n$ "gram" matrix G with element $G_{i,j} = K(p_i, p_j)$ is positive definite.

4.5.3 Normed Similarities between Sets

One can also extend these normed similarities to be between sets of vector. Now let $P = \{p_1, p_2, \ldots, p_n\} \subset \mathbb{R}^d$ and $Q = \{q_1, q_2, \ldots, q_m\} \subset \mathbb{R}^d$ be two sets of vectors. We use a kernel K to define a similarity κ between sets as a double sum over their kernel interactions

$$\kappa(P, Q) = \sum_{p \in P} \sum_{q \in Q} K(p, q).$$

This can again induce a generalization of the kernel distance between point sets as

$$\mathbf{d}_K(P, Q) = \sqrt{\kappa(P, P) + \kappa(Q, Q) - 2\kappa(P, Q)}.$$

Again, as long as the kernel is characteristic, this is a metric.

Geometry of Kernel Density Estimates

A kernel density estimate is a powerful modeling tool that allows one to take a finite set of points $P \subset \mathbb{R}^d$ and a smoothing operator, a kernel K, and transform them into a continuous function $\text{KDE}_P : \mathbb{R}^d \to \mathbb{R}$.

The most common kernel is the (normalized) Gaussian kernel

$$K(p, x) = \frac{1}{\sigma^d \sqrt{(2\pi)^d}} \exp\left(-\frac{\|x - p\|^2}{2\sigma^2}\right),$$

which is precisely the evaluation of the Gaussian distribution $\mathcal{G}_d(x)$ with mean p and standard deviation σ evaluated at x. The coefficient ensures the integral $\int_{x \in \mathbb{R}^d} K(p, x) \mathrm{d}x = 1$ for all choices of p. Thus it can be interpreted as taking the "mass" of a point p and spreading it out according to this probability distribution.

Now, a *kernel density estimate* is defined at a point $x \in \mathbb{R}^d$ as

$$\text{KDE}_P(x) = \frac{1}{|P|} \sum_{p \in P} K(p, x) = \frac{1}{|P|} \kappa(P, \{x\}).$$

That is, it gives each point $p \in P$ a mass of $\frac{1}{|P|}$ (each weighted uniformly), then spreads each of these $|P|$ masses out using the Gaussian kernels, and sums them up. While P is discrete in nature, the kernel density estimate KDE_P allows one to interpolate between these points in a smooth natural way.

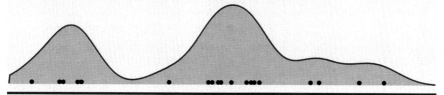

These set of vector similarities, and related structures, play a large role in non-linear data analysis. There are two ways to conceptualize this. First, each use of the Euclidean dot product (which will be quite ubiquitous in this book) is simply replaced with a kernel, which is not a linear operator on $\mathbb{R}^d$; this is often called the "kernel trick". Second, one can think of the functions $K(p, \cdot)$ or more generally $\kappa(P, \cdot)$ as an element of a function space, which acts like an infinite-dimensional vector space. For positive definite kernels, this space is called a *reproducing kernel Hilbert space* (RKHS) where $\kappa(P, Q)$ serves as the inner product, and thus $\sqrt{\kappa(P, P)}$ is the norm. Although one cannot represent this data exactly as vectors in this space, the structure and properties of this RKHS provide the mathematical validation for replacing the Euclidean inner product $\langle p, x \rangle$ with $K(p, x)$.

4.6 Locality Sensitive Hashing

This chapter so far has surveyed numerous different distances and similarities, which are themselves a small subset of all distances and similarities one may consider when modeling data and an analysis task. While metric properties and other nice mathematical properties are useful, another key concern is the computational cost associated with using the distances and similarities. This is also not just the case of a single evaluation, but when comparing a very large set of objects (say of $n = 100$ million objects). Then how can one quickly determine which ones are close, or given a query object q, which ones in the large set are close to this query?

In 1-dimensional Euclidean data, these problems are relatively simple to address. Start by sorting all objects, and storing them in sorted order. Then the nearby objects are the adjacent ones in the sorted order. And if this ordering induces a balanced binary tree, then on a query, nearby objects can be found in logarithmic running time (i.e., time proportional to $\log n$).

Next we introduce *locality sensitive hashing* to address these question for higher-dimensional vectorized objects and for metrics on set-based representations. The main idea is for a family of objects $\mathcal{B}$, and a similarity $\mathbf{s} : \mathcal{B} \times \mathcal{B} \to [0, 1]$, to define a family of random hash functions $\mathcal{H}$ with roughly the following property.

▷ Locality Sensitive Hash Family ◁

For a similarity function $\mathbf{s}$, a locality sensitive hash family $\mathcal{H}$ is a set of hash functions so for any two objects $p, q \in \mathcal{B}$ that

$$\mathbf{Pr}_{h \in \mathcal{H}}[h(p) = h(q)] \approx \mathbf{s}(p, q).$$

That is for any $p, q \in \mathcal{B}$, a randomly chosen hash function $h \in \mathcal{H}$ will cause those two objects to collide with probability roughly the same as their similarity.

Once a family is defined, it will be useful to randomly select a specific hash function from that family (e.g., the $h \in \mathcal{H}$ is the random variable for the probability). However, once some function h is chosen, then it is a deterministic function.

Example: Simple Gridding LSH for Triangle Similarity

Consider a 1-dimensional data set $X \subset \mathbb{R}^1$, and the triangle similarity

$$\mathbf{s}_\triangle(p, q) = \max\{0, 1 - |p - q|\}.$$

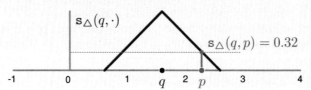

A simple hash family for this is a set of randomly shifted grids, where items in the same grid cell hash together. Define $\mathcal{H}_\Delta = \{h_\eta \mid \eta \in [0, 1)\}$ where $h_\eta(x) = \lceil x + \eta \rceil$. That is each h_η maps the input to an integer (representing the index of a hash bucket) where a grid defines consecutive sets (intervals) of numbers of length 1 which are all mapped to the same hash bucket. The parameter η randomly shifts these grid boundaries. So whether or not a pair of points are in the same grid cell depends on the random choice of η, and more similar points are more likely to be in the same grid cell, and hence the same hash bucket.

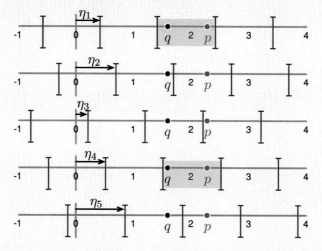

The example above shows 5 hash functions $h_{\eta_1}, h_{\eta_2}, h_{\eta_3}, h_{\eta_4}, h_{\eta_5} \in \mathcal{H}_\Delta$. In this example p and q are in the same hash bucket for 2 of the 5 functions (for h_{η_1} and h_{η_4}), so we would estimate the similarity between p and q as $2/5 = 0.4$.

Indeed, we can verify that for any $p, q \in \mathbb{R}$ that

$$\mathbf{Pr}_{h_\eta \in \mathcal{H}_\Delta}[h_\eta(p) = h_\eta(q)] = \mathbf{s}_\Delta(p, q).$$

For any p, q with $|p - q| > 1$, both the probability and the similarity is 0; and if $p = q$, then both the probability and similarity is 1. The probability that p and q are *not* hashed together is the probability a randomly shifted grid boundary falls between them, which is precisely $|p - q|$ given that $|p - q| \le 1$. Hence the probability and similarity in this case are both $1 - |p - q|$, as desired.

A common and important data structure is a *hash table* that relies on a different type of hash function, which to distinguish it from an LSH, we call a *separating hash function*. These hash functions $h : \mathcal{B} \rightarrow \mathcal{U}$ maps an object $b \in \mathcal{B}$ into a fixed universe $\mathcal{U}$; typically $\mathcal{U}$ can be represented as a set of integers $\{0, 1, 2, \ldots, u - 1\}$, representing indexes of an array of size u. Hash tables are again defined with respect to a family $\mathcal{H}$, and we consider a random choice $h \in \mathcal{H}$. Given this random choice, then a *perfect* separating hash function guarantees that for any two $b, b' \in \mathcal{B}$ that $\mathbf{Pr}_{h \in \mathcal{H}}[h(b) = h(b')] = 1/u$.

It is important to distinguish these two data structures and types of hash functions. Separating hash functions are powerful and useful, but are *not* locality sensitive hash functions.

4.6.1 Properties of Locality Sensitive Hashing

More generally, locality sensitive hash families are defined with respect to distance **d**. A hash family $\mathcal{H}$ is $(\gamma, \phi, \alpha, \beta)$-*sensitive* with respect to **d** when it has the following properties:

- $\mathbf{Pr}_{h \in \mathcal{H}}[h(p) = h(q)] > \alpha$ if $\mathbf{d}(p, q) < \gamma$
- $\mathbf{Pr}_{h \in \mathcal{H}}[h(p) = h(q)] < \beta$ if $\mathbf{d}(p, q) > \phi$

For this to make sense we need $\alpha > 0$ and $\beta < 1$ for $\gamma \le \phi$. Ideally we want α to be large, β to be small, and $\phi - \gamma$ to be small. Then we can repeat this with more random hash functions to *amplify* the effect using concentration of measure bounds; this can be used to make α larger and β smaller while fixing γ and ϕ. Ultimately, if α becomes close to 1, then we almost always hash items closer than γ in the same bucket, and as β becomes close to 0, then we almost never hash items further than ϕ in the same bucket. Thus to define similar items (within a $\phi - \gamma$ tolerance) we simply need to look at which ones are hashed together.

Example: Sensitivity for Triangle Similarity

Revisit the triangle similarity $\mathbf{s}_\triangle$ and its associated grid-based hash family $\mathcal{H}_\triangle$. Recall for any objects p, q we have $\mathbf{Pr}_{h \in \mathcal{H}_\triangle}[h(p) = h(q)] = \mathbf{s}_\triangle(p, q)$. Then for some similarity threshold $\tau = \mathbf{d}_\triangle(p, q) = 1 - \mathbf{s}_\triangle(p, q)$ we can set $\tau = \gamma = \phi$ and have $\alpha = 1 - \tau$ and $\beta = \tau$.

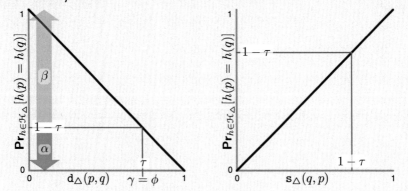

This same setting (and picture) can be used for any threshold $\tau \in (0, 1)$ where the similarity, hash family pair $(\mathbf{s}, \mathcal{H})$ satisfies $\mathbf{Pr}_{h \in \mathcal{H}}[h(p) = h(q)] = \mathbf{s}(p, q)$.

This $(\gamma, \phi, \alpha, \beta)$-sensitivity condition is indeed more complicated than the $\mathbf{Pr}[h(a) = h(b)] = \mathbf{s}(a, b)$ condition. But it is more general, and any such separation will be all that is required to allow the LSH data structure to work, as elaborated on next. Moreover, this flexibility is required for certain important distances like Euclidean distance $\mathbf{d}_2$.

4.6.2 Prototypical Tasks for LSH

This LSH machinery is useful to answer three types of tasks given a large data set P (which could have size $|P|$ in the hundreds of millions):

- For a threshold τ, determine all pairs $p, p' \in P$ so that $\mathbf{d}(p, p') \leq \tau$
  (or symmetrically, so that $\mathbf{s}(p, p') \geq \tau'$, e.g., with $\tau' = 1 - \tau$).
- For a threshold τ and a query q, return all $p \in P$ so that $\mathbf{d}(p, q) \leq \tau$.
- For a query q, return point $\tilde{p} \in P$ so that it is approximately $\mathrm{argmin}_{p \in P} \, \mathbf{d}(p, q)$.

The first two tasks are important in deduplication and plagiarism detection. For instance, this task occurs when a search engine wants to avoid returning two webpages which are very similar in content, or an instructor wants to quickly check if two students have turned in very similar assignments to each other, or very similar to one from an online solution repository.

In each case, we desire these tasks to take time roughly proportional to the number of items within the distance threshold τ. In very high-dimensional Euclidean space, or in the space of sets, a brute force checking of all distances would require time proportional to $|P|^2$ for the first task, and proportional to $|P|$ time for the second task. This can be untenable when for instance the size of $|P|$ is hundreds of millions.

The third task, as we will see in the classification section, is an essential tool for data analysis. If we know how the objects $p \in P$ in the database P behave, then we can guess that q will behave similarly to the closest known example $p^* = \mathrm{argmin}_{p \in P} \, \mathbf{d}(p, q)$. We can solve for p^* by invoking the second task multiple times while doing a binary-like search on τ; if the threshold is too large, we short cut the operation of the second task as soon as we find more than a constant number of τ-close items, and compare those directly.

The first task can also essentially be reduced to the second one, but just running the second task with each $p \in P$ as a query. So our description below will focus on the second task.

In all of these tasks, we will be able to solve them by allowing some approximation. We may not return the closest item, but one that is not too much further than the closest item. We may return some items slightly beyond our distance threshold and may miss some which are barely closer than it. Given that the choice of distance or similarity used is a modeling choice (why not use a different one?), we should not take its exact value too seriously.

4.6.3 Banding to Amplify LSH

If we only use a single hash function $h \in \mathcal{H}$ where $h : \mathcal{B} \to U$, then we can design a hash table over the universe $U = \{0, 1, 2, \ldots, u - 1\}$. That is, if item $h(p) = j$, then we store it in a linked list L_j stored in location j in a length u array. We can then, given a query item q so $h(q) = j$, access all items which hash to j in time proportional to however many items satisfy this condition.

While this is efficient, it is not selective, and has many false positives. There are two simple ways to retain the efficiency while either increasing the selectivity or decreasing false positives.

To increase the selectivity, label items $p \in P$ as not close to q unless multiple hash functions $h_1, h_2, \ldots, h_b$ all report that $h_i(p) = h_i(q)$. This can be handled efficiently with a hash table by creating a banded-hash $H(p) : \mathcal{B} \to U^b$, that is, it maps each object p to a b-dimensional vector, each element in U, where the ith coordinate is determined by the ith hash function. We can again create a hash table over these u^b possible entries (using standard tricks from data structures to avoid u^b space). However, this is too selective, and reduces the probability that any two items hash to the same bucket in the banded-hash table.

To reduce false negatives, again consider multiple hash functions $h_1, h_2, \ldots, h_r$, but now report two items similar if they collide in *any* hash function. If we keep r separate hash tables which store the contents of P, then we can report the union of the items which collide with q. However, this approach is not selective enough, it increases the probability that any two items hash to the same bucket.

Banding puts these together. We create r banded-hash functions H, with each composed of b hash functions. This requires a total of $r \cdot b$ hash functions, and a query to be hashed by all of these. Then if q intersects with *any* p in any banded-hash table (which requires *all* hash functions to collide), we return it as similar.

Example: Banding

Consider $r = 3$ banded-hash functions H_1, H_2, and H_3, where each banded-hash function H_i has $b = 2$ hash functions $(h_{i,1}, h_{i,2})$; each maps to a universe $U = \{0, 1, 2, \ldots 5\}$. We apply these hash functions to 5 points in a set P and a query point q, and the results are shown in the table below. We return objects close to q if and only if they have an exact match in any banded-hash function.

hash	$h_{1,1}$	$h_{1,2}$	$h_{2,1}$	$h_{2,2}$	$h_{3,1}$	$h_{3,2}$	close?
q	3	5	1	2	3	2	
p_1	[3,	5]	0	2	4	3	yes
p_2	3	4	1	0	4	2	no
p_3	0	2	4	3	5	1	no
p_4	0	1	[1,	2]	0	2	yes
p_5	4	5	3	5	4	1	no

We observe that only p_1 and p_4 are designated as close. Object p_1 collides on H_1 since both p_1 and q have $h_{1,1} = 3$ and $h_{1,2} = 5$. Object p_4 collides on H_2 since both p_4 and q have $h_{2,1} = 1$ and $h_{2,2} = 2$. Note that although p_2 collides with q on 3 individual hash functions ($h_{1,1} = 3$, $h_{2,1} = 1$, and $h_{3,2} = 2$) it never has an entire banded-hash function collide, so it is not marked as close.

Analysis of Banding

We will analyze the simple case where there is a hash family $\mathcal{H}$ so that we have $\mathbf{Pr}_{h \in \mathcal{H}}[h(p) = h(q)] = \mathbf{s}(p, q)$. Let $s = \mathbf{s}(p, q)$, and we will analyze the case where we use r bands with b hash functions each.

$$
\begin{aligned}
s &= \text{probability of collision on any one hash function} \\
s^b &= \text{probability all hash functions collide in 1 band} \\
(1 - s^b) &= \text{probability not all collide in 1 band} \\
(1 - s^b)^r &= \text{probability that in no band, do all hashes collide} \\
f_{b,r}(s) = 1 - (1 - s^b)^r &= \text{probability all hashes collide in at least 1 band}
\end{aligned}
$$

So this function $f_{b,r}(s) = 1 - (1 - s^b)^r$ describes the probability that two objects of similarity s are marked as similar in the banding approach. A similar, but slightly messier, analysis can be applied for any $(\gamma, \phi, \alpha, \beta)$-sensitive hash family.

Example: Plotting the LSH similarity function

We plot $f_{b,r}$ as an S-curve where on the x-axis $s = \mathbf{s}(p, q)$ and the y-axis represents the probability that the pair p, q is detected as close. In this example we have $t = r \cdot b = 15$ with $r = 5$ and $b = 3$.

s	0.1	0.2	0.3	0.4	0.5	0.6	0.7	0.8	0.9
$f_{3,5}(s)$	0.005	0.04	0.13	0.28	0.48	0.70	0.88	0.97	0.998

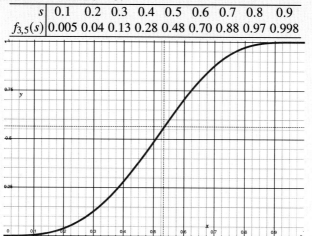

Next we change the r and b parameters and observe what happens to the plot of the S-curve $f_{r,b}(s)$. We show combinations of values $r \in \{4, 8, 16, 32\}$ and $b \in \{2, 4, 6, 8\}$. The value b is fixed in each row, and increases as the rows go down. The value r is fixed in each column, and increases as the columns go from left to right.

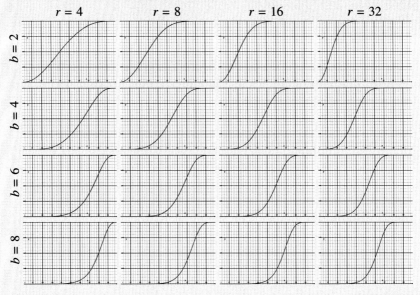

We see that as $r \cdot b$ increases, the curve becomes steeper. As r increases (move to the right), the curve shifts to the left, that is a smaller similarity is likely designated as close. As b increases (move down), the curve shifts to the right, that is a larger similarity is likely designated as close.

Choice of r and b

Usually there is a budget of t hash function one is willing to use. Perhaps it is $t = (1/2\varepsilon^2) \ln(2/\delta)$; e.g., the number required by a Chernoff-Hoeffding bound to estimate the similarity within ε-error, with probability at least $1 - \delta$. Then how does one divvy them up among r and b to $t = r \cdot b$?

The threshold τ' where f has the steepest slope is about $\tau' \approx (1/r)^{1/b}$. So given a similarity value $s = \mathbf{s}(p, q)$, if we want to return objects more similar to τ' (e.g., $\tau' = s = \alpha = 1 - \beta$) we can solve for $b = t/r$ in $\tau' = (b/t)^{1/b}$ to (very roughly) yield $b \approx -\log_{\tau'}(t)$ and set $r = t/b = t/(-\log_{\tau'}(t))$. This is only a rule-of-thumb. It is often best to plot the f function and play with values of b and r to find your desired trade-off, but this gives a good place to start searching.

If there is no budget on r and b, as they increase, the S-curve produced by $f(s)$ gets sharper and sharper and provides a more accurate retrieval of all and only the close data points.

4.6.4 LSH for Angular Distance

Locality sensitive hash families can be developed for more than just the simplistic 1-dimensional triangle similarity. Here we describe such a family for the angular distance $\mathbf{d}_{ang}$ and similarity $\mathbf{s}_{ang}$. Recall these functions operate on vectors $a, b \in \mathbb{S}^{d-1}$, which are unit vectors in $\mathbb{R}^d$.

A hash h_u from hash family $\mathcal{H}_{ang}$ is defined as

$$\mathcal{H}_{ang} = \left\{ h_u(\cdot) = \text{sign}(\langle \cdot, u \rangle) \mid u \sim \text{Unif}(\mathbb{S}^{d-1}) \right\}.$$

Let us unpack this. First choose a uniform random unit vector $u \in \mathbb{S}^{d-1}$, that is, so any unit vector in $\mathbb{R}^d$ is equally likely to be chosen ($u \sim \text{Unif}(\mathbb{S}^{d-1})$). Then $h_u(p) = \text{sign}(\langle a, u \rangle)$, that is, if the dot product of a and u is positive, it returns $+1$, and if it is negative, it returns -1.

We will show at $\mathbf{Pr}_{h_u \in \mathcal{H}_{ang}}[h_u(a) = h_u(b)] = \mathbf{s}_{ang}(a, b)/\pi$ for any $a, b \in \mathbb{S}^{d-1}$. Thus, after scaling by π, the amplification analysis for this similarity and hash family pair is exactly the same as with the simple triangle similarity example.

Geometry of Angular LSH

Here we see why $\mathbf{Pr}_{h_u \in \mathcal{H}_{ang}}[h_u(a) = h_u(b)] = \mathbf{s}_{ang}(a, b)/\pi$.

First consider data points $a, b \in \mathbb{S}^1$, the circle. Recall they induce a distance $\mathbf{d}_{ang}(a, b) \in [0, \pi]$ which is the minimum arclength between them. Each point corresponds to a set $F_{a,\perp} \subset \mathbb{S}^1$ of all points $u \in \mathbb{S}^1$ so $\langle a, u \rangle \geq 0$; it corresponds to a halfspace with boundary through the origin, and with normal a, when embedded in $\mathbb{R}^2$. A vector u results in $h_u(a) \neq h_u(b)$ only if it is in exactly one of $F_{a,\perp}$ or $F_{b,\perp}$. The arclength of the xor of these regions is $2 \cdot \mathbf{d}_{ang}(a, b)$; it includes example vector u' but not u.

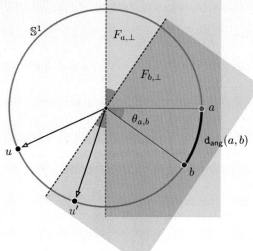

A random unit vector $u \in \mathbb{S}^1$ can be generated with a uniform value $\xi \in [0, 2\pi]$ and walking along the circle clockwise from $(1, 0)$ an arclength of ξ. Hence, the fraction of such u for which $h_u(a) \neq h_u(b)$ is $(2\mathbf{d}_{\text{ang}}(a, b))/(2\pi)$. And since $\mathbf{s}_{\text{ang}}(a, b)/\pi = 1 - \mathbf{d}_{\text{ang}}(a, b)/\pi$, then as desired, the probability $h_u(a) = h_u(b)$ is $\mathbf{s}_{\text{ang}}(a, b)/\pi$.

The more general case for points $p, q \in \mathbb{S}^{d-1}$ reduces to this analysis on $\mathbb{S}^1$. The great circle which passes through $p, q \in \mathbb{S}^{d-1}$ is an instance of $\mathbb{S}^1$. The hash signs of $h_u(a)$ and $h_u(b)$ are still induced by halfspaces $F_{a,\perp}$ and $F_{b,\perp}$ in $\mathbb{R}^d$ mapped to halfspaces in $\mathbb{R}^2$. And while there are now several ways to generate the random unit vectors in $\mathbb{S}^{d-1}$, a sufficient way is to first generate the "coordinate" as $\xi \in [0, 2\pi]$ as it pertains to the great circle through p, q on $\mathbb{S}^1$, and then generate the remaining "coordinates" which do not affect the inclusion in $F_{a,\perp}$ and $F_{b,\perp}$.

Generating Random Unit Vectors

A missing algorithmic element of this hash family is generating a random unit vector. While generating a random uniform scalar is assumed standard, random unit vectors require the following observations. For instance, generating d uniform vectors $u_1, u_2, \ldots, u_d \sim \text{Unif}[-1, 1]$ and then normalizing $u = (u_1, u_2, \ldots, u_d) \rightarrow u/\|u\|$ does not give the desired result. It places too much probability near vector $(1/\sqrt{d}, 1/\sqrt{d}, \ldots, 1/\sqrt{d})$ and similar vectors with signs flipped (i.e., the corners).

The easiest way to generate a random unit vector is through Gaussian random variables. A d-dimensional Gaussian distribution $\mathcal{G}_d$ is defined as

$$\mathcal{G}_d(x) = \frac{1}{(2\pi)^{d/2}} e^{-\|x\|_2^2/2}.$$

If we have two uniform random numbers $u_1, u_2 \sim \text{Unif}[0, 1]$ then we can generate two independent 1-dimensional Gaussian random variables as (using the *Box-Muller transform*)

$$y_1 = \sqrt{-2 \ln(u_1)} \cos(2\pi u_2)$$
$$y_2 = \sqrt{-2 \ln(u_1)} \sin(2\pi u_2).$$

A uniform Gaussian has the (*amazing!*) property that all coordinates (in any orthogonal basis) are independent of each other. Thus to generated a point $x \in \mathbb{R}^d$ from a d-dimensional Gaussian, for each coordinate i we assign it the value of an independent 1-dimensional Gaussian random variable.

Moreover, this amazing property for a d-dimensional Gaussian random variable $x \sim \mathcal{G}_d$ implies that if it is projected to any 1-dimensional subspace, the result is a 1-dimensional Gaussian distribution. That is for any unit vector v and $x \sim \mathcal{G}_d$, then $\langle x, v \rangle \sim \mathcal{G}_1$. This means it does not favor any directions (as the d-variate uniform distribution $(\text{Unif}[-1, 1])^d$ favors the "corners"), and hence if we normalize $u \leftarrow x/\|x\|_2$ for $x \sim \mathcal{G}_d$, then $u \sim \text{Unif}(\mathbb{S}^{d-1})$, as desired.

4.6.5 LSH for Euclidean Distance

The LSH hash family for Euclidean distance $\mathbf{d}_2$ is a combination of the ideas for angular distance and for the triangle similarity. However, the probability does not work out quite as nicely so we require the more general notion of $(\gamma, \phi, \alpha, \beta)$-sensitivity.

The hash family $\mathcal{H}_{L_2, \tau}$ requires a desired similarity parameter τ, which was implicit in the scaling of the triangle similarity, but not present in the angular similarity. The set is defined as

$$\mathcal{H}_{L_2, \tau} = \left\{ h_{u,\eta}(\cdot) = \lceil \langle \cdot, u \rangle + \eta \rceil \mid u \sim \mathsf{Unif}(\mathbb{S}^{d-1}), \eta \sim \mathsf{Unif}[0, \tau] \right\}.$$

That is, it relies on a random unit vector u, and a random offset η. The hash operation $h_{u,\eta}(x) = \lceil \langle x, u \rangle + \eta \rceil$ first projects onto direction u, and then offsets by η in this directions, and rounds up to an integer. For a large enough integer t (e.g., if there are at most n data points, then using $t = n^3$), it is also common to instead map to a finite domain $0, 1, 2, \ldots, t - 1$ as $h_{u,\eta}(x) \mod t$.

The hash family $\mathcal{H}_{L_2, \tau}$ is $(\tau/2, 2\tau, 1/2, 1/3)$-sensitive with respect to $\mathbf{d}_2$.

Geometry of Euclidean LSH Sensitivity

We show here that $\mathcal{H}_{L_2, \tau}$ is $(\tau/2 = \gamma, 2\tau = \phi, 1/2 = \alpha, 1/3 = \beta)$-sensitive with respect to $\mathbf{d}_2$.

To see that we do not miss too many real collisions, observe that the projection onto u is contractive. So for $a, b \in \mathbb{R}^d$, then $|\langle a, u \rangle - \langle b, u \rangle| \leq \|a - b\|$. Thus as with the 1-dimensional triangle similarity, if $\|a - b\| \leq \tau/2 = \gamma$ then $|\langle a, u \rangle - \langle b, u \rangle| \leq \tau/2$ and a, b fall in the same bin with probability at least $1/2 = \alpha$.

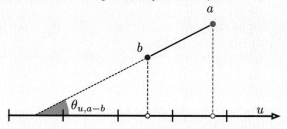

To see that we do not observe too many spurious collisions, we can use an argument similar to that for angular distance. If $\|a - b\| > 2\tau = \phi$ and they collide, then the vector u must be sufficiently orthogonal to $(a - b)$ so the norm is reduced by half; otherwise if $|\langle a, u \rangle - \langle b, u \rangle| \geq \tau$, they cannot be in the same bin. Formally

$$|\langle a, u \rangle - \langle b, u \rangle| = |\langle a - b, u \rangle| = \|a - b\| \|u\| |\cos \theta_{u, a-b}|,$$

where $\theta_{u, a-b}$ is the angle (in radians) between u and $a - b$. We know $|\cos(\theta_{u, a-b})| \leq 1/2$ only if $\theta_{u, a-b} \geq \pi/3$. Fixing the 2-dimensional subspace which includes $0, u$, and $a - b$, then we can see that the probability $\theta_{u,a-b} \geq \pi/3$ is at most $1/3 = \beta$ (that is there are $(2\pi/3)$ options for u out of 2π total).

---

p-Stable Random Variables and LSH for L_p-Norms

There exists a beautiful extension to create similar LSH hash families $\mathcal{H}_{L_p,\tau}$ for any L_p distance with $p \in (0, 2]$. The main difference is to replace the $u \sim \text{Unif}(\mathbb{S}^{d-1})$ which is really some $x \sim \mathcal{G}_d$ with another d-dimensional random variable from what is known as a p-stable distribution.

A distribution μ over $\mathbb{R}$ is p-*stable* (for $p \geq 0$) if the following holds. Consider $d + 1$ random variables $X_0, X_1, \ldots, X_d \sim \mu$. Then considering any d real values $\{v_1, \ldots, v_d\}$, the random variable $\sum_i v_i X_i$ has the same distribution as $(\sum_i |v_i|^p)^{1/p} X_0$. Such p-stable distributions exist for $p \in (0, 2]$. Special cases are

- The Gaussian distribution $\mathcal{G}(x) = \frac{1}{\sqrt{2\pi}} \exp(-x^2/2)$ is 2-stable.
- The Cauchy distribution $\mathcal{C}(x) = \frac{1}{\pi} \frac{1}{1+x^2}$ is 1-stable.

Intuitively, this allows us to replace a sum of random variables with a single random variable by adjusting coefficients carefully. But actually, it is the composition of the coefficients which is interesting.

In particular, for $p = 2$ and a vector $v = (v_1, v_2, \ldots, v_d)$ where we want to estimate $\|v\|_2 = (\sum_i v_i^2)^{1/2}$, we can consider each coordinate v_i individually as an estimate by using a p-stable random variable. That is, we can estimate $\|v\|_2$ by choosing $d + 1$ random Gaussian vectors $g_0, g_1, \ldots, g_d \sim \mathcal{G}$ and calculating $(1/g_0) \sum_i g_i v_i = (1/g_0)\langle g, v \rangle$, where $g = (g_1, \ldots, g_d)$. Note that by dividing by g_0, we are approximately correctly normalizing the d-dimensional Gaussian random variable. This turns out to be sufficient for Euclidean LSH and other similar methods for high-dimensional data we will see later.

Using the Cauchy random variables $c_0, c_1, \ldots, c_d \sim \mathcal{C}$ in place of the Gaussian ones allows us to estimate $\|v\|_1 = \sum_i |v_i|$ as $(1/c_0)\langle c, v \rangle$ with $c = (c_1, \ldots c_d)$.

4.6.6 Min Hashing as LSH for Jaccard Distance

The final example of LSH we will provide is for the Jaccard distance, which is defined on sets. This hash family is $\mathcal{H}_J$ again and has the nice property that $\mathbf{Pr}_{h_\sigma \in \mathcal{H}_J}[h_\sigma(A) = h_\sigma(B)] = \mathbf{s}_J(A, B)$, so the previous amplification analysis applies directly. Consider sets defined over a domain Ω of size n, and define Π_n as the set of all permutations from Ω to distinct integers in $[n] = \{1, 2, 3, \ldots, n\}$. So for each element $\omega \in \Omega$ then each $\sigma \in \Pi_n$ maps $\sigma(\omega) = i$ for some $i \in [n]$, and for $\omega \neq \omega'$ then $\sigma(\omega) \neq \sigma(\omega')$. In practice we can relax this $\neq$ restriction as we discuss below. Then we define

$$\mathcal{H}_J = \left\{ h_\sigma(A) = \min_{\omega \in A} \sigma(\omega) \mid \sigma \sim \text{Unif}(\Pi_n) \right\}.$$

That is, the hash function h_σ applies σ to each ω in A, and then returns the smallest value out of all of them. Due to the return of the smallest value, this type of approach is often called a *min hash*.

Example: Min Hashing

Consider two sets $A = \{a, b, e, f, h\}$ and $B = \{b, c, e, g, j\}$ over the domain $\Omega = \{a, b, c, d, e, f, g, h, i, j\}$. We can now choose two permutation functions $\sigma_1, \sigma_2 \in \mathcal{H}_J$:

	a	b	c	d	e	f	g	h	i	j
σ_1	4	6	3	1	2	8	9	5	10	7
σ_2	3	7	8	5	9	10	1	4	6	2

These permutations induce hash functions $h_{\sigma_1}, h_{\sigma_2} \in \mathcal{H}_J$. These can then be applied to sets A and B as

	A	B
h_{σ_1}	2	2
h_{σ_2}	3	1

For instance

$$h_{\sigma_1}(A) = \min\{\sigma_1(a), \sigma_1(b), \sigma_1(e), \sigma_1(f), \sigma_1(h)\} = \min\{4, 6, 2, 8, 5\} = 2.$$

Observe that the Jaccard similarity $\mathbf{s}_J(A, B) = \frac{2}{8}$. On the hash functions there is a collision for $h_{\sigma_1}(A) = h_{\sigma_1}(B)$, but not for h_{σ_2}, so they would estimate the similarity as $\frac{1}{2}$.

It is useful to understand why $\mathbf{Pr}_{h_\sigma \in \mathcal{H}_J}[h_\sigma(A) = h_\sigma(B)] = \mathbf{s}_J(A, B)$. We think of three types of elements from Ω: the objects $\Omega_{A \cap B}$ which are in both A and B; the objects $\Omega_{A \vee B}$ which are in either A or B, but not both; and the objects $\Omega_{\notin A, B}$ which are in neither A nor B. Note that the Jaccard similarity is precisely $\mathbf{s}_J(A, B) = \frac{|\Omega_{A \cap B}|}{|\Omega_{A \cap B}| + |\Omega_{A \vee B}|}$. Recall, the similarity has no dependence on the set $\Omega_{\notin A, B}$ hence we actually do not need to know all of Ω. But we can also see that the probability of a hash collision is precisely $\frac{|\Omega_{A \cap B}|}{|\Omega_{A \cap B}| + |\Omega_{A \vee B}|}$. Any value returned by h_σ can only be an element from $\Omega_{A \cap B} \cup \Omega_{A \vee B}$, these are equally likely, and it is only a collision if it is one from $\Omega_{A \cap B}$.

Fast Min Hash Functions

So this definition of min hash functions is easy to analyze but it is suboptimal. It needs to define and store a hash function from Ω to $[|\Omega|]$. But we would like to use these ideas over domains such as IP addresses or words or strings where there is an enormous implicit domain Ω that is often too big to store. An alternative is to replace σ with another separating hash function $h : \Omega \to [N]$ (see Section 11.1) where N is a sufficiently large value (such as the size of the largest set we expect to encounter to the power 3). There will now be some chance of collisions between objects in any set

A, but it will be small enough to have insignificant effect on the analysis. Moreover, most programming languages include built-in such hash functions that can operate on strings or large integers or double precisions values (e.g., using the SHA-1 hash with a salt). Let f be a built-in hash function that operates on strings Σ^k, and maps to some range $[N]$. Let $\oplus$ be a concatenation function so for input string $\omega \in \Sigma^k$ and S another string, then $\omega \oplus S$ is their concatenation, also a string. Then we can define a hash function f_S as

$$f_S(\omega) = f(\omega \oplus S).$$

Now we can define our hash family for Jaccard similarity using min hashing, defined on a set of objects A as

$$\mathcal{H}'_J = \{h_{f,S}(A) = \min_{\omega \in A} f_S(\omega) \mid f : \Omega \to [N], S \sim \text{random salt}\}.$$

This formulation leads to a simple algorithm, where given a set A, we can scan over the set and maintain the value of the element that is the smallest. When using several hash functions $h_1, \ldots, h_k \in \mathcal{H}'_J$, this leads to a *minhash signature*, a vector $v(A) \in [N]^k$; that is, the jth coordinate $v(A)_j = h_j(A)$. Given just a single built-in separating hash function f, and a set of k salts $S_1, S_2, \ldots, S_k$, we can still make a single scanning pass over A, and maintain all k coordinates.

≫ Efficient Min Hashing Signature

It is possible to maintain the minhash signature $v(A)$ in a single pass over the set A, updating each of the k coordinates, while processing each $\omega \in A$.

Algorithm 4.6.1 Efficient MinHash Signature$(A, \{S_1, \ldots, S_k\})$

Initialize $v = (v_1 = \infty, v_2 = \infty, \ldots, v_k = \infty)$
for $\omega \in A$ **do**
    for $j = 1$ to k **do**
        if $f_{S_j}(\omega) < v_j$ **then**
            $v_j \leftarrow f_{S_j}(\omega)$
return $v(A) = v$

If the set A is the k-gram representation of a text document, then this process can be even more streamlined. Rather than first creating set A, and then using this approach to calculate the minhash signature, one can scan the document as the first **for** loop above. Each k-gram ω generated on the fly can be hashed by the k hash functions, and compared against the maintained minhash signature. Duplicate k-grams will not affect the minhash value (since they are the same or greater, and never less than the existing value).

Exercises

4.1 Consider two vectors $a = (1, 2, -4, 3, -6)$ and $b = (1, 2, 5, -2, 0) \in \mathbb{R}^5$.

1. Calculate the $\mathbf{d}_1(a, b)$, $\mathbf{d}_2(a, b)$, $\mathbf{d}_0(a, b)$, and $\mathbf{d}_\infty(a, b)$, and sort the answers.
2. Without calculating $\mathbf{d}_3(a, b)$, explain where it will fall in the sorted order.
3. Does it make sense to normalize these vectors so they lie in Δ^4 (i.e., by dividing by $\|a\|_1$ and $\|b\|_1$, respectively), and use the Kullback-Liebler divergence $\mathbf{d}_{KL}$ on them? Why or why not?
4. Normalize the data sets to live on $\mathbb{S}^4$, and compute the cosine distance between them.

4.2 Consider sets $A = \{1, 2, 4, 8\}$ and $B = \{1, 2, 3\}$.

1. Calculate the Jaccard distance $\mathbf{d}_J(A, B)$.
2. Compute the following similarities if they are well-defined.

- Jaccard similarity $\mathbf{s}_J(A, B)$
- Hamming similarity $\mathbf{s}_{Ham}(A, B)$
- Andberg similarity $\mathbf{s}_{Andb}(A, B)$
- Rogers-Tanimoto similarity $\mathbf{s}_{RT}(A, B)$
- Soerensen-Dice similarity $\mathbf{s}_{Dice}(A, B)$

4.3 Find a set of 3 points in $\mathbb{R}^2$ so that $\mathbf{d}_{1/2}$ (the L_p distance with $p = 1/2$) does not satisfy the triangle inequality. Show that $\mathbf{d}_{1/2}$ still satisfies the other properties of a metric.

4.4 Find a set of 3 points in $\mathbb{S}^1$ so that $\mathbf{d}_{cos}$ does not satisfy the triangle inequality.

4.5 Find a set of 3 points $a, b, c \in \mathbb{R}^2$ so that if we sort the pairwise distances measured by $\mathbf{d}_2$ and $\mathbf{d}_1$ they are different.

4.6 Consider the two pieces of text:

T1   how much wood could a wood chuck chuck if a wood chuck
    could chuck wood
T2   a wood chuck could chuck a chuck of wood if a wood chuck
    could chuck wood

For both pieces of text create and report the k-grams.

1. With $k = 2$ and words.
2. With $k = 2$ and characters (treat a space as a character).
3. With $k = 3$ and characters (treat a space as a character).

For each pair or representations, calculate the Jaccard distance.

4.7 Find 4 news articles or websites on the internet (each at least 500 words) so that the k-grams of their text have significantly different pairwise Jaccard distances. Two should be very similar (almost duplicates), one similar to the duplicates, and one fairly different.

1. Report the texts used, the type of k-gram used, and the resulting Jaccard distances.
2. Build Min Hash signatures of length t (using t hash functions) to estimate the Jaccard distance between these sets of k-grams. Try various sizes t (between 10 and 500), and report a value sufficient to reliably empirically distinguish the three classes of distances (almost-duplicate, similar, and different).

4.8 Consider computing an LSH using $t = 216$ hash functions. We want to find all object pairs which have Jaccard similarity above threshold $\tau = .75$.

1. Estimate the best values of hash functions b within each of r bands to provide the S-curve
$$f(s) = 1 - (1 - s^b)^r$$
with good separation at τ. Report these values b, r.
2. Consider the 4 objects A, B, C, D, with the following pairwise similarities:

	A	B	C	D
A	1	0.75	0.25	0.35
B	0.75	1	0.1	0.45
C	0.25	0.1	1	0.92
D	0.35	0.45	0.92	1

Use your choice of r and b and $f(\cdot)$ designed to find pairs of objects with similarity greater than τ. What is the probability, for each pair of the four objects, of being estimated as similar (i.e., similarity greater than $\tau = 0.75$)? Report 6 numbers.

4.9 Explore random unit vectors: their generation and their average similarity.

1. Provide the code to generate a single random unit vector in $d = 10$ dimensions using only the operation $u \leftarrow \text{Unif}(0, 1)$ which generates a uniform random variable between 0 and 1. *(This $u \leftarrow \text{Unif}(0, 1)$ operation will be called multiple times.)* Other operations (linear algebraic, trigonometric, etc) are allowed, but no other sources of randomness.
2. Generate $t = 150$ unit vectors in $\mathbb{R}^d$ for $d = 10$. Plot cdf of their pairwise dot products (yes, calculate $\binom{t}{2}$ dot products).
3. Repeat the above experiment to generate a cdf of pairwise dot products in $d = 100$.
4. Now repeat the above step with $d = 100$ using Angular Similarity instead of just the dot product.

4.10 Consider 5 data points in one dimension:

x_1	x_2	x_3	x_4	x_5
1.1	1.2	1.5	2.0	2.3

1. Report the triangle similarity for each pair (report 10 numbers).
2. Use $t = 20$ hash functions from the hash family $\mathcal{H}_\Delta$ to estimate the triangle similarity. Report the 10 estimated values.
3. Report how large t needs to be to estimate the triangle similarity with error at most 0.05 on *all 10* pairs. Report the t and the 10 estimates and their errors.

Chapter 5
Linear Regression

Abstract This chapter introduces the model of linear regression and some simple algorithms to solve for it. In the simplest form it builds a linear model to predict one variable from one other variable or from a set of other variables. We will demonstrate how this simple technique can extend to building potentially much more complex polynomial models. To harness this model complexity, we introduce the central and extremely powerful idea of cross-validation; this method fundamentally changes the statistical goal of validating a model, to characterizing how it interacts with the data. Next we introduce basic techniques for regularized regression including ridge regression and lasso, which can improve the performance, for instances as measured under cross-validation, by introducing bias into the derived model. Some of these approaches, like lasso, also bias toward sparse solutions which can be essential for high-dimensional and over-parameterized data sets. We provide simple algorithms toward these goals using matching pursuit, and explain how these can be used toward compressed sensing.

5.1 Simple Linear Regression

We will begin with the simplest form of linear regression. The input is a set of n 2-dimensional data points $(X, y) = \{(x_1, y_1), (x_2, y_2), \ldots, (x_n, y_n)\}$. The ultimate goal will be to predict the y values using only the x values. In this case x is the *explanatory variable* and y is the *dependent variable*.

The notation (X, y) with an uppercase X and lowercase y will become clear later. This problem commonly generalizes where the x-part (explanatory variables) becomes multidimensional, but the y-part (the dependent variable) remains 1-dimensional.

In order to do this, we will "fit" a line through the data of the form

$$y = \ell(x) = ax + b,$$

© Springer Nature Switzerland AG 2021
J. M. Phillips, *Mathematical Foundations for Data Analysis*,
Springer Series in the Data Sciences,
https://doi.org/10.1007/978-3-030-62341-8_5

where a (the slope) and b (the intercept) are parameters of this line. The line ℓ is our "model" for this input data.

Example: Fitting a line to height and weight

Consider the following data set that describes a set of heights and weights.

height (in)	weight (lbs)
66	160
68	170
60	110
70	178
65	155
61	120
74	223
73	215
75	235
67	164
69	?

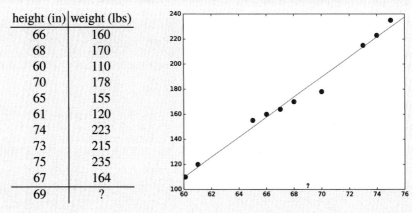

Note that in the last entry, we have a height of 69, but we do not have a weight. If we were to guess the weight in the last row, how should we do this?

We can draw a line (the red one) through the data points. Then we can guess the weight for a data point with height 69, by the value of the line at height 69 inches: about 182 pounds.

Measuring Error

The purpose of this line is not just to be close to all of the data (for this we will have to wait for PCA and dimensionality reduction). Rather, its goal is prediction; specifically, using the explanatory variable x to predict the dependent variable y.

In particular, for every value $x \in \mathbb{R}$, we can predict a value $\hat{y} = \ell(x)$. Then on our data set, we can examine for each x_i how close $\hat{y}_i$ is to y_i. This difference is called a *residual*:

$$r_i = y_i - \hat{y}_i = y_i - \ell(x_i).$$

Note that this residual is not the distance from y_i to the line ℓ, but the (signed) distance from y_i to the corresponding point on line ℓ with the same x value. Again, this is because our only goal is prediction of y. And this will be important as it allows techniques to be immune to the choice of units (e.g., inches or feet, pounds or kilograms)

So the residual measures the error of a single data point, but how should we measure the overall error of the entire data set? The common approach is the *sum of squared errors*:

$$\mathsf{SSE}((X, y), \ell) = \sum_{i=1}^{n} r_i^2 = \sum_{i=1}^{n} (y_i - \hat{y}_i)^2 = \sum_{i=1}^{n} (y_i - \ell(x_i))^2.$$

Why is this the most common measure? Here are 3 explanations:

- The sum of squared errors was the optimal result for a single point estimator under Gaussian noise using Bayesian reasoning, when there was assumed Gaussian noise (see Section 1.8). In that case the answer was simply the mean of the data.
- If you treat the residuals as a vector $r = (r_1, r_2, \ldots, r_n)$, then the standard way to measure total length of a vector r is through its norm $\|r\|$, which is most commonly its 2-norm $\|r\| = \|r\|_2 = \sqrt{\sum_{i=1}^{n} r_i^2}$. The square root part is not so important (it does not change which line ℓ minimizes this error), so removing this square root, we are left with SSE.
- For this specific formulation, there is a simple closed-form solution (which we will see next) for ℓ. And in fact, this solution will generalize to many more complex scenarios.

There are many other formulations of how best to measure error for the fit of a line (and other models), but we will not cover all of them in this text.

Solving for ℓ

To solve for the line which minimizes $\mathsf{SSE}((X, y), \ell)$ there is a very simply solution, in two steps. Calculate averages $\bar{x} = \frac{1}{n} \sum_{i=1}^{n} x_i$ and $\bar{y} = \frac{1}{n} \sum_{i=1}^{n} y_i$, and create centered n-dimension vectors $\bar{X} = (x_1 - \bar{x}, x_2 - \bar{x}, \ldots, x_n - \bar{x})$ for all x-coordinates and $\bar{Y} = (y_1 - \bar{y}, y_2 - \bar{y}, \ldots, y_n - \bar{y})$ for all y-coordinates.

1. Set $a = \langle \bar{Y}, \bar{X} \rangle / \|\bar{X}\|^2$
2. Set $b = \bar{y} - a\bar{x}$

This defines $\ell(x) = ax + b$.

We will provide the proof for why this is the optimal solution for the high-dimensional case (in short, it can be shown by expanding out the SSE expression, taking the derivative, and solving for 0). We will only provide some intuition here.

First let us examine the intercept

$$b = \frac{1}{n} \sum_{i=1}^{n} (y_i - ax_i) = \bar{y} - a\bar{x}.$$

This setting of b ensures that the line $y = \ell(x) = ax + b$ goes through the point $(\bar{x}, \bar{y})$ at the center of the data set since $\bar{y} = \ell(\bar{x}) = a\bar{x} + b$.

Second, to understand how the slope a is chosen, it is illustrative to reexamine the dot product as

$$a = \frac{\langle \bar{Y}, \bar{X} \rangle}{\|\bar{X}\|^2} = \frac{\|\bar{Y}\| \cdot \|\bar{X}\| \cdot \cos \theta}{\|\bar{X}\|^2} = \frac{\|\bar{Y}\|}{\|\bar{X}\|} \cos \theta,$$

where θ is the angle between the n-dimensional vectors $\bar{Y}$ and $\bar{X}$. Now in this expression, the $\|\bar{Y}\|/\|\bar{X}\|$ captures how much on (root-squared) average $\bar{Y}$ increases as $\bar{X}$ does (the rise-over-run interpretation of slope). However, we may want this to be negative if there is a negative correlation between $\bar{X}$ and $\bar{Y}$, or really this does not matter much if there is no correlation. So the $\cos\theta$ term captures the correlation after normalizing the units of $\bar{X}$ and $\bar{Y}$.

Example: Computing and Plotting Simple Linear Regression

This analysis is simple to carry out in Python. First we can plot the data, and its mean.

```python
import numpy as np
import matplotlib as mpl
import matplotlib.pyplot as plt
#  %matplotlib inline  # use this line in notebooks

x = np.array([66, 68, 60, 70, 65, 61, 74, 73, 75, 67])
y = np.array([160, 170, 110, 178, 155, 120, 233, 215, 235, 164])

ave_x = np.average(x)
ave_y = np.average(y)

plt.scatter(x,y, s=80, c="blue")
plt.scatter(ave_x,ave_y, s=100, c="red")
plt.show()
```

We can then compute the regression model as a, b directly

```python
xc = x - ave_x
yc = y - ave_y
a = xc.dot(yc)/xc.dot(xc)
b = ave_y - a*ave_x
print(a,b)
```

or using one of many built-in functions, including from NumPy

```python
(a,b)=np.polyfit(x,y,1)
print(a,b)
```

Finally, we make the prediction at unknown x value 69

```python
w = a*69 + b
print(w)
```

and plot the modeled line, and this predicted value

```python
h=np.linspace(60,76,17)
line=a*h+b
plt.scatter(x,y, s=80, c="blue")
plt.scatter(67.9,174, s=100, c="red")
plt.scatter(69,a*69+b, s=100, c="black")
plt.plot(h,line,'r-')
plt.show()
```

5.2 Linear Regression with Multiple Explanatory Variables

Magically, using linear algebra, everything extends gracefully to using more than one explanatory variables. Consider a data set $(X, y) = \{(x_1, y_1), (x_2, y_2), \ldots, (x_n, y_n)\}$ where each data point has $x_i = (x_{i,1}, x_{i,2}, \ldots, x_{i,d}) \in \mathbb{R}^d$ and $y_i \in \mathbb{R}$. That is there are d explanatory variables, as the coordinates of x_i, and one dependent variable in y_i. We would now like to use all of these variables at once to make a single (linear) prediction about the variable y_i. That is, we would like to create a model

$$\hat{y}_i = M_\alpha(x_i) = M_\alpha(x_{i,1}, x_{i,2}, \ldots, x_{i,d}) = \alpha_0 + \sum_{j=1}^{d} \alpha_j x_{i,j}$$

$$= \alpha_0 + \alpha_1 x_{i,1} + \alpha_2 x_{i,2} + \ldots + \alpha_d x_{i,d}.$$

$$= \langle \alpha, (1, x_{i,1}, x_{i,2}, \ldots, x_{i,d}) \rangle = \langle \alpha, (1, x_i) \rangle.$$

In the above equivalent notations α_0 serves the purpose of the intercept b, and all of the α_js replace the single coefficient a in the simple linear regression. Indeed, we can write this model as a dot product between the $(d + 1)$-dimensional vectors $\alpha = (\alpha_0, \alpha_1, \ldots, \alpha_d)$ and $(1, x_{i,1}, x_{i,2}, \ldots, x_{i,d}) = (1, x_i)$. As promised, the magic of linear algebra has allowed us to describe a more complex linear model M_α. Next we will see how to solve it.

Given a data point $x_i = (x_{i,1}, x_{i,2}, \ldots, x_{i,d})$, we can again evaluate our prediction $\hat{y}_i = M(x_i)$ using the *residual* value $r_i = y_i - \hat{y}_i = y_i - M(x_i)$. And to evaluate a set of n data points, it is standard to consider the sum of squared error as

$$\text{SSE}(X, y, M) = \sum_{i=1}^{n} r_i^2 = \sum_{i=1}^{n} (y_i - M(x_i))^2.$$

To obtain the coefficients which minimize this error, we can now do so with very simple linear algebra.

First we construct a $n \times (d + 1)$ data matrix $\tilde{X} = [\mathbf{1}, X_1, X_2, \ldots, X_d]$, where the first column $\mathbf{1}$ is the n-dimensional all ones column vector $[1; 1; \ldots; 1]$. Each of the next d columns is a column vector X_j, where $x_{i,j} = X_{i,j}$ is the ith entry of X_j and represents the jth coordinate of data point x_i. Then we let y be a n-dimensional column vector containing all the dependent variables. Now we can simply calculate the $(d + 1)$-dimensional column vector $\alpha = (\alpha_0, \alpha_1, \ldots, \alpha_d)$ as

$$\alpha = (\tilde{X}^T \tilde{X})^{-1} \tilde{X}^T y.$$

Let us compare to the simple case where we have 1 explanatory variable. The $(\tilde{X}^T \tilde{X})^{-1}$ term replaces the $\frac{1}{\|\tilde{X}\|^2}$ term. The $\tilde{X}^T y$ replaces the dot product $\langle \tilde{Y}, \tilde{X} \rangle$. And we do not need to separately solve for the intercept b, since we have created a new column in $\tilde{X}$ of all 1s. The α_0 replaces the intercept b; it is multiplied by a value 1 in $\tilde{X}$ equivalent to b just standing by itself.

Often the matrices X and $\tilde{X}$ are used interchangeably, and hence we drop the ~ from $\tilde{X}$ in most situations. We can either simply treat all data points x_i as one-dimension larger (with always a 1 in the first coordinate), or we can fit a model on the original matrix X and ignore the offset parameter α_0, which is then by default 0. The former approach, where each x_i is just assumed one dimension larger, is more common since it automatically handles the offset parameter.

Example: Predicting customer value

A website specializing in dongles (dongles-r-us.com) wants to predict the total dollar amount that visitors will spend on their site. It has installed some software that can track three variables:

*   time (the amount of time on the page in seconds): X_1,
*   jiggle (the amount of mouse movement in cm): X_2, and
*   scroll (how far they scroll the page down in cm): X_3.
    Also, for a set of past customers they have recorded the
*   sales (how much they spend on dongles in cents): y.

We see a portion of their data set here with $n = 11$ customers:

time: X_1	jiggle: X_2	scroll: X_3	sales: y
232	33	402	2201
10	22	160	0
6437	343	231	7650
512	101	17	5599
441	212	55	8900
453	53	99	1742
2	2	10	0
332	79	154	1215
182	20	89	699
123	223	12	2101
424	32	15	8789

To build a model, we recast the data as an 11×4 matrix $X = [\mathbf{1}, X_1, X_2, X_3]$. We let y be the 11-dimensional column vector.

$$X = \begin{bmatrix} 1 & 232 & 33 & 402 \\ 1 & 10 & 22 & 160 \\ 1 & 6437 & 343 & 231 \\ 1 & 512 & 101 & 17 \\ 1 & 441 & 212 & 55 \\ 1 & 453 & 53 & 99 \\ 1 & 2 & 2 & 10 \\ 1 & 332 & 79 & 154 \\ 1 & 182 & 20 & 89 \\ 1 & 123 & 223 & 12 \\ 1 & 424 & 32 & 15 \end{bmatrix} \quad y = \begin{bmatrix} 2201 \\ 0 \\ 7650 \\ 5599 \\ 8900 \\ 1742 \\ 0 \\ 1215 \\ 699 \\ 2101 \\ 8789 \end{bmatrix}$$

The goal is to learn the 4-dimensional column vector $\alpha = [\alpha_0; \alpha_1; \alpha_2; \alpha_3]$ so

$$y \approx X\alpha.$$

Setting $\alpha = (X^T X)^{-1} X^T y$ obtains (roughly) $\alpha_0 = 2626$, $\alpha_1 = 0.42$, $\alpha_2 = 12.72$, and $\alpha_3 = -6.50$. This implies an average customer who visits the site, but with no interaction on the site generates $\alpha_0 = \$2.62$. It is tempting to also make analysis specific to the other coefficients, but this is dangerous unless the variables (what each column represents) are independent. All values are proportional so this is unlikely the case. These coefficients offer a joint prediction of sales, and changing one may significantly affect the others.

Again, this is straight-forward to compute in Python

```
alpha = LA.inv(X.T @ X) @ X.T @ y.T    # or
alpha = LA.lstsq(X,y)[0]
```

Geometry of the Normal Equations

Why does $\alpha = (X^T X)^{-1} X^T y$ minimize the sum of squared errors:

$$\text{SSE}(X, y, M) = \sum_{i=1}^{n} r_i^2 = \sum_{i=1}^{n} (y_i - \langle \alpha, x_i \rangle)^2?$$

Fixing the data X and y, and representing M by its parameters α, we can consider a function $S(\alpha) = \text{SSE}(X, y, M)$. Then we observe that $S(\alpha)$ is a quadratic function in α, and it is convex (see Section 6.1), so its minimum is when the gradient is 0. This will be true when each partial derivative is 0. Let $X_{i,j}$ be the jth coordinate of x_i so residual $r_i = (y_i - \langle \alpha, x_i \rangle)$ has partial derivative $\frac{dr_i}{d\alpha_j} = -X_{i,j}$. Then set to 0 each partial derivative:

$$0 = \frac{dS(\alpha)}{d\alpha_j} = 2 \sum_{i=1}^{n} r_i \frac{dr_i}{d\alpha_j} = 2 \sum_{i=1}^{n} r_i(-X_{i,j}) = 2 \sum_{i=1}^{n} (y_i - \langle \alpha, x_i \rangle)(-X_{i,j})$$

We can rearrange this into the *normal equations*

$$\sum_{i=1}^{n} X_{i,j} \langle x_i, \alpha \rangle = \sum_{i=1}^{n} X_{i,j} y_i \qquad \text{for all } j \in \{1, 2, \ldots, d\}$$

equivalently $(X^T X)\alpha = X^T y$.

Multiplying by the inverse *gram matrix* $X^T X$ on both sides reveals the desired

$$\alpha = (X^T X)^{-1} X^T y.$$

Geometry: To see why these are called the normal equations, consider a form

$$0 = (y - X\alpha)^T X,$$

where $\mathbf{0}$ is the all-zeros vector in $\mathbb{R}^n$. Thus for any vector $v \in \mathbb{R}^n$, then

$$0 = (y - X\alpha)^T Xv = \langle y - X\alpha, Xv \rangle.$$

This includes when $v = \alpha$; under this setting then $Xv = X\alpha = \hat{y}$. Notice that $r = y - X\alpha$ is the vector in $\mathbb{R}^n$ that stores all residuals (so $y = \hat{y} + r$). Then the normal equations implies that

$$0 = \langle y - X\alpha, X\alpha \rangle = \langle r, \hat{y} \rangle;$$

that is, for the optimal α, the prediction $\hat{y}$ and the residual vector r are orthogonal. Since $\hat{y} = X\alpha$ is restricted to the $(d + 1)$-dimensional span of the columns of X, and α minimizes $\|r\|^2$, then this orthogonality implies that r is the *normal* vector to this $(d + 1)$-dimensional subspace.

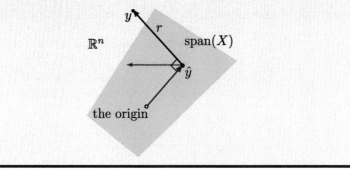

5.3 Polynomial Regression

Sometimes linear relations are not sufficient to capture the true pattern going on in the data with even a single dependent variable x. Instead we would like to build a model of the form:

$$\hat{y} = M_2(x) = \alpha_0 + \alpha_1 x + \alpha_2 x^2$$

or more generally for some polynomial of degree p

$$\hat{y} = M_p(x) = \alpha_0 + \alpha_1 x + \alpha_2 x^2 + \ldots + \alpha_p x^p$$

$$= \alpha_0 + \sum_{i=1}^{p} \alpha_i x^i.$$

Example: Predicting Height and Weight with Polynomials

We found more height and weight data, in addition to the ones in the height-weight example above.

height (in)	weight (lbs)
61.5	125
73.5	208
62.5	138
63	145
64	152
71	180
69	172
72.5	199
72	194
67.5	172

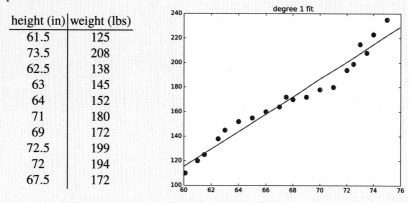

But can we do better if we fit with a polynomial?

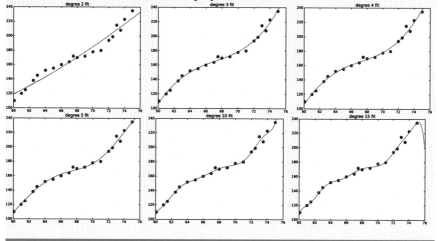

Again we can measure error for a single data point (x_i, y_i) as a residual as $r_i = \hat{y} - y_i = M_\alpha(x_i) - y_i$ and the error on n data points as the sum of squared residuals

$$\text{SSE}(P, M_\alpha) = \sum_{i=1}^{n} r_i^2 = \sum_{i=1}^{n} (M_p(x_i) - y_i)^2.$$

Under this error measure, it turns out we can again find a simple solution for the residuals $\alpha = (\alpha_0, \alpha_1, \ldots, \alpha_p)$. For each explanatory variable data value x we create a $(p + 1)$-dimensional vector

$$v = (1, x, x^2, \ldots, x^p).$$

And then for n data points $(x_1, y_1), \ldots, (x_n, y_n)$ we can create an $n \times (p + 1)$ data matrix

$$\tilde{X}_p = \begin{bmatrix} 1 & x_1 & x_1^2 & \ldots & x_1^p \\ 1 & x_2 & x_2^2 & \ldots & x_2^p \\ \vdots & \vdots & \vdots & \ddots & \vdots \\ 1 & x_n & x_n^2 & \ldots & x_n^p \end{bmatrix} \qquad y = \begin{bmatrix} y_1 \\ y_2 \\ \vdots \\ v_n \end{bmatrix}.$$

Example: Polynomial Expansion

Consider the input data set with $n = 3$ points (X, y)

$$X = \begin{bmatrix} 2 \\ 4 \\ 3 \end{bmatrix} \quad \text{and} \quad y = \begin{bmatrix} 1 \\ 6 \\ 5 \end{bmatrix}$$

The polynomial expansion of X generates with $p = 5$ a 3×6 matrix

$$\tilde{X}_p = \begin{bmatrix} 1 & 2 & 4 & 8 & 16 & 32 \\ 1 & 4 & 16 & 64 & 256 & 1024 \\ 1 & 3 & 9 & 27 & 81 & 243 \end{bmatrix}$$

while y is unchanged.

Then we can solve the same way as if each data value raised to a different power was a different explanatory variable. That is we can solve for the coefficients $\alpha = (\alpha_0, \alpha_1, \alpha_2, \ldots, \alpha_p)$ as

$$\alpha = (\tilde{X}_p^T \tilde{X}_p)^{-1} \tilde{X}_p^T y.$$

5.4 Cross-Validation

So it appears, as we increase p larger and larger, the data is fit better and better. The only downside appears to be the number of columns needed in the matrix $\tilde{X}_p$, right? Unfortunately, that is not right. Then how (and why?) do we choose the correct value of p, the degree of the polynomial fit?

A (very basic) statistical (hypothesis testing) approach may be to choose a model of the data (the best fit curve for some polynomial degree p, and assume Gaussian noise), then calculate the probability that the data fell outside the error bounds of that model. But maybe many different polynomials are a good fit?

In fact, if we choose p as $n - 1$ or greater, then the curve will *polynomially interpolate* all of the points. That is, it will pass through all points, so all points have a residual of exactly 0 (up to numerical precision). This is the basis of a lot of geometric modeling (e.g., for CAD), but it turns out bad for data modeling.

Example: Simple Polynomial Fitting

Consider the simple data set of 9 points

x	1	2	3	4	5	6	7	8	9
y	4	6	8.2	9	9.5	11	11.5	12	11.2

With the following polynomial fits for $p = \{1, 2, 3, 4, 5, 8\}$. Believe your eyes, for $p = 8$, the curve actually passes through each and every point exactly.

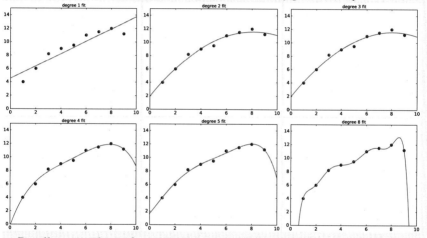

Recall, our goal was for a new data point with only an x value to predict its y value. Which do you think does the best job?

Generalization and Cross-Validation

Our ultimate goal in regression is *generalization* (how well do we predict on new data), not SSE! Using some error measure (SSE) to fit a line or curve is a good proxy for what we want, but in many cases (as with polynomial regression), it can be abused. We want to know how our model will generalize to new data. How would we measure this without new data?

The solution is *cross-validation*. In the simplest form, we **randomly** split our data into training data (on which we build a model) and testing data (on which we evaluate our model). The testing serves to estimate how well we would do on future data that we do not have.

- *Why randomly?* Because you do not want to bias the model to do better on some parts than others in how you choose the split. Also, since we assume the data elements are iid from some underlying distribution, then the test data is also iid from that distribution if chosen randomly.
- *How large should the test data be?* It depends on the data set. Both 10% and 33% are common. Intuitively, the goal is to estimate that the SSE, which after dividing by its size, is an average squared error. That is, the goal is to estimate an average of an unknown distribution, and it maps to the central limit theorem, and to concentration of measure bounds. We could use these bounds via the number of test samples and properties of the observed distributions to estimate the accuracy of the generalization prediction.

Let (X, y) be the full data set (with n rows of data), and we split it into data sets $(X_{\text{train}}, y_{\text{train}})$ and $(X_{\text{test}}, y_{\text{test}})$ with n_{train} and n_{test} rows, respectively, with $n = n_{\text{train}} + n_{\text{test}}$. Next we build a model with the training data, e.g.,

$$\alpha = (X_{\text{train}}^T X_{\text{train}})^{-1} X_{\text{train}}^T y_{\text{train}}.$$

Then we evaluate the model M_α on the test data X_{test}, often using $\text{SSE}(X_{\text{test}}, y_{\text{test}}, M_\alpha)$ as

$$\begin{aligned}
\text{SSE}(X_{\text{test}}, y_{\text{test}}, M_\alpha) &= \sum_{(x_i, y_i) \in (X_{\text{test}}, y_{\text{test}})} (y_i - M_\alpha(x_i))^2 \\
&= \sum_{(x_i, y_i) \in (X_{\text{test}}, y_{\text{test}})} (y_i - \langle (1; x_i), \alpha \rangle)^2.
\end{aligned}$$

We can use the testing data for two purposes:

- To estimate how well our model would perform on new data, yet unseen. That is the predicted residual of a new data point is precisely $\text{SSE}(X_{\text{test}}, y_{\text{test}}, M_\alpha)/n_{\text{test}}$.
- To choose the correct parameter for a model (e.g., which p to use), based on which one provides the least error on the test data.

Its important to not use the same $(X_{\text{test}}, y_{\text{test}})$ to do both tasks. If we choose a model with $(X_{\text{test}}, y_{\text{test}})$, then we should reserve even more data for predicting the generalization error. When using the test data to choose a model parameter, then it is being used to build the model; thus evaluating generalization with this same data can suffer the same fate as testing and training with the same data.

Specifically, how should we choose the best p? We calculate models M_{α_p} for each value p on the same training data. Then we calculate the model error $\text{SSE}(X_{\text{test}}, y_{\text{test}}, M_{\alpha_p})$ for each p, and see which has the smallest value. That is we train on $(X_{\text{train}}, y_{\text{train}})$ and test(evaluate) on $(X_{\text{test}}, y_{\text{test}})$.

Example: Simple polynomial example with Cross-Validation

Now split our data sets into a train set and a test set:

$$\text{train: } \frac{x\ 2\ 3\ 4\ 6\ 7\ 8}{y\ 6\ 8.2\ 9\ 11\ 11.5\ 12} \qquad \text{test: } \frac{x\ 1\ 5\ 9}{y\ 4\ 9.5\ 11.2}$$

with the following polynomial fits for $p = \{1, 2, 3, 4, 5, 8\}$ generating model M_{α_p} on the test data. We then calculate the $\text{SSE}(x_{\text{test}}, y_{\text{test}}, M_{\alpha_p})$ score for each (as shown):

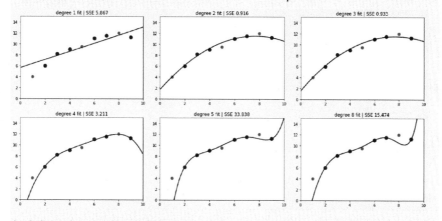

And the polynomial model with degree $p = 2$ has the lowest SSE score of 0.916. It is also the simplest model that does a very good job by the "eye-ball" test. So we would choose this as our model.

Using Python, it is easy to generate the model M on the training data, in this case with polynomial degree 3

```
xT = np.array([2, 3, 4, 6, 7, 8])
yT = np.array([6, 8.2, 9, 11, 11.5, 12])
xE = np.array([1, 5, 9])
yE = np.array(4, 9.5, 11.2)

p=3
coefs=np.polyfit(xT,yT,p)
M = np.poly1d(coefs)
```

We can then calculate the residual and error on the test data

```
resid = ffit(xE) - yE
print(resid)
SSE = LA.norm(resid)**2
print(SSE)
```

Finally, we can plot the result on the data

```
plt.axis([0,10,0,15])
s=np.linspace(0,10,101)
plt.plot(s,ffit(s),'r-',linewidth=2.0)

plt.scatter(xT,yT, s=80, c="blue")
plt.scatter(xE,yE, s=50, c="green")
plt.title('degree_%s_fit_|_SSE_%0.3f' % (p, SSE))
plt.show()
```

Leave-one-out Cross-Validation

But, not training on the test data means that you use less data, and your model is worse! If your data is limited, you may not want to "waste" this data!

If your data is very large, then leaving out 10% is not a big deal. But if you only have 9 data points it can be. The smallest the test set could be is 1 point. But then it is not a very good representation of the full data set.

The alternative, called *leave-one-out cross-validation*, is to create n different training sets, each of size $n - 1$ ($X_{1,\text{train}}, X_{2,\text{train}}, \ldots, X_{n,\text{train}}$), where $X_{i,\text{train}}$ contains all points except for x_i, which is a one-point test set. Then we build n different models $M_1, M_2, \ldots, M_n$, evaluate each model M_i on the one test point x_i to get an error $E_i = (y_i - M_i(x_i))^2$, and average their errors $E = \frac{1}{n} \sum_{i=1}^{n} E_i$. Again, the parameter with the smallest associated av erage error E is deemed the best. This allows you to build a model on as much data as possible, while still using all of the data to test.

However, this requires roughly n times as long to compute as the other techniques, so is often too slow for really big data sets.

5.4.1 Other ways to Evaluate Linear Regression Models

The statistics literature has developed numerous tests to evaluate the effectiveness of linear models for regression. For instance, an F-test evaluates if the residuals are normally distributed—as one would expect if minimizing SSE.

A popular statistic is the *coefficient of determination* R^2. It captures the proportion of variance in the data explained by the model. Again let $\bar{y} = \frac{1}{n} \sum_{i=1}^{n} y_i$ be the average dependent variable, and then the population variance in y is defined as

$$SS_{\text{tot}} = \sum_{i=1}^{n} (y_i - \bar{y})^2.$$

This is compared to the sum of squared residuals

$$SS_{\text{reg}} = \sum_{i=1}^{n} (y_i - \hat{y})^2 = \|r_i\|^2,$$

to provide the value

$$R^2 = 1 - \frac{SS_{\text{tot}}}{SS_{\text{reg}}}.$$

The best fitting regression will have R^2 closer to 1, and as the fit explains less variance, the value decreases. A value of 0 is attained by using a horizontal line at $\ell(x) = \bar{y}$.

While these mechanisms provide a score to assess the performance of linear regression on a particular data set, they cannot directly be used for two tasks of cross-validation. They do not directly predict how well the model will work on new data, and they do not seamlessly allow one to compare across any set of models (e.g., compare to a model M which might be something other than a linear regression model).

We can replace the L_2 measure in the SSE with other cost functions like *mean absolute error*

$$\mathsf{MAE}(M_\alpha, (X, y)) = \sum_{i=1}^n |y_i - M_\alpha(x_i)| = \sum_{i=1}^n |r_i|.$$

This is less sensitive to outliers and corresponds with a Laplace noise model. And while algorithms exist for minimizing this, they are more complicated, and far less common than SSE.

5.5 Regularized Regression

In linear regression with multiple explanatory variables, cross-validation can again be useful to help tune our model. This at first may seem contradictory, as there is one optimal model which minimizes SSE; but we will see this is useful when there are more dimensions d (the explanatory variables) than needed to build a model from n data points. Indeed these methods can extend the least squares regression framework when d is greater than n, which from a linear algebraic perspective seems impossible since then the SSE solution is under-defined.

Gauss-Markov Theorem

Linear regression is in some sense optimal, as formalized by the *Gauss-Markov Theorem*. This states that, for data (X, y), the linear model M_α derived from $\alpha = (X^T X)^{-1} X^T y$ is the best possible, conditioned on three things:

- best is defined as lowest variance in residuals (the SSE cost we study),
- the solution has 0 expected error, and
- all residuals $r_i = y_i - x_i \alpha$ are not known to be correlated.

It is important to examine these benign seeming conditions. There are reasonable alternatives to the first condition, but studying the SSE is ultimately sensible, it is the *minimum variance* solution. The third condition can in some cases be challenged as well, but requires more information than we typically assume and will involve ideas beyond the scope of this text. The second condition seems the most benign—why would we want to expect to have non-zero error? But this condition, it turns out, is the most ripe to exploit ways that improve upon ordinary linear squares regression.

5.5.1 Tikhonov Regularization for Ridge Regression

We first introduce Tikhonov regularization, which results in a parameterized linear regression known as *ridge regression*. Consider again a data set (X, y) with n data points. Now instead of minimizing α over

$$\mathsf{SSE}(X, y, M_\alpha) = \sum_{i=1}^{n}(y_i - M_\alpha(x_i))^2 = \|X\alpha - y\|_2^2$$

we introduce a new cost function for a parameter $s > 0$ as

$$W_\circ(X, y, \alpha, s) = \frac{1}{n}\|X\alpha - y\|_2^2 + s\|\alpha\|_2^2.$$

This cost function is similar to before, except it adds a *regularization* term $+s\|\alpha\|_2^2$. This term is adding a bias toward α, and in particular, it is biasing α to be smaller; if there are two vectors α_1 and α_2 such that $\|X\alpha_1 - y\|_2^2 = \|X\alpha_2 - y\|_2^2$, then this cost function will prefer whichever has a smaller norm. In fact this will also prefer some vectors (representing models) α which do not correspond to the smallest values $\|X\alpha - y\|_2^2$, since they will make up for it with an even smaller term $s\|\alpha\|_2^2$. Indeed the larger that parameter s, the more a small norm of α is favored versus its fit to the data.

The normalization of the $\|X\alpha - y\|_2^2$ by $1/n$ makes it easier to understand the $s\|\alpha\|_2^2$ as the number of data points (the value n) changes. This essentially compares the $s\|\alpha\|_2^2$ term with the effect of a single data point.

Geometry of Favoring a Small $\|\alpha\|$

Consider the simple example with one explanatory variable. In this case, the α_1 parameter corresponds to the slope of the line. Very large sloped models (like $\alpha^{(1)}$) are inherently unstable. A small change in the x value will lead to large changes in the y value (e.g., $\hat{y} = \alpha^{(1)}x$ then $\hat{y}_\delta = \alpha^{(1)}(x + \delta) = \hat{y} + \alpha^{(1)}\delta$). The figure shows two models $\alpha^{(1)}$ with large slope and model $\alpha^{(2)}$ with small slope, both with similar sum of squared residuals. However, changing x_3 to $x_3 + \delta$ results in a much larger change for the predicted value using models $\alpha^{(1)}$.

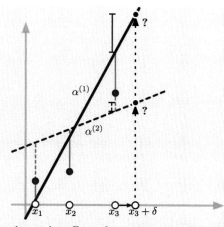

This picture only shows the effect of α with one explanatory variable, but this is most useful with many explanatory variables. In this higher-dimensional setting, it is more common to *extrapolate*: make predictions on new x' which are outside the region occupied by the data X. In higher dimensions, it is hard to cover the entire domain with observations. And in these settings where predictions are made past the extent of data, it is even more dangerous to have parameters α_j which seem to fit the data locally, but change rapidly beyond the extent of the data.

Hence, smaller α are simpler models that are more cautious in extrapolated predictions, and so the predictions $\hat{y}$ are less sensitive to changes in x.

Moreover, it turns out that solving for the solution α_s° which minimizes $W_o(X, y, \alpha, s)$ is as simple as for the ordinary least squares where $s = 0$. In particular

$$\alpha_s^\circ = \arg\min_\alpha W_o(X, y, \alpha, s) = (X^T X + (s/n)I)^{-1} X^T y,$$

where I is the $d \times d$ identity matrix.

Recall that the matrix inverse is not defined if the matrix is not full rank. So in the case where X has fewer data points (n) than dimensions (d), then ordinary least squares (with $s = 0$) will not be solvable this way. Moreover, even if $n > d$, but if some of the eigenvectors of $X^T X$ are very small, then this matrix inverse operation will be numerically unstable. Adding the $(s/n)I$ term explicitly adds (s/n) to all eigenvalues, hence making the operation more stable. Explicitly, it makes the regression biased but more stable with respect to directions within $\mathbb{R}^d$ where X has small covariance.

Improved Generalization

Indeed, it can be shown that appropriately setting some $s > 0$ can "get around" the Gauss-Markov Theorem. Formally, consider data drawn iid from a distribution

$(X, y) \sim \mu_\alpha$ where given some choice of x, then $y = \langle x, \alpha \rangle + \varepsilon$, where ε is unbiased noise. Then unless always $|\varepsilon| = 0$, there exists some $s > 0$ such that predictions on new data from the same μ_α will have smaller SSE using the optimum from $W_\circ$ than from ordinary least squares.

This improved error argument is non-constructive since we do not explicitly know the distribution μ_α from which our data is drawn (we only observe our data (X, y)). However, we can use cross-validation to select the best choice of s on some testing data.

Example: Improved Generalization with Ridge Regression

Consider a data set (X, y) with $X \subset \mathbb{R}^1$ split into a training set

$$\{(60, 110), (61, 120), (65, 180), (66, 160), (68, 170), (70, 225), (73, 215), (75, 235)\}$$

and test set
$$\{(64, 140), (72, 200), (74, 220)\}.$$

Building a model on the training set for ordinary least squares ($s = 0$) results in model parameters $\alpha^{(0)} = (-379.75, 8.27)$, shown in red. Instead using ridge regression with $s = 0.01$ results in model $\alpha^{(0.01)} = (-309.24, 7.23)$, shown in green.

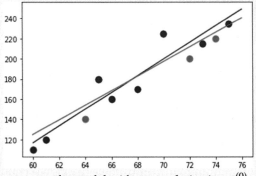

On the held out test set, the model without regularization $\alpha^{(0)}$ achieved a sum of squared errors of 524.05, and the model with regularization $\alpha^{(0.01)}$ achieved a sum of squared errors of 361.30.

5.5.2 Lasso

A surprisingly effective variant of this regularization is known as the *lasso*. It replaces the $W_\circ$ cost function, again using a parameter $s > 0$ as

$$W_\circ(X, y, \alpha, s) = \frac{1}{n}\|X\alpha - y\|_2^2 + s\|\alpha\|_1.$$

This has the same general form as ridge regression, but it makes the subtle change in the norm of the regularizer from $\| \cdot \|_2^2$ to $\| \cdot \|_1$.

This alternative form of regularization has all of the basic robustness and stronger generalization properties associated with biasing toward a smaller normed choice of α in ridge regression. However, it also has two additional effects. First, the function with L_1 norm is less simple to optimize. Although there are algorithmic and combinatorial approaches, there is not a simple closed form using the matrix inverse available anymore. We will discuss a couple of ways to optimize this object soon, and then in the more general context of the following chapter. Second, this formulation has the additional benefit that it also biases toward sparser solutions. That is, for large enough values of s, the optimal choice of

$$\alpha_s^\diamond = \arg\min_\alpha W_\diamond(X, y, \alpha, s)$$

will have multiple coordinates $\alpha_{s,j}^\diamond = 0$. And, in general, as s increases, the number of indexes j with $\alpha_{s,j}^\diamond = 0$ will increase.

Sparser models α, those with several coordinates $\alpha_j = 0$, are useful for various reasons. If the number of non-zero coordinates is small, it is simpler to understand and more efficient to use. That is, to make a prediction from a new data point x, only the coordinates which correspond to non-zero α_j values need to be considered. And while this does not definitively say which coordinates are actually meaningful toward understanding the dependent variable, it provides a small and viable set. The process of selecting such a subset is known as *variable selection*. It is known that under some often reasonable assumptions, if the data is indeed drawn from a sparse model with noise, then using lasso to recover $\alpha_s^\diamond$ for the proper s can recover the true model exactly. This striking result is relatively recent, and already has many important implications in statistical inference, signal processing, and machine learning.

5.5.3 Dual Constrained Formulation

There is an equivalent formulation to both ridge regression and lasso that will make it easier to understand their optimization properties, and to describe a common approach used to solve for their solution. That is, instead of solving the modified objective functions $W_\circ$ and $W_\diamond$, we can solve the original least squares objective $\|X\alpha - y\|_2^2$, but instead provide a hard constraint on the allowable values of α.

Specifically, we can reformulate ridge regression for some parameter $t > 0$ as

$$\alpha_t^\circ = \underset{\alpha}{\operatorname{argmin}} \|X\alpha - y\|_2^2 \quad \text{such that} \quad \|\alpha\|_2^2 \le t$$

and lasso again using some $t > 0$ as

$$\alpha_t^\diamond = \underset{\alpha}{\operatorname{argmin}} \|X\alpha - y\|_2^2 \quad \text{such that} \quad \|\alpha\|_1 \le t.$$

For ridge regression for each parameter $s > 0$ and solution α_s° there is a corresponding parameter $t > 0$ so $\alpha_t^\diamond = \alpha_s^\circ$. And respectively, for lasso for each s there is a t so $\alpha_s^\diamond = \alpha_t^\circ$. To see this, for a fixed s, find the solution α_s° (e.g., using the closed-form solution), then set $t = \|\alpha_s^\circ\|$. Now using this value of t, the solution α_t° will match that of α_s° since it satisfies the hard norm constraint, and if there was another solution α satisfying that norm constraint with a smaller $\|X\alpha - y\|$ value, this would be smaller in both terms of $W_\circ$ than α_s°, a contradiction to the optimality of α_s°.

Hence, we can focus on solving either the soft norm formulations $W_\circ$ and $W_\diamond$ or the hard norm formulations, for ridge regression or lasso. Ultimately, we will search over the choices of parameters s or t using cross-validation, so we do not need to know the correspondence ahead of time.

Geometry of Sparse Models with Lasso

Consider the hard constraint variants of lasso and ridge regression with some parameter t. We will visualize the geometry of this problem when there are $d = 2$ explanatory variables. The value of t constrains the allowable values of α to be within metric balls around the origin. These are shown below as an L_1-ball for lasso (in green) and an L_2-ball for ridge regression (in blue). Note that the constraint for ridge is actually $\|\alpha\|_2^2 \leq t$, where the norm is squared (this is the most common formulation, which we study), but we can always just consider the square root of that value t as the one we picture—the geometry is unchanged.

Then we can imagine the (unconstrained) ordinary least squares solution α^* as outside of both of these balls. The cost function $\|X\alpha - y\|_2^2$ is quadratic, so it must vary as a parabola as a function of α, with minimum at α^*. This is shown with red shading. Note that the values of α which have a fixed value in $\|X\alpha - y\|_2^2$ form concentric ellipses centered at α^*.

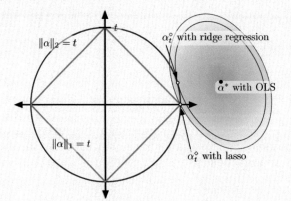

The key difference between ridge regression and lasso is now apparent in their solutions. Because the L_1-ball is "pointier," and specifically along the coordinate axes, it reaches out along these coordinate axes. Thus the innermost ellipse around α^* which the L_1-ball reaches tends to be along these coordinate axes—whereas there is no such property with the L_2-ball. These coordinate axes correspond with values of α which have 0 coordinates, thus the lasso solution with L_1 constraint explicitly biases toward sparse solutions.

The same phenomenon holds in higher dimensions (but which are harder to draw). However, in this setting, the L_1-ball is even pointier. The quadratic cost function now may not always first intersect the L_1-ball along a single coordinate axes (which has only a single non-zero coordinate), but it will typically intersect the L_1-ball along multidimensional ridges between coordinate axes, with only those coordinates as non-zero.

As t becomes smaller, it is more and more likely that the solution is found on a high-degree corner or ridge (with fewer non-zero coordinates), since it is harder for the convex ellipses to sneak past the pointy corners. As t becomes larger, the solution approaches the OLS solution a^*, and unless a^* itself has coordinates very close to 0, then it will tend to reach the L_1 ball away from those corners.

5.5.4 Matching Pursuit

We next describe a simple procedure called *matching pursuit* (MP) that provides a simple greedy alternative toward solving regression problems. This is particularly helpful for the lasso variants for which there is no closed-form solution. In particular, when MP is run with the lasso objective, then it is sometimes called *basis pursuit* or *forward subset selection* because it iteratively reveals a meaningful set of features or a basis within the data matrix X which captures the trends with respect to the dependent variable.

Ethical Questions with Feature Selection

Lasso and Matching Pursuit can be used for feature selection, that is choosing a limited number of dimensions (the features) with which predictions from regression can generalize nearly as well, or in some cases better, than if all of the features are used. However, these approaches can be unstable when two features (or sets of features) are correlated. That is, there could be multiple subsets of features of the same size which provide nearly as good generalization.

Now consider you are working in a large business that pays for various sorts of indicators or features for customers or products it is trying to model. You notice that by small changes in the regularization parameter, two different subsets of features are selected, with both providing approximately the same generalization estimates when cross-validating. Changing which subsets you choose to purchase and use will dramatically affect the business of one of the companies trying to sell these features. Do you have an obligation to keep the financial benefits of this company in mind as you select features?

Alternatively, while the two subsets of features provide similar generalization predictions when averaged across the entire data set, on many individual data sets, it changes the predictions drastically. For instance, this may update the prediction for a product line to go from profitable to unprofitable. How should you investigate this prognostication before acting on it?

The MP algorithm (sketched in Algorithm 5.5.1) uses a fixed value of s, initially sets $\alpha_j = 0$ in all coordinates, and iteratively chooses coordinates j which tend to minimize the objective $W_\diamond$ when made non-zero. It can also be run with $W_\circ$. As coordinates are chosen and assigned a value, a residual vector r of the part of y not yet captured is updated. Initially $r = y$, and as the algorithm is run and better fits are found $\|r\|_2^2$ decreases, as desired. In each successive round, the next best coordinate is selected using only r (not the original y); this is the key insight that makes this algorithm tractable. And in particular, this can be done by choosing the coordinate j which maximizes a linear operator (a dot product)

$$j^* = \underset{j}{\operatorname{argmax}} |\langle r, X_j \rangle|$$

where X_j is the n-dimensional, jth column of the data matrix X. It is then possible to solve for α_{j^*} as

$$\alpha_{j^*} = \underset{\gamma}{\operatorname{argmin}} \|r - X_j\gamma\|^2 + s|\gamma| = \frac{1}{\|X_j\|^2} \left(\langle r, X_j \rangle \pm \frac{s}{2} \right).$$

The choice of $\pm$ (either addition or subtraction of the $s/2$ term needs to be checked in the full expression.

Algorithm 5.5.1 Matching Pursuit

Set $r = y$; and $\alpha_j = 0$ for all $j \in [d]$
for $i = 1$ to k **do**
    Set $X_j = \operatorname{argmax}_{X_{j'} \in X} |\langle r, X_{j'} \rangle|$
    Set $\alpha_j = \operatorname{argmin}_\gamma \|r - X_j\gamma\|^2 + s|\gamma|$
    Set $r = r - X_j\alpha_j$
Return α

This algorithm is greedy, and may not result in the true optimal solution: it may initially choose a coordinate which is not in the true optimum, and it may assign it a value α_j which is not the true optimum value. However, there are situations when the noise is small enough that this approach will still find the true optimum. When using the $W_\circ$ objective for ridge regression, then at each step when solving for α_j, we can solve for all coordinates selected so far; this is more robust to local minimum and is often called *orthogonal matching pursuit*.

For either objective, this can be run until a fixed number k coordinates have been chosen (as in Algorithm 5.5.1), or until the residual's norm $\|r\|_2^2$ is below some threshold. For the feature selection goal, these coordinates are given an ordering where the more pertinent ones are deemed more relevant for the modeling problem.

Example: Matching Pursuit with $n < d$

Consider the data set (X, y) where X has $n = 5$ data points and $d = 7$ dimensions.

$$X = \begin{bmatrix} 1 & 8 & -3 & 5 & 4 & -9 & 4 \\ 1 & -2 & 4 & 8 & -2 & -3 & 2 \\ 1 & 9 & 6 & -7 & 4 & -5 & -5 \\ 1 & 6 & -14 & -5 & -3 & 9 & -2 \\ 1 & -2 & 11 & -6 & 3 & -5 & 1 \end{bmatrix} \quad y = \begin{bmatrix} 43.22 \\ 46.11 \\ -24.63 \\ -42.61 \\ -19.76 \end{bmatrix}$$

This was generated by applying a model $\alpha^T = [0, 0, 0, 5, 0, -2, 0]$ onto X and adding a small amount of noise to obtain y. Running ordinary least squares would fail since the system has more unknowns $(d = 7)$ than equations $(n = 5)$. However, we could run ridge regression; setting $s = 0.5$ fits a dense model $\alpha^\circ_{0.5} = [-0.30, 0.26, -0.17, 4.94, -0.22, -1.87, 0.31]$.

Running MP with the $W_\diamond$ objective, again using regularization parameter $s = 0.5$, does recover a sparse model. The first step identifies index $j = 4$ as having its column $X_6^T = [5, 8, -7, -5, -6]$ as being most correlated with $r = y$. Solving for $\alpha_4^* = 5.47$ which is larger than the optimal 5.0, but reasonably close, we then update

$$r = r - X_4\alpha_4^* = [-15.91, -2.41, -13.60, 15.29, -13.01]$$

This suboptimal choice of α_4^* is still enough to reduce the norm of r from $\|r\| = 82.50$ to $\|r\| = 29.10$.

The next step again correctly selects $j = 6$ for column $X_6^T = [-9, -3, -5, 9, -5]$ having the most remaining alignment with r. It solves for $\alpha_6^* = -1.93$, less (in absolute value) than the ideal -2.0, but again fairly close. This updates

$$r = r - X_6\alpha_6^* = [-1.26, -3.36, 4.14, 1.87, 3.54]$$

reducing the residual norm to $\|r\| = 6.79$.

If we run for another step, the algorithm will again choose $j = 4$, and solve for $\alpha_4^* = -0.46$. If we sum this with the previous choice, then the final value is 5.01, very close to the true model. After updating the residual norm goes down to $\|r\| = 1.68$.

The following Python code replicates this example.

```python
X = np.array([[1, 8, -3, 5, 4, -9, 4],[1,-2,4,8,-2,-3,2],
              [1,9,6,-7,4,-5,-5],[1,6,-14,-5,-3,9,-2],
              [1,-2,11,-6,3,-5,1]])
a = np.array([0,0,0,5,0,-2,0])
noise = np.random.normal(0,0.5,5)
y = X @ a + noise
s = 0.5   # regularization parameter
k = 3     # number of iterations

print("norm:␣", LA.norm(y), y)

r = y
for i in range(k):
```

```
# select column index most aligned with residual
j = np.argmax(abs(r @ X))

# find best model parameter at j to fit residual
ajp = (np.dot(r,X[:,j])+s/2) / (LA.norm(X[:,j])**2)
ajm = (np.dot(r,X[:,j])-s/2) / (LA.norm(X[:,j])**2)
if (LA.norm(r-X[:,j]*ajp) + s*abs(ajp) <
    LA.norm(r-X[:,j]*ajm) + s*abs(ajm)):
  aj = ajp
else:
  aj = ajm

# udpate residual
r = r - X[:,j]*aj
print("update:␣", j, aj, LA.norm(r))

# Ordinary Least Squares
print("OLS:␣", LA.inv(X.T @ X) @ X.T @ y.T)
# Ridge Regression
print("ridge:␣", LA.inv(X.T @ X + s*np.identity(7)) @ X.T @ y.T)
```

An alternative to matching pursuit is called *least angle regression*. Instead of explicitly adding one coordinate at a time for a fixed s, it follows the hard constraint objective. This method iteratively increases the constraint t while maintaining the optimal solution. With $t = 0$, then the solution $\alpha_t^\diamond = 0$. And as t increases, then initially a single coordinate is made non-zero in the optimal solution. Because the boundaries facets of the L_1-ball grow at a simple linear rate as t increases, then we can exactly describe the increases in the solution to $\alpha_t^\diamond$ as a linear equation of t. Moreover, solving another quadratic equation can be used to determine when the next coordinate is made non-zero, until eventually all coordinates are non-zero, and the OLS solution α^* is recovered. This also builds a model $M_{\alpha_t^\diamond}$ for every value t, so it can be cross-validated as it is built.

Non-Comparable Columns and Normalization

It is important to note that these regularized algorithms directly compare norms of the columns X_j of matrix X. This is inherently an issue with the regularization term (either $\|\alpha\|_2$ for $W_\diamond$ or $\|\alpha\|_1$ for $W_\diamond$) in the problem formulation. If these columns measure distinct non-comparable quantities (like height and weight) of each data point (say with different units), then these steps are arbitrary. Changing the measurement from feet to centimeters may completely change the result. Thus it is essential the input matrix X is a *data matrix* (see Section 7.1) where all coordinates have the same unit or are unit-less. As before, when this is not the case, as a stop-gap, all columns can be normalized; however such a process may be easily affected by outliers. In either case, even with a data matrix, it is recommended to first "center" all columns (other than the first offset column) by subtracting its mean before initiating the algorithms. Otherwise large constant effects (absorbed in the offset) appear as meaningful coefficients.

Compressed Sensing

A demonstration of what sort of data can be recovered exactly using lasso is shown in the compressed sensing problem. There are variants of this problem that show up in computer vision, astronomy, and medical imagining; we will examine a simple form.

First consider an unknown signal vector $s \in \mathbb{R}^d$. It is d-dimensional, but is known to have most of its coordinates as 0. In our case, we consider with $m \ll d$ non-zero coordinates, and for simplicity assume these have value 1. For example, imagine a telescope scanning the sky. For most snapshots, the telescope sees nothing (a 0), but occasionally it sees a star (a 1). This maps to our model when it takes d snapshots, and only sees m stars. Many other examples exist: a customer will buy m out of d products a store sells; an earthquake registers on only m out of d days on record; or only m out of d people test positive for a rare genetic marker. We will use as an example s as

$$s^T = [0\,1\,0\,0\,0\,1\,0\,0\,0\,0\,0\,0\,0\,0\,1\,0\,0\,0\,1\,1\,0\,1\,0\,0\,0\,1\,0\,0\,1\,0\,0].$$

But this problem is interesting because, as mentioned, s is unknown. Instead we "sense" s with a known compressed vector $x_i \in \{-1, 0, +1\}^d$. For example, let

$$x_i^T = [-1\,0\,1\,0\,1\,1\,-1\,1\,0\,-1\,0\,0\,1\,-1\,-1\,1\,0\,1\,0\,1\,-1\,-1\,-1\,0\,1\,0\,0\,-1\,0\,1\,0\,0].$$

The sensing of s with x_i is recorded as $y_i = \langle x_i, s \rangle$, which is a single scalar value (in our setting an integer). In our example

$$y_i = \langle x_i, s \rangle = 0+0+0+0+1+0+0+0+0+0+0+0+0+0+1+0+0+0+1-1+0-1$$
$$+0+0+0+0+0+0+1+0+0 = 2.$$

Moreover, in general we will have n measurements values $y_1, y_2, \ldots, y_n$ using n known measurement vectors $x_1, x_2, \ldots, x_n$. Stacking the measurement values $y = [y_1; y_2; \ldots; y_n]$ and vectors into a matrix $X = [x_1; x_2; \ldots; x_n]$ results in the familiar form

$$y = Xs.$$

And the goal is to recover s using only y and X. This appears to be precisely a regression problem! However, we are interested in the case where $d \gg n$. In particular, this can be solvable when $n = C \cdot m \log(d/m)$, for the constant $C \in (4, 20)$ depending on the recovery approach. Given our assumption that $m \ll d$ (e.g., $m = 50$ and $d = 10{,}000$) this is a remarkable compression, since for instance just storing the location of each 1 bit uses $\log d$ bits (its index in binary), for a total of about $m \log d$.

The simplest recovery approach is with Matching Pursuit; this (actually the OMP variant) requires the measurement size constant C closer to 20. This modified version of MP is sketched in Algorithm 5.5.2. The key step is modified to choose the dimension j where the column in the measurement matrix X_j has the largest dot product with r. We guess this has a 1 bit, and then factor out the effect of the measurement matrix witnessing a 1 bit at that location in the residual r. This is repeated either for a fixed (m) number of steps, or until the residual becomes all 0.

Algorithm 5.5.2 Matching Pursuit for Compressed Sensing

Set $r = y$; and $s = [0; 0; \ldots, 0]$
for $i = 1$ **to** m **do**
    Set $j = \text{argmax}_{j'} \langle r, X_{j'} \rangle$
    Set $s_j = 1$
    Set $r = r - X_j$
Return s

> ### Example: Recovery in Compressed Sensing

Consider a specific example for running Matching Pursuit, this has $d = 10$, $m = 3$, and $n = 6$. Let the (unknown) input signal be

$$s = [0, 0, 1, 0, 0, 1, 0, 0, 1, 0].$$

Let the known measurement matrix be

$$X = \begin{bmatrix} 0 & 1 & 1 & -1 & -1 & 0 & -1 & 0 & -1 & 0 \\ -1 & -1 & 0 & 1 & -1 & 0 & 0 & -1 & 0 & 1 \\ 1 & -1 & 1 & -1 & 0 & -1 & 1 & 1 & 0 & 0 \\ 1 & 0 & -1 & 0 & 0 & 1 & -1 & -1 & 1 & 1 \\ -1 & 0 & 0 & 0 & 1 & 0 & 1 & 0 & 1 & -1 \\ 0 & 0 & -1 & -1 & -1 & 0 & -1 & 1 & -1 & 0 \end{bmatrix}$$

so for instance the first row $x_1 = (0, 1, 1, -1, -1, 0, -1, 0, -1, 0)$ yields measurement

$$\langle x_1, s \rangle = 0 + 0 + 1 + 0 + 0 + 0 + 0 + 0 + (-1) + 0 = 0.$$

The observed measurement vector is

$$y = Xs^T = [0, 0, 0, 1, 1, -2]^T.$$

Column 9 has the most explanatory power toward y, based on X. We let $j = 9$ so $X_j = X_9 = (-1, 0, 0, 1, 1, -1)^T$. Then $s_9 = 1$ and $r = y - X_9 = (1, 0, 0, 0, 0, -1)$.

Next, we observe that column 3 has the most explanatory power for the new r. We set $s_3 = 1$ and update $r = r - X_3 = (0, 0, -1, 1, 0, 0)$. *Note: This progress seemed sideways at best. It increased our non-zero s values, but did not decrease $\|r - y\|$.*

Finally, we observe column 6 has the most explanatory power of the new r. We set $s_6 = 1$ and update $r = r - X_6 \gamma_3 = (0, 0, 0, 0, 0, 0)$. We have now completely explained y using only 3 data elements.

This will not always work so cleanly on a small example. Using MP typically needs something like $n = 20m \log d$ measurements (instead of $n = 6$). Larger measurement sets act like concentrating random variables, and the larger the sets the more likely that at each step we chose the correct index j as most explanatory.

Exercises

We will use data sets found here:
https://mathfordata.github.io/data/D.csv
https://mathfordata.github.io/data/x.csv
https://mathfordata.github.io/data/y.csv
https://mathfordata.github.io/data/XX.csv
https://mathfordata.github.io/data/YY.csv

5.1 Let the first column of the data set D be the explanatory variable x, and let the fourth column be the dependent variable y. [That is: ignore columns 2 and 3 for now]

1. Run simple linear regression to predict y from x. Report the linear model you found. Predict the value of y for new x values of 0.3, 0.5, and 0.8.
2. Use cross-validation to predict generalization error, with error of a single data point $(\mathbf{x}, \mathbf{y})$ from a model M as $(M(x) - y)^2$. Describe how you did this, and which data was used for what.
3. On the same data, run polynomial regression for $p = 2, 3, 4, 5$. Report polynomial models for each. With each of these models, predict the value of y for a new x values of 0.3, 0.5, and 0.8.
4. Cross-validate to choose the best model. Describe how you did this, and which data was used for what.

5.2 Now let the first three columns of the data set D be separate explanatory variables x_1, x_2, x_3. Again let the fourth column be the dependent variable y.

- Run linear regression simultaneously using all three explanatory variables. Report the linear model you found. Predict the value of y for new (x_1, x_2, x_3) values of $(0.3, 0.4, 0.1)$, $(0.5, 0.2, 0.4)$, and $(0.8, 0.2, 0.7)$.
- Use cross-validation to predict generalization error; as usual define the error of a single data point (x_1, x_2, x_3, y) from a model M as $(M(x_1, x_2, x_3) - y)^2$. Describe how you did this, and which data was used for what.

5.3 Let data set $\mathbf{x} \in \mathbb{R}^n$ hold the data for an explanatory variable, and data set $\mathbf{y} \in \mathbb{R}^n$ be the data for the dependent variable. Here $n = 100$.

1. Run simple linear regression to predict $\mathbf{y}$ from $\mathbf{x}$. Report the linear model you find. Predict the value of y for the new x values of 0.4 and 0.7.
2. Split the data into a training set (the first 70 values) and the test set (the last 30 values). Run simple linear regression on the training set, and report the linear model. Again predict the y value at x value of 0.4 and of 0.7.
3. Using the testing data, report the residual vector (it should be 30-dimensional, use the absolute value for each entry) for the model built on the full data, and another one using the model built just from the training data. Report the 2 norm of each vector.

Also compute the 2-norm of the residual vector for the training data (a 70-dimensional vector) for the model build on the full data, and also for the model built on the training data.

4. Expand data set x into a $n \times (p+1)$ matrix $\tilde{X}_p$ using standard polynomial expansion for $p = 5$. Report the first 5 rows of this matrix.

   Build and report the degree-5 polynomial model using this matrix on the training data.

   Report the 2 norm of the residual vector built for the testing data (from a 30-dimensional vector) and for the training data (from a 70-dimensional vector).

5.4 Consider a data set (X, y) where $X \in \mathbb{R}^{n \times 3}$, and its decomposition into a test $(X_{\text{test}}, y_{\text{test}})$ and a training data set $(X_{\text{train}}, y_{\text{train}})$. Assume that X_{train} is not just a subset of X, but also prepends a columns of all 1s. We build a linear model

$$\alpha = (X_{\text{train}}^T X_{\text{train}})^{-1} X_{\text{train}}^T y_{\text{train}},$$

where $\alpha \in \mathbb{R}^4$. The test data $(X_{\text{test}}, y_{\text{test}})$ consists of two data points: (x_1, y_1) and (x_2, y_2), where $x_1, x_2 \in \mathbb{R}^3$. Explain how to use (write a mathematical expression) this test data to estimate the generalization error. That is, if one new data point x arrives, how much squared error would we expect the model α to have compared to the unknown true value y?

5.5 Instead of prepending column vectors of all 1s from X to $\tilde{X}$, consider prepending a vector of all 2s. Explain if this approach is more powerful, less powerful, or the same as the standard approach. How would the coefficients be interpreted?

5.6 Consider input data (X, y) where $X \in \mathbb{R}^{n \times d}$, and assume the rows are drawn iid from some fixed and unknown distribution.

1. Describe three computational techniques to solve for a model $\alpha \in \mathbb{R}^{d+1}$ which minimizes $\|X\alpha - y\|$ (so no regularization, you may use ideas from Chapter 6). For each, briefly describe what the methods are —do not just list different commands in Python—and explain potential advantages of each; these advantages may depend on the values of n and d or the variant in the model being optimized.
2. Now contrast these above model (no regularization) with two regularized models (ridge regression and lasso). Explain the advantages of each scenario.

5.7 Assume that for a regression problem the dependent variable y values are in $[0, 1]$, and so we can assume the residuals of any reasonable model built will also be in $[0, 1]$. Use a Chernoff-Hoeffding bound to calculate how many test set points are required to ensure, we estimate the generalization error of our model on a single new point within $\varepsilon = 0.02$ with probability of at least $1 - \delta = 0.99$.

5.8 Using data sets **XX** (which we refer to as $X \in \mathbb{R}^{100 \times 15}$) and **YY** (referred to as $y \in \mathbb{R}^{100}$), we will compare ordinary least squares and ridge regression under cross-validation. We will use both of the direct, closed-form solutions.

Least Squares $\quad \alpha = (X^T X)^{-1} X^T y$

Ridge Regression $\alpha_s = (X^T X + s^2 I)^{-1} X^T y$

1. Solve for the coefficients α or α_s using Least Squares and Ridge Regression with $s = \{1, 5, 10, 15, 20, 25, 30\}$; that is generate the 1 least squares solution (implicitly $s = 0$), and 7 different ridge regression solutions. For each set of coefficients, report the error in the estimate $\hat{y}$ of y as $\|y - X\alpha_s\|_2 = \|y - \hat{y}\|_2$.
2. Create three row-subsets of X and y

   - X_1 and y_1: rows 1...66 of both data sets.
   - X_2 and y_2: rows 34...100 of both data sets.
   - X_3 and y_3: rows 1...33 and 67...100 of both data sets.

   Repeat the above procedure to build models α_s on these subsets and *cross-validate* the solution on the remainder of X and y that was held out. Specifically, learn the coefficients α_s using, say, X_1 and y_1, and then measure the norm on rows 67...100 as $\|\bar{y}_1 - \bar{X}_1 \alpha_s\|$ where $\bar{X}_1$ and $\bar{y}_1$ are rows 67...100.
3. Which approach works best (averaging the results from the three subsets): Least Squares, or for which value of s using Ridge Regression?

5.9 Consider a new regularized linear regression formulation

$$W_\star(X, y, \alpha, s) = \frac{1}{n}\|X\alpha - y\|_2^2 + s\|\alpha\|_{1/2}.$$

that regularizes the solution with the $L_{1/2}$ norm on the coefficients α. For appropriate value of s, will the solution to this problem favor a sparse solution (like lasso) or a dense solution (like ridge regression)? Explain why.

5.10 Consider the measurement matrix $M \in \mathbb{R}^{8 \times 10}$ (which was populated with random values from $\{-1, 0, +1\}$) and observation vector $y \in \mathbb{R}^8$, generated as $y = Ms$ using a sparse signal $s \in \mathbb{R}^{10}$.

$$
M = \begin{bmatrix}
0 & 0 & -1 & -1 & 0 & 0 & 1 & 0 & 1 & -1 \\
0 & 1 & 1 & 1 & -1 & 0 & 1 & -1 & 0 & 1 \\
-1 & -1 & 0 & 1 & -1 & 1 & 1 & 0 & 0 & 0 \\
-1 & 1 & 0 & -1 & 1 & 1 & 1 & 1 & 0 & -1 \\
-1 & -1 & 1 & 1 & 1 & 0 & 0 & 1 & 1 & 1 \\
-1 & 1 & 0 & -1 & 1 & -1 & 1 & -1 & 1 & -1 \\
0 & 1 & 0 & 1 & -1 & 1 & 1 & 0 & 1 & 0 \\
-1 & 0 & 1 & 0 & 0 & -1 & 1 & -1 & -1 & 1
\end{bmatrix}
\quad \text{and} \quad
y = \begin{bmatrix} 0 \\ 1 \\ 0 \\ 2 \\ -1 \\ 0 \\ 2 \\ -1 \end{bmatrix}.
$$

Use Matching Pursuit (as described in Algorithm 5.5.2) to recover the non-zero entries from s. Record the order in which you find each entry and the residual vector after each step.

Chapter 6
Gradient Descent

Abstract In this topic, we will discuss optimizing over general functions f. Typically, the function is defined as $f : \mathbb{R}^d \rightarrow \mathbb{R}$; that is its domain is multidimensional (in this case d-dimensional point α) and output is a real scalar ($\mathbb{R}$). This often arises to describe the "cost" of a model which has d parameters which describe the model (e.g., degree $(d-1)$-polynomial regression) and the goal is to find the parameters $\alpha = (\alpha_1, \alpha_2, \ldots, \alpha_d)$ with minimum cost. Although there are special cases where we can solve for these optimal parameters exactly, there are many cases where we cannot. What remains in these cases is to analyze the function f, and try to find its minimum point. The most common solution for this is gradient descent where we try to "walk" in a direction so the function decreases until we no longer can.

6.1 Functions

We review some basic properties of a function $f : \mathbb{R}^d \rightarrow \mathbb{R}$. Again, the goal will be to unveil abstract tools that are often easy to imagine in low dimensions, but automatically generalize to high-dimensional data. We will first provide definitions without any calculus.

Let $B_r(\alpha)$ define a Euclidean ball around a point $\alpha \in \mathbb{R}^d$ of radius r. That is, it includes all points $\{p \in \mathbb{R}^d \mid \|\alpha - p\| \leq r\}$, within a Euclidean distance of r from α. We will use $B_r(\alpha)$ to define a *local neighborhood* around a point α. The idea of "local" is quite flexible, and we can use *any* value of $r > 0$, basically it can be as small as we need it to be, as long as it is strictly greater than 0.

Minima and Maxima

A *local maximum* of f is a point $\alpha \in \mathbb{R}^d$ so for some neighborhood $B_r(\alpha)$, all points $p \in B_r(\alpha)$ have smaller (or equal) function value than at α: $f(p) \leq f(\alpha)$. A

© Springer Nature Switzerland AG 2021
J. M. Phillips, *Mathematical Foundations for Data Analysis*,
Springer Series in the Data Sciences,
https://doi.org/10.1007/978-3-030-62341-8_6

local minimum of f is a point $\alpha \in \mathbb{R}^d$ so for some neighborhood $B_r(\alpha)$, all points $p \in B_r(\alpha)$ have larger (or equal) function value than at α: $f(p) \geq f(\alpha)$. If we remove the "or equal" condition for both definitions for $p \in B_r(\alpha), p \neq \alpha$, we say the maximum or minimum points are *strict*.

A point $\alpha \in \mathbb{R}^d$ is a *global maximum* of f if for all $p \in \mathbb{R}^d$, then $f(p) \leq f(\alpha)$. Likewise, a point $\alpha \in \mathbb{R}^d$ is a *global minimum* if for all $p \in \mathbb{R}^d$, then $f(p) \geq f(\alpha)$. There may be multiple global minimum and maximum. If there is exactly one point $\alpha \in \mathbb{R}^d$ that is a global minimum or global maximum, we again say it is *strict*.

When we just use the term *minimum* or *maximum* (without local or global) it implies a local minimum or maximum.

Focusing on a function with domain restricted to a closed and bounded subset $S \subset \mathbb{R}^d$, if the function is *continuous* (that is, for any sufficiently small value $\delta > 0$, for all $\alpha \in \mathbb{R}^d$, there exists a radius r_δ such that if $p \in B_{r_\delta}(\alpha)$, then $|f(\alpha) - f(p)| \leq \delta$), then the function must have a global minimum and a global maximum. It may occur on the boundary of S.

A *saddle point* $\alpha \in \mathbb{R}^d$ (for $d > 1$) has within any neighborhood $B_r(\alpha)$ points $p \in B_r(\alpha)$ with $f(p) < f(\alpha)$ (the lower points) and points $p' \in B_r(\alpha)$ with $f(p') > f(\alpha)$ (the upper points). In particular, it is a saddle if within $B_r(\alpha)$ there are disconnected regions of upper points (and of lower points). The notion of saddle point is defined differently for $d = 1$; it can also be defined using gradients (see definitions below; then a saddle has a gradient of 0, but is not a minimum or maximum).

If these lower and upper regions are each connected, and it is not a minimum or maximum, then it is a *regular point*. For an arbitrary (or randomly) chosen point α, it is usually a regular point (except for examples you are unlikely to encounter, the set of strict minimum, maximum, and saddle points are finite, while the set of regular points is infinite).

Example: Continuous Functions

Here, we show some example functions, where the α-value on the x-axis represents a d-dimensional space. The first function has local minimum and maximum. The second function has a constant value, so every point is a global minimum and a global maximum. The third function f is convex, which is demonstrated with the line segment between points $(p, f(p))$ and $(q, f(q))$ is always above the function f, and hence satisfies $f(\lambda p + (1 - \lambda)q) \leq \lambda f(p) + (1 - \lambda)f(q)$ for all $\lambda \in [0, 1]$.

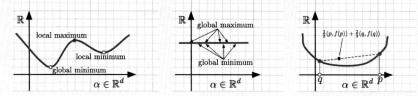

In higher dimensions, the categorization of function neighborhoods should be viewed differently, especially for saddle points. The following image shows the domain of functions in $\mathbb{R}^2$ as the neighborhood ball around a point (in red). The function value is illustrated by color relative to the function value at the red point: where it is blue indicates the function value is above that of the red point, and where it is green it is below the red point.

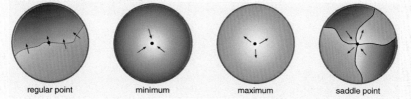

regular point   minimum   maximum   saddle point

Minimum and maximum are special points where the function value are all the same color in some neighborhood ball (either all blue for a minimum or all green for a maximum). The saddle points are special points where the neighborhood colors are not connected.

Convex Functions

In many cases we will assume (or at least desire) that our function is convex.

To define this it will be useful to define a line $\ell \subset \mathbb{R}^d$ as follows with any two points $p, q \in \mathbb{R}^d$. Then, a line $\ell_{p,q}$ is the set of all points defined by any scalar $\lambda \in \mathbb{R}$ as

$$\ell_{p,q} = \{x = \lambda p + (1 - \lambda)q \mid \lambda \in \mathbb{R}\}.$$

When $\lambda \in [0, 1]$, then this defines the line segment between p and q.

A function is *convex* if for any two points $p, q \in \mathbb{R}^d$, on the line segment between them has value less than (or equal) to the values at the weighted average of p and q. That is, it is convex if

For all $p, q \in \mathbb{R}$ and for all $\lambda \in [0, 1]$ $f(\lambda p + (1 - \lambda)q) \leq \lambda f(p) + (1 - \lambda)f(q)$.

Removing the "or equal" condition, the function becomes *strictly convex*.

There are many very cool properties of convex functions. For instance, for two convex functions f and g, then $h(\alpha) = f(\alpha) + g(\alpha)$ is convex and so is $h(\alpha) = \max\{f(\alpha), g(\alpha)\}$. But one will be most important for us:

▷ **Convexity and Global Minimums** ◁

Any local minimum of a convex function will also be a global minimum. A strictly convex function will have at most a single minimum: the global minimum.

This means if we find a minimum, then we must have also found a global minimum (our goal).

6.2 Gradients

For a function $f(\alpha) = f(\alpha_1, \alpha_2, \ldots, \alpha_d)$, and a unit vector $u = (u_1, u_2, \ldots, u_d)$ which represents a direction, then the *directional derivative* is defined as

$$\nabla_u f(\alpha) = \lim_{h \to 0} \frac{f(\alpha + hu) - f(\alpha)}{h}.$$

We are interested in functions f which are *differentiable*; this implies that $\nabla_u f(\alpha)$ is well-defined for all α and u.

Let $e_1, e_2, \ldots, e_d \in \mathbb{R}^d$ be a specific set of unit vectors defined so that $e_i = (0, 0, \ldots, 0, 1, 0, \ldots, 0)$ where for e_i the 1 is in the ith coordinate.

Then define

$$\nabla_i f(\alpha) = \nabla_{e_i} f(\alpha) = \frac{\mathrm{d}}{\mathrm{d}\alpha_i} f(\alpha).$$

It is the derivative in the ith coordinate, treating all other coordinates as constants.

We can now, for a differentiable function f, define the *gradient of f* as

$$\nabla f = \frac{\mathrm{d}f}{\mathrm{d}\alpha_1} e_1 + \frac{\mathrm{d}f}{\mathrm{d}\alpha_2} e_2 + \ldots + \frac{\mathrm{d}f}{\mathrm{d}\alpha_d} e_d = \left(\frac{\mathrm{d}f}{\mathrm{d}\alpha_1}, \frac{\mathrm{d}f}{\mathrm{d}\alpha_2}, \ldots, \frac{\mathrm{d}f}{\mathrm{d}\alpha_d} \right).$$

Note that ∇f is a function from $\mathbb{R}^d \to \mathbb{R}^d$, which we can evaluate at any point $\alpha \in \mathbb{R}^d$.

Example: Gradient

For $\alpha = (x, y, z) \in \mathbb{R}^3$, consider the function $f(x, y, z) = 3x^2 - 2y^3 - 2xe^z$. Then $\nabla f = (6x - 2e^z, -6y^2, -2xe^z)$ and $\nabla f(3, -2, 1) = (18 - 2e, -24, -6e)$.

Linear Approximation

From the gradient we can easily recover the directional derivative of f at point α, for any direction (unit vector) u as

$$\nabla_u f(\alpha) = \langle \nabla f(\alpha), u \rangle.$$

This implies the gradient describes the linear approximation of f at a point α. The slope of the tangent plane of f at α in any direction u is provided by $\nabla_u f(\alpha)$.

Hence, the direction which f is increasing the most at a point α is the unit vector u where $\nabla_u f(\alpha) = \langle \nabla f(\alpha), u \rangle$ is the largest. This occurs at $\overline{\nabla f}(\alpha) = \nabla f(\alpha)/\|\nabla f(\alpha)\|$, the normalized gradient vector.

To find the minimum of a function f, we then typically want to move from any point α in the direction $-\overline{\nabla f}(\alpha)$; at regular points this is the direction of steepest descent.

6.3 Gradient Descent

Gradient descent is a family of techniques that, for a differentiable function $f : \mathbb{R}^d \to \mathbb{R}$, try to identify either

$$\min_{\alpha \in \mathbb{R}^d} f(\alpha) \quad \text{and/or} \quad \alpha^* = \operatorname*{argmin}_{\alpha \in \mathbb{R}^d} f(\alpha).$$

This is effective when f is convex, and we do not have a "closed-form" solution α^*. The algorithm is iterative, in that it may never reach the completely optimal α^*, but it keeps getting closer and closer.

Algorithm 6.3.1 Gradient Descent(f, α_{start})

Initialize $\alpha^{(0)} = \alpha_{\text{start}} \in \mathbb{R}^d$; $k = 0$.
repeat
　$\alpha^{(k+1)} := \alpha^{(k)} - \gamma_k \nabla f(\alpha^{(k)})$; increment $k = k + 1$.
until ($\|\nabla f(\alpha^{(k)})\| \leq \tau$)
return $\alpha^{(k)}$

Basically, for any starting point $\alpha^{(0)}$ the algorithm moves to another point in the direction opposite to the gradient, in the direction that locally decreases f the fastest. How fast it moves depends on the scalar learning rate γ_k and the magnitude of the gradient vector $\nabla f(\alpha^{(k)})$.

Stopping Condition

The parameter τ is the tolerance of the algorithm. If we assume the function is differentiable, then at the minimum α^*, we must have that $\nabla f(\alpha) = (0, 0, \ldots, 0)$. So for α close to the minimum, $\nabla f(\alpha)$ should also have a small norm. The algorithm may never reach the true minimum (and we do not know what it is, so we cannot directly compare against the function value). So it is common to use $\|\nabla f\|$ as a proxy.

In other settings, we may run for a fixed number T steps. Although this does not automatically tune the algorithm to the input, as using a tolerance τ may, it is easier to describe and compare. Hence, most examples in this text will use this fixed number of step T approach.

6.3.1 Learning Rate

The most critical parameter of gradient descent is γ, the *learning rate*. In many cases the algorithm will keep $\gamma_k = \gamma$ fixed for all k. It controls how fast the algorithm works. But if it is too large, as we approach the minimum, then the algorithm may go too far, and overshoot it.

How should we choose γ? There is no consensus to this answer, and often in practice it is tuned in ad-hoc ways. In the following, we will describe some mathematically described scenarios where something formal can be said about how to choose γ. In some cases, these analyses show that if the function satisfies a mild property, then many prescribed choices of γ will result in accuracy guarantees. We will also show methods where it helps to adjust the learning parameter γ_k adaptively.

Lipschitz Bound

We say a function $g : \mathbb{R}^d \to \mathbb{R}^k$ is *L-Lipschitz* if for all $p, q \in \mathbb{R}^d$ that

$$\|g(p) - g(q)\| \le L\|p - q\|.$$

This property is useful when $g = \nabla f$ is describing the gradient of a cost function f. If ∇f is L-Lipschitz, and we set $\gamma \le \frac{1}{L}$, then gradient descent will converge to a stationary point. Moreover, if f is convex with global minimum α^*, then after at most $k = C/\varepsilon$ steps, for some constant C, we can guarantee that

$$f(\alpha^{(k)}) - f(\alpha^*) \le \varepsilon.$$

For the $k = C/\varepsilon$ claim (and others stated below), we assume that $f(\alpha^{(0)}) - f(\alpha^*)$ is less than some absolute constant, which relates to the constant C. Intuitively, the closer we start to the optimum, the fewer steps it will take.

For a convex quadratic function f (e.g., most cost functions derived by sum of squared errors), the gradient ∇f is L-Lipschitz.

Strongly Convex Functions

A function $f : \mathbb{R}^d \to \mathbb{R}$ is *η-strongly convex* with parameter $\eta > 0$ if for all $\alpha, p \in \mathbb{R}^d$ then

$$f(p) \ge f(\alpha) + \langle \nabla f(\alpha), p - \alpha \rangle + \frac{\eta}{2}\|p - \alpha\|^2.$$

Intuitively, this implies that f is lower bounded by a quadratic function. That is, along any direction $u = \frac{p-\alpha}{\|p-\alpha\|}$ the function $\langle f, u \rangle$ is one-dimensional, and then its second derivative $\frac{d^2}{du^2}\langle f, u \rangle$ is strictly positive; it is at least $\eta > 0$. Similarly, saying a function f has an L-Lipschitz gradient is equivalent to the condition that $\frac{d^2}{du^2}\langle f(\alpha), u \rangle \le L$ for all $\alpha, p \in \mathbb{R}^d$ where $u = \frac{p-\alpha}{\|p-\alpha\|}$.

For an η-strongly convex function f, that has an L-Lipschitz gradient, with global minimum α^*, then gradient descent with learning rate $\gamma \le 2/(\eta + L)$ after only $k = C \log(1/\varepsilon)$ steps, again for a constant C, will achieve

$$f(\alpha^{(k)}) - f(\alpha^*) \le \varepsilon.$$

The constant C in $k = C \log(1/\varepsilon)$ also depends on the condition number L/η. The conditions of this bound, imply that f is sandwiched between two convex quadratic functions; specifically for any $\alpha, p \in \mathbb{R}^d$ that we can bound

$$f(\alpha) + \langle \nabla f(\alpha), p - \alpha \rangle + \frac{\eta}{2}\|p - \alpha\|^2 \le f(p) \le f(\alpha) + \langle \nabla f(\alpha), p - \alpha \rangle + \frac{L}{2}\|p - \alpha\|^2.$$

Geometry of Strongly Convex Function

We show an example η-strongly convex function f, in blue. At any point p, it is sandwiched between two convex quadratic functions in green. The convex quadratic function which lower bounds f has an L-Lipschitz gradient.

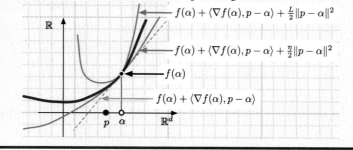

When an algorithm converges at such a rate (takes at most $C \log(1/\varepsilon)$ steps to obtain ε error), it is known as *linear convergence* since the log-error $\log(f(\alpha^{(k)}) - f(\alpha^*))$ looks like a linear function of k.

In practice, since many functions we consider will be convex quadratic functions (e.g., are derived from sum of squared error cost functions), then the error will decreases exponentially fast in terms of the number of steps of gradient descent, if the learning rate is set sufficiently small. That is, only a constant number of steps are required to resolve each bit (or digit) of precision in the function value at the optimum! However, if the learning rate is set too small, then this constant (number of steps to resolve one bit) will increase.

So at this point, we have explained that for many situations there is a learning rate for which gradient descent will work extremely well. If we can analytically bound various properties of the second derivative of the function, then we can use these bounds to choose such a rate. However, we have not yet explained a formal way to find such a rate in general if we can only evaluate the gradient at any point $\alpha \in \mathbb{R}^d$.

Line Search

An alternative, referred to as *line search* is to solve for the (approximately) optimal γ_k at each step. Once we have computed the gradient $\nabla f(\alpha^{(k)})$ then we have reduced the high-dimensional minimization problem to a one-dimensional problem. Note if f is convex, then f restricted to this one-dimensional search is also convex. We still need to find the minimum of an unknown function, but we can perform some procedure akin to binary search. We first find a value γ' such that

$$f\left(\alpha^{(k)} - \gamma' \nabla f(\alpha^{(k)})\right) > f(\alpha^{(k)})$$

then we keep subdividing the region $[0, \gamma']$ into pieces, and excluding ones which cannot contain the minimum.

For instance, the *golden section search* divides a range $[b, t]$ containing the optimal γ_k into three sub-intervals (based on the golden ratio) so $[b, t] = [b, b') \cup [b', t'] \cup (t', t]$. And each step, we can determine that either $\gamma_k \notin [b, b')$ if $f(t') < f(b')$, or $\gamma_k \notin (t', t]$ if $f(b') < f(t')$. This reduces the range to $[b', t]$ or $[b, t']$, respectively, and we recurse.

Example: Golden Section Search

A function f is shown in blue in the domain $[b, t]$. Observe that in the first step, $f(b') < f(t')$ so the domain containing the minimum shrinks to $[b, t']$. In the next step, $f(t') < f(b')$ so the domain containing the minimum shrinks to $[b', t]$.

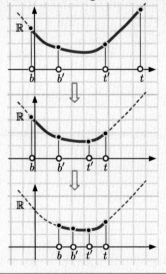

In other situations, we can solve for the optimal γ_k exactly at each step. This is the case if we can again analytically take the derivative $\frac{d}{d\gamma}\left(f(\alpha^{(k)}) - \gamma \nabla f(\alpha^{(k)})\right)$ and solve for the γ where it is equal to 0.

Adjustable Rate

In practice, line search is often slow. Also, we may not have a Lipschitz bound. It is often better to try a few fixed γ values, probably being a bit conservative. As long as $f(\alpha^{(k)})$ keep decreasing, it works well. This also may alert us if there is more than one local minimum if the algorithm converges to different locations.

An algorithm called *backtracking line search* automatically tunes the parameter γ. It uses a fixed parameter $\beta \in (0, 1)$ (preferably in $(0.1, 0.8)$; for instance, use $\beta = 3/4$). Start with a large step size γ (e.g., $\gamma = 1$, but "large" very much depends on the scale of f). Then at each step of gradient descent at location α, if

$$f(\alpha - \gamma \nabla f(\alpha)) > f(\alpha) - \frac{\gamma}{2} \|\nabla f(\alpha)\|^2$$

then update $\gamma = \beta \gamma$. This shrinks γ over the course of the algorithm, and if f is strongly convex, it will eventually decrease γ until it satisfies the condition for linear convergence.

Geometry of Backtracking and Adjustable Rates

The figure shows a function $f : \mathbb{R}^d \to \mathbb{R}$ along the gradient at $\alpha \in \mathbb{R}^d$ as $f(\alpha - \gamma \nabla f(\alpha))$. That is, the value of f is plotted after the next step, as a function of the current learning rate γ. In relation, we also plot $f(\alpha) - \frac{\gamma}{2} \|\nabla f(\alpha)\|^2$, which is linear in γ, and approximates what the upper bound of f is supposed to be at the current learning rate. If γ is too large, as in the figure, then a step of gradient descent will be too far, and may overshoot the minimum. Note at a decreased value of γ the linear function is above f, and so by decreasing γ by a fixed fraction until this is satisfied will ensure the learning rate is not too large, and also still sufficiently large so the process converges quickly.

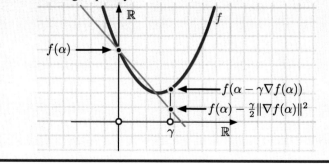

Example: Gradient Descent with Fixed Learning Rate

Consider the function f where $\alpha = (x, y) \in \mathbb{R}^2$ is defined

$$f(x, y) = \left(\frac{3}{4}x - \frac{3}{2}\right)^2 + (y - 2)^2 + \frac{1}{4}xy$$

and has gradient

$$\nabla f(x, y) = \left(\frac{9}{8}x - \frac{9}{4} + \frac{1}{4}y , \quad 2y - 4 + \frac{1}{4}x\right).$$

We run gradient descent for 10 iterations within initial position $(5, 4)$, while varying the learning rate in the range $\gamma = \{0.01, 0.1, 0.2, 0.3, 0.5, 0.75\}$.

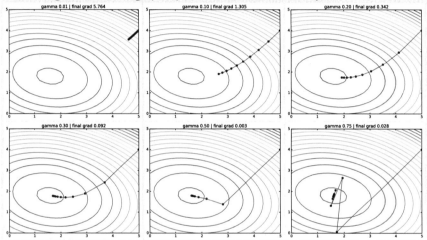

We see that with γ very small, the algorithm does not get close to the minimum. When γ is too large, then the algorithm jumps around a lot, and is in danger of not converging. But at a learning rate of $\gamma = 0.3$, it converges fairly smoothly and reaches a point where $\|\nabla f(x, y)\|$ is very small. Using $\gamma = 0.5$ almost overshoots in the first step; $\gamma = 0.3$ is smoother, and it is probably best to use a curve that looks smooth like that one, but with a few more iterations.

These contour plots can be easily generated in Python:

```python
import numpy as np
from numpy import linalg as LA

import matplotlib as mpl
import matplotlib.pyplot as plt
%matplotlib inline

def func(x,y):
  return (0.75*x-1.5)**2 + (y-2.0)**2 + 0.25*x*y

def func_grad(vx,vy):
  dfdx = 1.125*vx - 2.25 + 0.25*vy
  dfdy = 2.0*vy - 4.0 + 0.25*vx
  return np.array([dfdx,dfdy])
```

```
xlist = np.linspace(0, 5, 26)
ylist = np.linspace(0, 5, 26)
x, y = np.meshgrid(xlist, ylist)
z = func(x,y)
lev = np.linspace(0,20,21)

plt.contour(x,y,z,levels=lev)
plt.plot(values[:,0],values[:,1],'r-')
plt.plot(values[:,0],values[:,1],'bo')
```

After initializing some settings, in this case with $\gamma = 0.3$

```
v_init = np.array([5,4])
num_iter = 10
values = np.zeros([num_iter,2])
values[0,:] = v_init
v = v_init
gamma = 0.3
```

The gradient descent algorithm is quite simple

```
for i in range(1,num_iter):
    v = v - gamma * func_grad(v[0],v[1])
    values[i,:] = v
```

And, then we can add its trace onto the plots

```
grad_norm = LA.norm(func_grad(v[0],v[1]))
title = "gamma_%0.2f_|_final_grad_%0.3f" % (gamma,grad_norm)
plt.title(title)
plt.show()
```

6.4 Fitting a Model to Data

For data analysis, the most common use of gradient descent is to fit a model to data. In this setting, we have a data set $(X, y) = \{(x_1, y_1), (x_2, y_2), \ldots, (x_n, y_n)\} \in \mathbb{R}^d \times \mathbb{R}$ and a family of models $\mathcal{M}$ so each possible model M_α is defined by a k-dimensional vector $\alpha = \{\alpha_1, \alpha_2, \ldots, \alpha_k\}$ with k parameters.

Next we define a *loss function* $L((X, y), M_\alpha)$ which measures the difference between what the model predicts and what the data values are. To choose which parameters generate the best model, we let $f(\alpha) : \mathbb{R}^k \to \mathbb{R}$ be our function of interest, defined $f(\alpha) = L((X, y), M_\alpha)$. Then, we can run gradient descent to find our model M_{α^*}. For instance, we can set

$$f(\alpha) = L((X, y), M_\alpha) = \mathsf{SSE}((X, y), M_\alpha) = \sum_{(x_i,y_i)\in(X,y)} (y_i - M_\alpha(x_i))^2. \qquad (6.1)$$

This is used for example in maximum likelihood (or maximum log-likelihood) estimators from Bayesian inference. This includes finding a single point estimator assuming Gaussian noise (where we had a closed-form solution), but also many other variants (often where there is no known closed-form solution). It also includes least squares regression and its many variants; we will see this in much more detail next. And will include other topics (including clustering, PCA, classification) we will see later in the book.

6.4.1 Least Mean Squares Updates for Regression

Now, we will work through how to use gradient descent for simple quadratic regression on one-dimensional explanatory variables. That is, this will specify the function $f(\alpha)$ in equation (6.1) to have $k = 3$ parameters as $\alpha = (\alpha_0, \alpha_1, \alpha_2)$, and so for each $(x_i, y_i) \in (X, y)$ we have $x_i \in \mathbb{R}$. Then, we can write the model again as a dot product

$$M_\alpha(x_i) = \langle \alpha, (1, x_i, x_i^2) \rangle = \alpha_0 + \alpha_1 x_i + \alpha_2 x_i^2.$$

It is straightforward to generalize to linear regression, multiple explanatory variable linear regression, or general polynomial regression from here. For instance, in this case, we can represent x_i as the three-dimensional feature vector $q = (q_0 = x_i^0 = 1, q_1 = x_i^1, q_2 = x_i^2)$ of explanatory variables. Then the model is $M_\alpha(x_i) = \langle \alpha, q \rangle$. For simple linear regression, we simply omit the quadratic term $\alpha_2 x_i^2$ from the above model. For other polynomial forms of regression, the expansion simply includes more terms and powers. And for multiple explanatory variables they are each given a corresponding feature coordinate, possibly to some power, or multiplied with other explanatory variables in the polynomial models. We will continue using one-dimensional quadratic regression as an example.

Now to specify the gradient descent step

$$\alpha := \alpha - \gamma \nabla f(\alpha)$$

we only need to define $\nabla f(\alpha)$. We will first show this for the case where $n = 1$, that is when there is a single data point (x_1, y_1). For quadratic regression, the cost function $f(\alpha) = f_1(\alpha) = (\alpha_0 + \alpha_1 x_1 + \alpha_2 x_1^2 - y_1)^2$ is convex. Next derive

$$\frac{d}{d\alpha_j} f(\alpha) = \frac{d}{d\alpha_j} f_1(\alpha) = \frac{d}{d\alpha_j} (M_\alpha(x_1) - y_1)^2$$

$$= 2(M_\alpha(x_1) - y_1) \frac{d}{d\alpha_j} (M_\alpha(x_1) - y_1)$$

$$= 2(M_\alpha(x_1) - y_1) \frac{d}{d\alpha_j} (\sum_{j=0}^{2} \alpha_j x_1^j - y_1)$$

$$= 2(M_\alpha(x_1) - y_1) x_1^j.$$

Using this convenient form (which generalizes to any polynomial model), we define

$$\nabla f(\alpha) = \left(\frac{d}{d\alpha_0} f(\alpha), \frac{d}{d\alpha_1} f(\alpha), \frac{d}{d\alpha_2} f(\alpha) \right)$$

$$= 2 \left((M_\alpha(x_1) - y_1), \; (M_\alpha(x_1) - y_1)x_1, \; (M_\alpha(x_1) - y_1)x_1^2 \right).$$

Applying $\alpha := \alpha - \gamma \nabla f(\alpha)$ according to this specification is known as the *least mean squares (LMS) update rule* or the *Widrow-Hoff learning rule*. Quite intuitively, the magnitude of the update is proportional to the residual norm $(M_\alpha(x_1) - y_1)$. So if we have a lot of error in our guess of α, then we take a large step; if we do not have a lot of error, we take a small step.

6.4.2 Decomposable Functions

To generalize this to multiple data points ($n > 1$), there are two standard ways. Both of these take strong advantage of the cost function $f(\alpha)$ being *decomposable*. That is, we can write

$$f(\alpha) = \sum_{i=1}^{n} f_i(\alpha),$$

where each f_i depends only on the ith data point $(x_i, y_i) \in (X, y)$. In particular, for quadratic regression

$$f_i(\alpha) = (M_\alpha(x_i) - y_i)^2 = (\alpha_0 + \alpha_1 x_i + \alpha_2 x_i^2 - y_i)^2.$$

First notice that since f is the sum of f_is, where each is convex, then f must also be convex; in fact the sum of these usually becomes strongly convex (it is strongly convex as long as the corresponding feature vectors are full rank).

Geometry of Full Rank Decomposable Functions

Consider now where each $x_i \in \mathbb{R}$ so $(1, x) \in \mathbb{R}^2$ and the corresponding model $\alpha = (\alpha_0, \alpha_1) \in \mathbb{R}^2$. We can now visualize a function $f_i(\alpha) = (\langle \alpha, (1, x) \rangle - y_i)^2$ as a "halfpipe" in the two-dimensional space parameterized by α_0 and α_1. That is f_i is a described by a set of parabolas that have a minimum of exactly 0 on the line $\langle \alpha, (1, x) \rangle = y_i$ in this two-dimensional space.

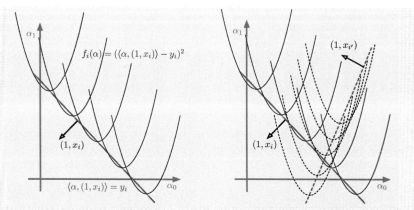

Thus while each f_i is convex, it is not strongly convex since along the line (shown in green) it has a 0 directional derivative. If we add a second function $f_{i'}$ (the dashed one) defined by $(x_{i'}, y_{i'})$, then $f(\alpha) = f_i(\alpha) + f_{i'}(\alpha)$. That it is the sum of the blue halfpipe and the dashed-purple halfpipe. As long as these are not parallel, then f becomes strongly convex. Linear algebraically, this happens as long as the matrix $\begin{bmatrix} 1 & x_i \\ 1 & x_{i'} \end{bmatrix}$ is full rank.

Two approaches toward gradient descent will take advantage of this decomposition in slightly different ways. This decomposable property holds for most loss functions for fitting a model to data.

Batch Gradient Descent

The first technique, called *batch gradient descent*, simply extends the definition of $\nabla f(\alpha)$ to the case with multiple data points. Since f is decomposable, then we use the linearity of the derivative to define

$$\frac{d}{d\alpha_j} f(\alpha) = \sum_{i=1}^{n} \frac{d}{d\alpha_j} f_i(\alpha) = \sum_{i=1}^{n} 2(M_\alpha(x_i) - y_i)x_i^j$$

and thus

$$\nabla f(\alpha) = \left(2\sum_{i=1}^{n}(M_\alpha(x_i) - y_i),\ 2\sum_{i=1}^{n}(M_\alpha(x_i) - y_i)x_i,\ 2\sum_{i=1}^{n}(M_\alpha(x_i) - y_i)x_i^2\right)$$

$$= \sum_{i=1}^{n}\left(2(M_\alpha(x_i) - y_i),\ 2(M_\alpha(x_i) - y_i)x_i,\ 2(M_\alpha(x_i) - y_i)x_i^2\right)$$

$$= 2\sum_{i=1}^{n}(M_\alpha(x_i) - y_i)\left(1, x_i, x_i^2\right).$$

That is, the step is now just the sum of the terms from each data point. Since f is (strongly) convex, then we can apply all of the nice convergence results discussed about (strongly) convex f before. However, computing $\nabla f(\alpha)$ each step takes time proportional to n, which can be slow for large n (i.e., for large data sets).

Stochastic and Incremental Gradient Descent

The second technique is called *incremental gradient descent*; see Algorithm 6.4.1. It avoids computing the full gradient each step, and only computes $\nabla f_i(\alpha)$ for a single data point $(x_i, y_i) \in (X, y)$. For quadratic regression with $M_\alpha(x) = \langle \alpha, (1, x, x^2)\rangle$, it is

$$\nabla f_i(\alpha) = 2(M_\alpha(x_i) - y_i)(1, x_i, x_i^2).$$

Algorithm 6.4.1 Incremental Gradient Descent(f, α_{start})

Initialize $\alpha^{(0)} = \alpha_{\text{start}} \in \mathbb{R}^d$; $i = 1$.
repeat
    $\alpha^{(k+1)} := \alpha^{(k)} - \gamma_k \nabla f_i(\alpha^{(k)})$
    $i := (i + 1) \mod n$
until $(\|\nabla f(\alpha^{(k)})\| \le \tau)$ $(\star)$
return $\alpha^{(k)}$

≫ Implementation Hint on Convergence

$(\star)$ The norm of a single gradient is not stable. Instead a better stopping condition averages the gradient norm over several (let us say B steps). The condition may then be $\frac{1}{B}\sum_{b=0}^{B-1}\|\nabla f(\alpha^{(k-b)})\| \le \tau$, and is only checked after B steps are taken.

A more common variant of this is called *stochastic gradient descent*; see Algorithm 6.4.2. Instead of choosing the data points in order, it selects a data point (x_i, y_i) at random each iteration (the term "stochastic" refers to this randomness).

On very large data sets (i.e., big data!), these algorithms are often much faster than the batch version since each iteration now takes constant time. However, it does not automatically inherit all of the nice convergence results from what is known

---

Algorithm 6.4.2 Stochastic Gradient Descent(f, α_{start})

Initialize $\alpha^{(0)} = \alpha_{\text{start}} \in \mathbb{R}^d$
repeat
    Randomly choose $i \in \{1, 2, \ldots, n\}$
    $\alpha^{(k+1)} := \alpha^{(k)} - \gamma_k \nabla f_i(\alpha^{(k)})$
until $(\|\nabla f(\alpha^{(k)})\| \leq \tau)$ ($\star$)
return $\alpha^{(k)}$

---

about (strongly) convex functions. Yet in many settings, there is an abundance of data points described by the same model. They should have a roughly similar effect. In practice when one is far from the optimal model, these steps converge about as well as the batch version (but much much faster in runtime). When one is close to the optimal model, then the incremental/stochastic variants may not exactly converge. However, if one is satisfied to reach a point that is close enough to optimal, there are some randomized (PAC-style) guarantees possible for the stochastic variant. And in fact, for very large data sets (i.e., n is very big) they typically converge before the algorithm even uses all (or even most) of the data points.

Exercises

We will use a dataset `https://mathfordata.github.io/data/G.csv`

6.1 Consider a function $f(x, y)$ with gradient $\nabla f(x, y) = (x - 1, 2y + x)$. Starting at a value $(x = 1, y = 2)$, and a learning rate of $\gamma = 1$, execute one step of gradient descent.

6.2 Consider running gradient descent with a fixed learning rate γ. For each of the following, we plot the function value over 10 steps (the function is different each time). Decide whether the learning rate is probably **too high**, **too low**, or **about right**.

1. f_1: 100, 99, 98, 97, 96, 95, 94, 93, 92, 91
2. f_2: 100, 50, 75, 60, 65, 45, 75, 110, 90, 85
3. f_3: 100, 80, 65, 50, 40, 35, 31, 29, 28, 27.5, 27.3
4. f_4: 100, 80, 60, 40, 20, 0, –20, –40, –60, –80, –100

6.3 Consider two functions

$$f_1(x, y) = (x-5)^2 + (y+2)^2 - 2xy \qquad f_2(x, y) = (1-(y-4))^2 + 20((x+6)-(y-4)^2)^2$$

Starting with $(x, y) = (0, 2)$ run the gradient descent algorithm for each function. Run for T iterations, and report the function value at the end of each step.

1. First, run with a fixed learning rate of $\gamma = 0.05$ for f_1 and $\gamma = 0.0015$ for f_2.
2. Second, run with any variant of gradient descent you want. Try to get the smallest function value after T steps.

For f_1 you are allowed only $T = 10$ steps. For f_2 you are allowed $T = 100$ steps.

6.4 In the first `G.csv` dataset provided, use the first three columns as explanatory variables x_1, x_2, x_3, and the fourth as the dependent variable y. Run gradient descent on $\alpha \in \mathbb{R}^4$, using the dataset provided to find a linear model

$$\hat{y} = \alpha_0 + \alpha_1 x_1 + \alpha_2 x_2 + \alpha_3 x_3$$

minimizing the sum of squared errors. Run for as many steps as you feel necessary. On each step of your run, print on a single line: (1) the model parameters $\alpha^{(i)} = [\alpha_0^{(i)}, \alpha_1^{(i)}, \alpha_2^{(i)}, \alpha_3^{(i)}]$, (2) the value of a function $f(\alpha^{(i)})$, estimating the sum of squared errors, and (3) the gradient $\nabla f(\alpha^{(i)})$. *(These are the sort of things you would do to check/debug a gradient descent algorithm; you may also want to plot some of these.)*

1. First run batch gradient descent.
2. Second run incremental gradient descent.

Choose one method which you preferred (either is ok to choose), and explain why you preferred it to the other methods.

6.5 Explain what parts of the above procedures on data set `G.csv` would change if you instead are minimizing the sum of absolute value of residuals, not the sum of squared residuals?

- Is the function still convex?
- Does the gradient always exist?

6.6 Consider the quadratic (polynomial of degree 2) regression on a data set (X, y) where each of n data points (x_i, y_i) has $x_i \in \mathbb{R}^2$ and $y_i \in \mathbb{R}$. To simplify notation, let each $x_i = (a_i, b_i)$.

1. Expand $x_i = (a_i, b_i)$ and write the model $M_\alpha(x_i)$ as a single dot product of the form

$$M_\alpha(x_i) = \langle \alpha, (?, ?, \ldots, ?) \rangle$$

   where α is a vector, and you need to fill in the appropriate ?s.
2. Write the batch (of size n) gradient $\nabla f(\alpha)$ for this problem, where

$$f(\alpha) = \sum_{i=1}^{n} (M_\alpha(x_i) - y_i)^2.$$

Your expression for $\nabla f(\alpha)$ should use the term $(M_\alpha(x_i) - y_i)$ as part of its solution.

Chapter 7
Dimensionality Reduction

Abstract This chapter will build a series of techniques to deal with high-dimensional data. Unlike regression problems, our goal is not to predict a value (the y-coordinate), it is to understand the "shape" of the data, for instance a low-dimensional representation that captures most of the meaning of the high-dimensional data. This is sometimes referred to as *unsupervised learning* (as opposed to regression and classification, where the data has labels, known as supervised learning). Like most unsupervised scenarios (as most parents know), you can have a lot of fun, but it is easy to get yourself into trouble if you are not careful. We will cover many interconnected tools, including the singular value decomposition (SVD), eigenvectors and eigenvalues, the power method, principal component analysis (PCA), multidimensional scaling, as well as several other use cases of the SVD. The chapter will conclude by contrasting PCA with the main alternative method for dimensionality reduction: random projections.

7.1 Data Matrices

We will start with data in a matrix $A \in \mathbb{R}^{n \times d}$, and will call upon linear algebra to rescue us. It is useful to think of each row a_i of A as a data point in $\mathbb{R}^d$, so there are n data points. Each dimension $j \in 1, 2, \ldots, d$ corresponds with an attribute of the data points.

Example: Data Matrices

There are many situations where data matrices arise.

- Consider a set of n weather stations reporting temperature over d points in time. Then each row a_i corresponds to a single weather station, and each coordinate $A_{i,j}$ of that row is the temperature at station i at time j.

© Springer Nature Switzerland AG 2021
J. M. Phillips, *Mathematical Foundations for Data Analysis*,
Springer Series in the Data Sciences,
https://doi.org/10.1007/978-3-030-62341-8_7

- In movie ratings, we may consider n users who have rated each of d movies on a score of $1 - 5$. Then each row a_i represents a user, and the jth entry of that user is the score given to the j movie.
- Consider the price of a stock measured over time (say the closing price each day). Many time-series models consider some number of days (d days, for instance 25 days) to capture the pattern of the stock at any given time. So for a given closing day, we consider the d previous days. If we have data on the stock for 4 years (about 1000 days the stock market is open), then we can create a d-dimensional data points (the previous $d = 25$ days) for each day (except the first 25 or so). The data matrix is then comprised of n data points a_i, where each corresponds to the closing day, and the previous d days. The jth entry of a_i is the value on $(j - 1)$ days before the closing day i.
- Finally, consider a series of pictures of a shape (say the Utah teapot). The camera position is fixed as is the background, but we vary two things: the rotation of the teapot, and the amount of light. Here each pictures is a set of say d pixels (say 10,000 if it is 100×100), and there are n pictures. Each picture is a row of length d, and each pixel corresponds to a column of the matrix. Similar, but more complicated scenarios frequently occur with pictures of a person's face, or $3d$-imaging of an organ.

In each of these scenarios, there are many (n) data points, each with d attributes.

▷ Data Matrices and Units ◁

For a *data matrix*, to consider dimensionality reduction

- **all coordinates should have the same units!**

If this "same units" property does not hold, then when we measure a distance between data points in $\mathbb{R}^d$, usually using the L_2-norm, then the distance is nonsensical. While there exists "normalization" heuristics to address these issues (see discussion in Section 4.2), these each may cause their own unintended consequences.

The next goal is to uncover a pattern, or a model M. In this case, the model will be a low-dimensional subspace B. It will represent a k-dimensional space, where $k \ll d$. For instance, in the example with images, there are only two parameters which are changing (rotation and lighting), so despite having $d = 10{,}000$ dimensions of data, 2 dimension should be enough to represent everything about the data in that controlled environment.

7.1.1 Projections

Different than in linear regression this family of techniques will measure error as a projection from $a_i \in \mathbb{R}^d$ to the closest point $\pi_B(a_i)$ on B. We next define these concepts more precisely using linear algebra.

First recall, that given a unit vector $v \in \mathbb{R}^d$ and any data point $p \in \mathbb{R}^d$, then the dot product

$$\langle v, p \rangle$$

is the norm of p after it is projected onto the line through v. If we multiply this scalar by v then

$$\pi_v(p) = \langle v, p \rangle v;$$

it results in the point on the line through unit vector v that is closest to data point p. This is a *projection of p onto vector v*.

To understand this for a subspace B, we will need to define an orthogonal basis. *For now we will assume that B contains the origin* $(0, 0, 0, \ldots, 0)$ *(as did the line through v)*. Then if B is k-dimensional, there is a k-dimensional *orthogonal basis* $V_B = \{v_1, v_2, \ldots, v_k\}$ so that

- For each $v_i \in V_B$ we have $\|v_j\| = 1$, that is v_j is a unit vector.
- For each pair $v_i, v_j \in V_B$ with $i \neq j$ we have $\langle v_i, v_j \rangle = 0$; the pairs are orthogonal.
- For any point $x \in B$ we can write $x = \sum_{j=1}^{k} \alpha_j v_j$; in particular $\alpha_j = \langle x, v_j \rangle$.

Given such a basis, then the projection on to B of a point $p \in \mathbb{R}^d$ is simply

$$\pi_B(p) = \sum_{j=1}^{k} \langle v_j, p \rangle v_j = \sum_{j=1}^{k} \pi_{v_j}(p).$$

Thus if p happens to be exactly in B, then this recovers p exactly.

Geometry of Projection

Consider a point $p \in \mathbb{R}^3$, and a two-dimensional subspace B. Given a basis $V_B = \{v_1, v_2\}$, the projection of p onto B is represented as

$$\pi_B(p) = \langle p, v_1 \rangle v_1 + \langle p, v_2 \rangle v_2 = \alpha_1(p)v_1 + \alpha_2(p)v_2$$

where $\alpha_1(p) = \langle p, v_1 \rangle$ and $\alpha_2(p) = \langle p, v_2 \rangle$. Thus, we can also represent $\pi_B(p)$ in the basis $\{v_1, v_2\}$ with the two coordinates $(\alpha_1(p), \alpha_2(p))$.

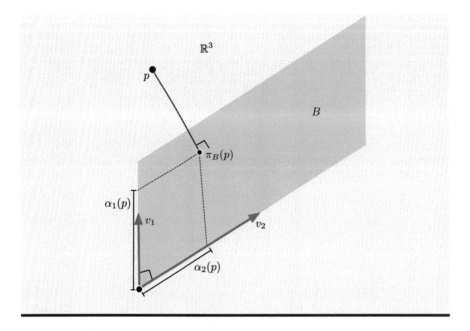

The other powerful part of the basis V_B is that it defines a *new coordinate system*. Instead of using the d original coordinates, we can use new coordinates $(\alpha_1(p), \alpha_2(p), \ldots, \alpha_k(p))$ where $\alpha_i(p) = \langle v_i, p \rangle$. To be clear $\pi_B(p)$ is still in $\mathbb{R}^d$, but there is a k-dimensional representation if we restrict to B.

When B is d-dimensional, this operation can still be interesting. The basis we choose $V_B = \{v_1, v_2, \ldots, v_d\}$ could be the same as the original coordinate axis, that is we could have $v_i = e_i = (0, 0, \ldots, 0, 1, 0, \ldots, 0)$ where only the ith coordinate is 1. But if it is another basis, then this acts as a rotation (with possibility of also a mirror flip). The first coordinate is rotated to be along v_1; the second along v_2; and so on. In $\pi_B(p)$, the point p does not change, just its representation.

7.1.2 Sum of Squared Errors Goal

As usual our goal will be to minimize the sum of squared errors (SSE). In this case, we define this objective as

$$\mathsf{SSE}(A, B) = \sum_{a_i \in A} \|a_i - \pi_B(a_i)\|^2,$$

and our desired k-dimensional subspace B is

$$B^* = \underset{B}{\operatorname{argmin}} \, \mathsf{SSE}(A, B).$$

As compared to linear regression, this is much less a "proxy goal" where the true goal was prediction. Now, we have no labels (the y_i values), so we simply try to fit a model through all of the data.

Example: SSE for Projection

The standard goal for a projection subspace B is to minimize the sum of squared errors between each a_i to its projected approximation $\pi_B(a_i)$.

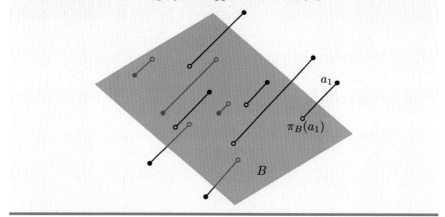

How do we solve for this? Our methods for linear regression do not solve the correct problem. The cost function is different in that it only measures error in one coordinate (the dependent variable), not the projection cost. It is also challenging to use the standard versions of gradient descent. The restriction that each $v_i \in V_B$ is a unit vector is a non-convex constraint. So ... linear algebra will come to the rescue—in the form of the SVD.

7.2 Singular Value Decomposition

A really powerful and useful linear algebra operation is called the *singular value decomposition* (the SVD). It extracts an enormous amount of information about a matrix A. This section will define it and discuss many of its uses. Then, we will describe one algorithm (the power method) which constructs the SVD. But in general, one simply calls the procedure in your favorite programming language and it calls the same highly optimized back-end from the Fortran LAPACK library.

The SVD takes in a matrix $A \in \mathbb{R}^{n \times d}$ and outputs three matrices $U \in \mathbb{R}^{n \times n}$, $S \in \mathbb{R}^{n \times d}$ and $V \in \mathbb{R}^{d \times d}$, so that $A = USV^T$.

Example: SVD in Rectangles

It is useful to visualize the SVD by representing the matrices as rectangles. A "tall and thin" matrix A has more rows than columns ($n > d$). Given its decomposition

$$[U, S, V] = \mathsf{svd}(A)$$

then U and V are square matrices of dimension $n \times n$ and $d \times d$, respectively, so they are represented as squares. S is a tall and thin rectangle, since its dimensions are the same as A; to show it is diagonal, we can leave most of the matrix empty where it has 0s.

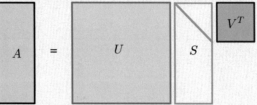

For a "short and fat" matrix A with more columns that rows ($d > n$), the picture is similar. In fact, we can recover the tall and thin view by in this case considering A^T and swapping the roles of $U \rightarrow V^T$ and $V^T \rightarrow U$.

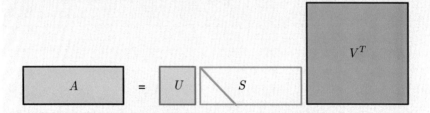

The Structure that Lurks Beneath

The matrix S is mainly all zeros; it only has non-zero elements along its diagonal. So $S_{i,j} = 0$ if $i \neq j$. The remaining values $\sigma_1 = S_{1,1}$, $\sigma_2 = S_{2,2}$, ..., $\sigma_r = S_{r,r}$ are known as the *singular values*. They are non-decreasing so

$$\sigma_1 \geq \sigma_2 \geq \ldots \geq \sigma_r \geq 0$$

where $r \leq \min\{n, d\}$ is the rank of the matrix A. That is the number of non-zero singular values reports the rank (this is a numerical way of computing the rank of a matrix).

The matrices U and V are orthogonal. Thus, their columns are all unit vectors and orthogonal to each other (within each matrix). The columns of U, written $u_1, u_2, \ldots, u_n$, are called the *left singular vectors*; and the columns of V (i.e., rows of V^T), written $v_1, v_2, \ldots, v_d$, are called the *right singular vectors*.

This means for any vector $x \in \mathbb{R}^d$, the columns of V (the right singular vectors) provide a basis. That is, we can write $x = \sum_{i=1}^{d} \alpha_i v_i$ for $\alpha_i = \langle x, v_i \rangle$. Similarly for any vector $y \in \mathbb{R}^n$, the columns of U (the left singular vectors) provide a basis. This also implies that $\|x\| = \|V^T x\|$ and $\|y\| = \|yU\|$. So these matrices and their vectors do not capture any "scale" information about A, that is all in S. Rather they provide the bases to describe the scale in a special form.

Example: Singular Vectors and Values

The left singular vectors are columns in U shown as thin vertical boxes, and right singular vectors are columns in V and thus thin horizontal boxes in V^T. The singular values are shown as small boxes along the diagonal of S, in decreasing order. The grey boxes in S and U illustrate the parts which are unimportant (aside from keeping the orthogonality in U); indeed in some programming languages (like Python), the standard representation of S is as a square matrix to avoid maintaining all of the irrelevant 0s in the grey box in the lower part.

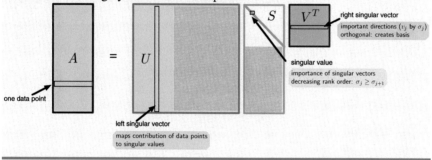

Geometry of SVD as the Aggregate Shape of a Matrix

One can picture the geometry of the SVD by how a unit vector $x \in \mathbb{R}^d$ interacts with the matrix A as a whole. The *aggregate shape* of A is a contour, the set of points

$$\{x(\|Ax\|) \mid \|x\| = 1\}$$

induced by unit vectors x as a movement away from the origin in direction x by amount $\|Ax\|$. When A is a (scaled) identity matrix, this defines a circle. When A is a diagonal matrix with values $\lambda_1, \lambda_2, \ldots, \lambda_d$ on the diagonal, this is an ellipse stretched along the coordinate axes, so the jth coordinate axis is stretched by λ_j.

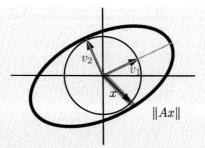

However, when A is general matrix, as $\|Ax\|$ is still a (square-root of a) quadratic equation, it must still have a similar form. But since it is not diagonal, it is not restricted to stretch along the coordinate axes. Using the SVD so $A = USV^T$, however, we can rewrite $\|Ax\| = \|USV^T x\| = \|SV^T x\|$ (since U does not affect the norm). Note that $p = V^T x$ must still be a unit vector, and S is diagonal. So after rotating x by V^T, then it again maps out an ellipse defined by the singular values $\sigma_1, \sigma_2, \ldots$. The stretching basis of the ellipse is now precisely the right singular value basis $v_1, v_2, \ldots$.

Tracing the Path of a Vector

To illustrate what this decomposition demonstrates, a useful exercise is to trace what happens to a vector $x \in \mathbb{R}^d$ as it is left-multiplied by A, that is $Ax = USV^T x$.

First $V^T x$ produces a new vector $p \in \mathbb{R}^d$. It essentially changes no scale information, it only changes the basis to that described by the right singular values. For instance, the new jth coordinate $p_j = \langle v_j, x \rangle$.

Next $q \in \mathbb{R}^n$ is the result of $SV^T x = Sp$. It scales p by the singular values of S. Note that if $d < n$ (the case we will focus on), then the last $n - d$ coordinates of q are 0. In fact, for $j > r$ (where $r = \text{rank}(A)$) then $q_j = 0$. For $j \le r$, then the vector q is stretched longer in the first coordinates since these are multiplied by larger singular values.

The final result is a vector $y \in \mathbb{R}^n$, the result of $Ax = USV^T x = Uq$. This again just changes the basis of q so that it aligns with the left singular vectors. In the setting $n > d$, the last $n - d$ left singular vectors are meaningless since the corresponding entries in q are 0.

Working backwards ... this final U matrix can be thought of mapping the effect of q onto each of the data points of A. The q vector, in turn, can be thought of as scaling by the content of the data matrix A (the U and V^T matrices contain no scaling information). And the p vector arises via the special rotation matrix V^T that puts the starting point x into the correct basis to perform coordinate-wise scaling (from the original d-dimensional coordinates to one that is aligned with the shape of the data, in its aggregate shape).

Example: Tracing through the SVD

Consider a matrix

$$A = \begin{pmatrix} 4 & 3 \\ 2 & 2 \\ -1 & -3 \\ -5 & -2 \end{pmatrix},$$

and its SVD $[U, S, V] = \mathsf{svd}(A)$:

$$U = \begin{pmatrix} -0.6122 & 0.0523 & 0.0642 & 0.7864 \\ -0.3415 & 0.2026 & 0.8489 & -0.3487 \\ 0.3130 & -0.8070 & 0.4264 & 0.2625 \\ 0.6408 & 0.5522 & 0.3057 & 0.4371 \end{pmatrix}, \quad S = \begin{pmatrix} 8.1655 & 0 \\ 0 & 2.3074 \\ 0 & 0 \\ 0 & 0 \end{pmatrix}, \quad V = \begin{pmatrix} -0.8142 & -0.5805 \\ -0.5805 & 0.8142 \end{pmatrix}.$$

Now consider a vector $x = (0.243, 0.97)$ (scaled very slightly so it is a unit vector, $\|x\| = 1$). Multiplying by V^T rotates (and flips) x to $p = V^T x$; still $\|p\| = 1$.

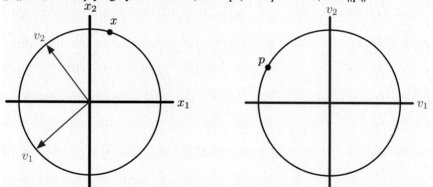

Next multiplying by S scales p to $q = Sp$. Notice there are an imaginary third and fourth coordinates now; they are both coming out of the page! Don't worry, they won't poke you since their magnitude is 0.

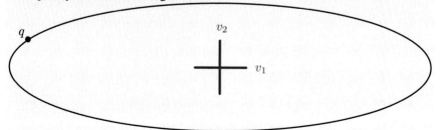

Finally, $y = Uq = Ax$ is again another rotation of q in this four dimensional space.

The SVD can be easily calculated in Python:

```
import numpy as np
from SciPy import linalg as LA
```

```
A = np.array([[4.0,3.0], [2.0,2.0], [-1.0,-3.0], [-5.0,-2.0]])
U, s, Vt = LA.svd(A)
print(U)
print(s)
print(Vt)
```

Then for a vector x we can observe its progress as it moves through the components

```
x = np.array([0.243,0.97])
x = x/LA.norm(x)

p = Vt @ x
print(p)

S = LA.diagsvd(s,2,2)
q = S @ p
print(q)

y = U @ q
print(y)
print(A @ x)
```

7.2.1 Best Rank-k Approximation of a Matrix

So how does this help solve the initial problem of finding B^*, which minimized the SSE? The singular values hold the key.

It turns out that there is a *unique* singular value decomposition, up to ties in the singular values. This means, there is exactly one (up to singular value ties) set of right singular vectors which rotate into a basis so that $\|Ax\| = \|SV^T x\|$ for all $x \in \mathbb{R}^d$ (recall that U is orthogonal, so it does not change the norm, $\|Uq\| = \|q\|$).

Next we realize that the singular values come in sorted order $\sigma_1 \geq \sigma_2 \geq \ldots \geq \sigma_r$. In fact, they are defined so that we choose v_1 so it maximizes $\|Av_1\|$ and $\|Av_1\| = \sigma_1$ (and thus $\|A\|_2 = \sigma_1$). Then, we find the next singular vector v_2 which is orthogonal to v_1 and maximizes $\|Av_2\|$, as $\|Av_2\| = \sigma_2$, and so on. In general $\sigma_j = \|Av_j\|$.

If we define B with the basis $V_B = \{v_1, v_2, \ldots, v_k\}$, then for any vector a

$$\|a - \pi_B(a)\|^2 = \left\| \sum_{j=1}^{d} v_j \langle a, v_j \rangle - \sum_{j=1}^{k} v_j \langle a, v_j \rangle \right\|^2 = \sum_{j=k+1}^{d} \langle a, v_j \rangle^2.$$

So the projection error of this V_B is that part of a in the last $(d - k)$ right singular vectors.

But we are not trying to directly predict new data here (like in regression). Rather, we are trying to approximate the data we have. We want to minimize $\sum_i \|a_i - \pi_B(a_i)\|^2$. But for any vector v, we recall now that

$$\|Av\|^2 = \left\| \begin{bmatrix} \langle a_1, v \rangle \\ \langle a_2, v \rangle \\ \vdots \\ \langle a_n, v \rangle \end{bmatrix} \right\|^2 = \sum_{i=1}^{n} \langle a_i, v \rangle^2.$$

Thus, the projection error can be measured with a set of orthonormal vectors $w_1, w_2, \ldots, w_{d-k}$ which are each orthogonal to B, as $\sum_{j'=1}^{d-k} \|Aw_{j'}\|^2$. When defining B as the first k right singular values, then these orthogonal vectors can be the remaining $(d-k)$ right singular vectors (that is $w_{j'} = w_{j-k} = v_j$), so the projection error is

$$\sum_{i=1}^{n} \|a_i - \pi_B(a_i)\|^2 = \sum_{i=1}^{n} \left(\sum_{j=k+1}^{d} \langle a_i, v_j \rangle^2 \right)$$

$$= \sum_{j=k+1}^{d} \left(\sum_{i=1}^{n} \langle a_i, v_j \rangle^2 \right) = \sum_{j=k+1}^{d} \|Av_j\|^2 = \sum_{j=k+1}^{d} \sigma_j^2.$$

And thus by how the right singular vectors are defined, the above expression is equivalent to the sum of squared errors

$$\mathsf{SSE}(A, B) = \sum_{a_i \in A} \|a_i - \pi_B(a_i)\|^2.$$

This is minimized (restricting that B contains the origin) when B is defined as the span of the first k singular vectors.

Matrix A_k, the Best Rank-k Approximation of Matrix A

A similar goal is to find the *best rank-k approximation* of A. That is a matrix $A_k \in \mathbb{R}^{n \times d}$ so that $\mathsf{rank}(A_k) = k$ and it minimizes both

$$\|A - A_k\|_2 \quad \text{and} \quad \|A - A_k\|_F.$$

In fact, the solution A_k will satisfy $\|A - A_k\|_2 = \sigma_{k+1}$ and $\|A - A_k\|_F^2 = \sum_{j=k+1}^{d} \sigma_j^2$.

This A_k matrix also comes from the SVD. If we set S_k as the matrix S in the decomposition so that all except the first k singular values are 0, then it has rank k. Hence, $A_k = US_kV^T$ also has rank k and is our solution. But we can notice that when we set most of S_k to 0, then the last $(d-k)$ columns of V are meaningless since they are only multiplied by 0s in US_kV^T, so we can also set those to all 0s, or remove them entirely (along with the last $(d-k)$ columns of S_k). Similarly, we can make 0 or remove the last $(n-k)$ columns of U. These matrices are referred to as V_k and U_k respectively, and also $A_k = U_k S_k V_k^T$.

Another way to understand this construction is to rewrite A as

$$A = USV^T = \sum_{j=1}^{r} \sigma_j u_j v_j^T$$

where each $u_j v_j^T$ is an $n \times d$ matrix with rank 1, the result of an outer product. We can ignore terms for $j > r$ since then $\sigma_j = 0$. This makes clear how the rank of A is at most r since it is the sum of r rank-1 matrices. Extending this, then A_k is simply

$$A_k = U_k S_k V_k^T = \sum_{j=1}^{k} \sigma_j u_j v_j^T.$$

This implies that $\|A\|_F^2 = \sum_{j=1}^{d} \sigma_j^2$ and $\|A_k\|_F^2 = \sum_{j=1}^{k} \sigma_j^2$. We can see this since orthonormal matrices U and V satisfy $\|U\|_F = 1$ and $\|V\|_F = 1$; so

$$\|A\|_F^2 = \|USV^T\|_F^2 = \|S\|_F^2 = \sum_{j=1}^{d} \sigma_j^2$$

$$\|A_k\|_F^2 = \|US_k V^T\|_F^2 = \|S_k\|_F^2 = \sum_{j=1}^{k} \sigma_j^2.$$

Example: Best Rank-k Approximation

For a matrix $A \in \mathbb{R}^{n \times d}$, the best rank-$k$ approximation is a matrix $A_k \in \mathbb{R}^{n \times d}$ derived from the SVD. We can zero-out all but the first k singular values in S to create a matrix S_k. If we similarly zero out all but the first k columns of U (the top k left singular vectors) and the first k rows of V^T (the top k right singular vectors), these represent matrices U_k and V_k. The resulting product $U_k S_k V_k^T$ is A_k.

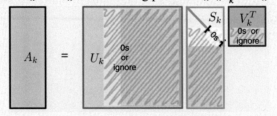

Geometry of Projection on to top k

The rank-k matrix V_k also encodes the projection onto the best k-dimensional subspace. Simply left-multiply any data point $x \in \mathbb{R}^d$ with V_k^T and the resulting vector $V_k^T x \in \mathbb{R}^k$ is the k-dimensional representation of the best k-dimensional subspace (that contains the origin) in terms of maximizing the sum of squared variation among the remaining points.

Each row of V_k^T, the right singular vector v_j results in the jth coordinate of the new vector as $\langle v_j, x \rangle$.

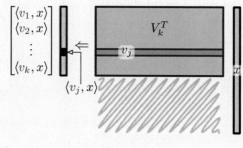

Low-Rank and Heavy-Tailed Distributions

For many data matrices $A \in \mathbb{R}^{n \times d}$, most of the variation in the data will occur within a low rank approximation. That is, for a small $k \ll d$ it will be that $\|A - A_k\|_F^2 / \|A\|_F^2$ will be very small. For instance for $n = 1000$, $d = 50$ and $k = 3$ it may be that $\|A - A_k\|_F^2 / \|A\|_F^2 \leq 0.05$ or for $k = 10$ that $\|A - A_k\|_F^2 / \|A\|_F^2 \leq 0.001$ (less than 0.1% of all variation is removed). Such matrices have *light-tailed distributions* of singular values.

However, for many other internet-scale data sets this is very much not the case. And even for large values of k (say $k = 100$), then $\|A - A_k\|_F^2 / \|A\|_F^2$ is large (say > 0.4). These are known as *heavy-tailed distributions*, and discussed more in Section 11.2.

7.3 Eigenvalues and Eigenvectors

A related matrix decomposition to the SVD is the *eigendecomposition*. This is only defined for a square matrix $M \in \mathbb{R}^{n \times n}$. Recall an *eigenvector* of M is a vector v such that there is some scalar λ that satisfies $Mv = \lambda v$, and when $\|v\| = 1$, then λ is the associated *eigenvalue*. We can think of the n eigenvectors as a matrix $V \in \mathbb{R}^{n \times n}$ and the associated n eigenvalues as a vector $l \in \mathbb{R}^n$.

To avoid issues of complex eigenvalues, we will focus on positive semidefinite matrices. How do we get positive semidefinite matrices? Let us start with a data matrix $A \in \mathbb{R}^{n \times d}$. Then, we can construct two positive semidefinite matrices

$$M_R = A^T A \quad \text{and} \quad M_L = A A^T.$$

Matrix M_R is $d \times d$ and M_L is $n \times n$. If the rank of A is d, then M_R is positive definite. If the rank of A is n, then M_L is positive definite.

Example: Eigendecomposition

Consider the matrix $M = A^T A \in \mathbb{R}^{2 \times 2}$, using the same $A \in \mathbb{R}^{3 \times 2}$ as above. We get symmetric, square matrix

$$M = \begin{bmatrix} 46 & 29 \\ 29 & 26 \end{bmatrix}.$$

Its eigenvectors are the columns of the matrix

$$V = \begin{bmatrix} 0.814 & 0.581 \\ -0.581 & 0.814 \end{bmatrix},$$

and eigenvalues are

$$\lambda_1 = 66.68 \quad \text{and} \quad \lambda_2 = 5.32.$$

The following Python code achieves these values

```
M = A.T @ A
l, V = LA.eig(M)
```

Eigenvectors and Eigenvalues Relation to SVD

Next consider the SVD of A so that $[U, S, V] = \mathsf{svd}(A)$. Then we can write

$$M_R V = A^T A V = (V S U^T)(U S V^T) V = V S^2.$$

Note that the last step follows because for orthogonal matrices U and V, then $U^T U = I$ and $V^T V = I$, where I is the identity matrix, which has no effect in the product. The matrix S is a diagonal square[1] matrix $S = \mathsf{diag}(\sigma_1, \sigma_2, \ldots, \sigma_d)$. Then $S^2 = SS$ (the product of S with S) is again diagonal with entries $S^2 = \mathsf{diag}(\sigma_1^2, \sigma_2^2, \ldots, \sigma_d^2)$.

Now consider a single column v_j of V (which is the jth right singular vector of A). Then extracting this column's role in the linear system $M_R V = V S^2$ we obtain

$$M_R v_j = v_j \sigma_i^2.$$

This means that v_j, the jth right singular vector of A, is an eigenvector (in fact the jth eigenvector) of $M_R = A^T A$. Moreover, the jth eigenvalue λ_j of M_R is the jth singular value of A squared: $\lambda_j = \sigma_j^2$.

Similarly, we can derive

---

[1] Technically, the S matrix from the svd has dimensions $S \in \mathbb{R}^{n \times d}$. To make this simple argument work with this technicality, let us first assume w.l.o.g. (without loss of generality) that $d \leq n$. Then the bottom $n - d$ rows of S are all zeros, which mean the right $n - d$ rows of U do not matter. So we can ignore both these $n - d$ rows and columns. Then S is square. This makes U no longer orthogonal, so $U^T U$ is then a projection, not identity; but it turns out this is a projection to the span of A, so the argument still works.

$$M_L U = A A^T U = (U S V^T)(V S U^T) U = U S^2,$$

and hence the left singular vectors of A are the eigenvectors of $M_L = A A^T$ and the eigenvalues of M_L are also the squared singular values of A.

Eigendecomposition

In general, the eigenvectors provide a basis for a matrix $M \in \mathbb{R}^{n \times n}$ in the same way that the right V or left singular vectors U provide a basis for matrix $A \in \mathbb{R}^{n \times d}$. In fact, it is again a very special basis, and is unique up to the multiplicity of eigenvalues. This implies that all eigenvectors are orthogonal to each other.

Let $V = [v_1, v_2, \ldots, v_n]$ be the eigenvectors of the matrix $M \in \mathbb{R}^{n \times n}$, as columns in the matrix V. Also let $L = \text{diag}(\lambda_1, \lambda_2, \ldots, \lambda_n)$ be the eigenvalues of M stored on the diagonal of matrix L. Then, we can decompose M as

$$M = V L V^{-1}.$$

The inverse of L is $L^{-1} = \text{diag}(1/\lambda_1, 1/\lambda_2, \ldots, 1/\lambda_n)$. Hence, we can write

$$M^{-1} = V L^{-1} V^{-1}.$$

When M is positive definite, it has n positive eigenvectors and eigenvalues; hence V is orthogonal, so $V^{-1} = V^T$. Thus in this situation, given the eigendecomposition, we now have a way to compute the inverse

$$M^{-1} = V L^{-1} V^T,$$

which was required in our almost closed-form solution for linear regression. Now we just need to compute the eigendecomposition, which we will discuss next.

7.4 The Power Method

The *power method* refers to what is probably the simplest algorithm to compute the first eigenvector and value of a matrix; shown in Algorithm 7.4.1. By factoring out the effect of the first eigenvector, we can then recursively repeat the process on the remainder until we have found all eigenvectors and values. Moreover, this implies we can also reconstruct the singular value decomposition as well.

We will consider $M \in \mathbb{R}^{n \times n}$, a positive semidefinite matrix: $M = A^T A$.

We can unroll the for loop to reveal another interpretation. Directly set $u^{(q)} = M^q u^{(0)}$, so all iterations are incorporated into one matrix-vector multiplication. Recall that $M^q = M \cdot M \cdot M \cdot \ldots \cdot M$, for q times. However, these q matrix multiplications are much more computationally expensive then q matrix-vector multiplications.

Algorithm 7.4.1 PowerMethod(M, q)

Initialize $u^{(0)}$ as a random unit vector.
for $i = 1$ **to** q **do**
  $u^{(i)} := M u^{(i-1)}$
return $v := u^{(q)} / \|u^{(q)}\|$

Alternatively we are provided only the matrix A (where $M = A^T A$) then we can run the algorithm without explicitly constructing M (since for instance if $d > n$ and $A \in \mathbb{R}^{n \times d}$, then the size of M (d^2) may be much larger than A (nd)). Then, we simply replace the inside of the for loop with

$$u^{(i)} := A^T (A u^{(i-1)})$$

where we first multiply $\tilde{u} = A u^{(i-1)}$ and then complete $u^{(i)} = A^T \tilde{u}$.

Recovering All Eigenvalues

The output of PowerMethod($M = A^T A, q$) is a single unit vector v, which we will argue is arbitrarily close to the first eigenvector v_1. Clearly, we can recover the first eigenvalue as $\lambda_1 = \|M v_1\|$. Since we know the eigenvectors form a basis for M, they are orthogonal. Hence, after we have constructed the first eigenvector v_1, we can factor it out from M as follows:

$$A_1 := A - A v_1 v_1^T$$
$$M_1 := A_1^T A_1$$

Then we run PowerMethod($M_1 = A_1^T A_1, q$) to recover v_2, and λ_2; factor them out of M_1 to obtain M_2, and iterate.

Geometry of Why the Power Method Works?

To understand why the power method works, assume we know the eigenvectors $v_1, v_2, \ldots, v_n$ and eigenvalues $\lambda_1, \lambda_2, \ldots, \lambda_n$ of $M \in \mathbb{R}^{n \times n}$.

Since the eigenvectors form a basis for M, and assuming it is full rank, then it is also a basis for all of $\mathbb{R}^n$ (if not, then it does not have n eigenvalues, and we can fill out the rest of the basis of $\mathbb{R}^n$ arbitrarily). Hence any vector, including the initialization random vector $u^{(0)}$, can be written as

$$u^{(0)} = \sum_{j=1}^{n} \alpha_j v_j.$$

Recall that $\alpha_j = \langle u^{(0)}, v_j \rangle$, and since it is random, it is possible to claim[*] that with probability at least $1/2$ that for any α_j we have that $|\alpha_j| \geq \frac{1}{2\sqrt{n}}$. We will now assume that this holds for $j = 1$, so $\alpha_1 > 1/(2\sqrt{n})$.

[(*) Since $u^{(0)}$ is a unit vector, its norm is 1, and because $\{v_1, \ldots, v_n\}$ is a basis, then $1 = \|u^{(0)}\|^2 = \sum_{j=1}^{n} \alpha_j^2$. Since it is random, then $\mathbf{E}[\alpha_j^2] = 1/n$ for each j. Applying a concentration of measure (almost a Markov Inequality, but need to be more careful), we can argue that with probability $1/2$ any $\alpha_j^2 > (1/4) \cdot (1/n)$, and hence $\alpha_j > (1/2) \cdot (1/\sqrt{n})$.]

Geometrically, as the algorithm is run, the aggregated shape of M^q stretches, and as long as $u^{(0)}$ is not too close to orthogonal to v_1, then it quickly stretches as $u^{(q)} = M^q u^{(0)}$ along with M^q toward v_1.

Next toward formalizing this, since we can interpret that algorithm as $v = M^q u^{(0)}$, we analyze M^q. If M has jth eigenvector v_j and eigenvalue λ_j, that is, $Mv_j = \lambda_j v_j$, then M^q has jth eigenvalue λ_j^q since

$$M^q v_j = M \cdot M \cdot \ldots \cdot M v_j = M^{q-1}(v_j \lambda_j) = M^{q-2}(v_j \lambda_j)\lambda_j = \ldots = v_j \lambda_j^q.$$

This holds for each eigenvalue of M^q. Hence, we can rewrite the output by summing over the terms in the eigenbasis as

$$v = \frac{\sum_{j=1}^{n} \alpha_j \lambda_j^q v_j}{\sqrt{\sum_{j=1}^{n} (\alpha_j \lambda_j^q)^2}}.$$

Finally, we would like to show our output v is close to the first eigenvector v_1. We can measure closeness with the dot product (actually we will need to use its absolute value since we might find something close to $-v_1$). If they are almost the same, the dot product will be close to 1 (or -1).

$$|\langle M^q u^{(0)}, v_1 \rangle| = \frac{\alpha_1 \lambda_1^q}{\sqrt{\sum_{j=1}^{n} (\alpha_j \lambda_j^q)^2}}$$

$$\geq \frac{\alpha_1 \lambda_1^q}{\sqrt{\alpha_1^2 \lambda_1^{2q} + n\lambda_2^{2q}}} \geq \frac{\alpha_1 \lambda_1^q}{\alpha_1 \lambda_1^q + \lambda_2^q \sqrt{n}} = 1 - \frac{\lambda_2^q \sqrt{n}}{\alpha_1 \lambda_1^q + \lambda_2^q \sqrt{n}}$$

$$\geq 1 - 2n \left(\frac{\lambda_2}{\lambda_1} \right)^q.$$

The first inequality holds because $\lambda_1 \geq \lambda_2 \geq \lambda_j$ for all $j > 2$. The third inequality (going to third line) holds by dropping the $\lambda_2^q \sqrt{n}$ term in the denominator, and since $\alpha_1 > 1/(2\sqrt{n})$.

Thus if there is "gap" between the first two eigenvalues (λ_1/λ_2 is large), then this algorithm converges quickly (at an exponential rate) to where $|\langle v, v_1 \rangle| = 1$.

7.5 Principal Component Analysis

Recall that the original goal of this topic was to find the k-dimensional subspace B to minimize

$$\|A - \pi_B(A)\|_F^2 = \sum_{a_i \in A} \|a_i - \pi_B(a_i)\|^2.$$

We have not actually solved this problem yet. The top k right singular values V_k of A only provided this bound assuming that B contains the origin: $(0, 0, \ldots, 0)$. However, this might not be the case!

Geometry of Principal Components vs. Right Singular Vectors

The right singular vectors directly from the SVD (without centering) and the principal components can be very different. And they can represent very different amounts of variation of the data in the top components. For instance, in the example, the 1 top principal component (in green) does a significantly better job of representing the shape of the data than the 1 top right singular vector (in blue).

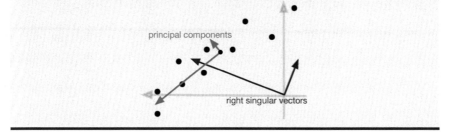

Principal component analysis (PCA) is an extension of the SVD when we do not restrict that the subspace V_k must go through the origin. It turns out, like with simple linear regression, that the optimal B must go through the mean of all of the data. So we can still use the SVD, after a simple preprocessing step called centering to shift the data matrix so its mean is exactly at the origin.

Specifically, *centering* is adjusting the original input data matrix $A \in \mathbb{R}^{n \times d}$ so that each column (each dimension) has an average value of 0. This is easier than it seems. Define $\bar{a}_j = \frac{1}{n} \sum_{i=1}^{n} A_{i,j}$ (the average of each column j). Then set each $\tilde{A}_{i,j} = A_{i,j} - \bar{a}_j$ to represent the entry in the ith row and jth column of centered matrix $\tilde{A}$.

There is a *centering matrix* $C_n = I_n - \frac{1}{n}\mathbf{1}\mathbf{1}^T$ where I_n is the $n \times n$ identity matrix, $\mathbf{1}$ is the all-ones column vector (of length n) and thus $\mathbf{1}\mathbf{1}^T$ is the all-ones $n \times n$ matrix. Then, we can also just write $\tilde{A} = C_n A$. The centering matrix C_n can be interpreted as a projection operator which projects to the $(d - 1)$-dimensional space in the direction along the vector from the origin to the mean.

Now to perform PCA on a data set A, we compute $[U, S, V] = \mathrm{svd}(C_n A) = \mathrm{svd}(\tilde{A})$. Then, the resulting singular values $\mathrm{diag}(S) = \{\sigma_1, \sigma_2, \ldots, \sigma_r\}$ are known as the *principal values*, and the top k right singular vectors $V_k = [v_1 \; v_2 \; \ldots \; v_k]$ are known as the top-k *principal directions*.

This often gives a better fitting to the data than just SVD. The SVD finds the best rank-k approximation of A, which is the best k-dimensional subspace (under Frobenius and spectral norms) **which passes through the origin**. If all of the data is far from the origin, this can essentially "waste" a dimension to pass through the origin. However, PCA may also need to store the shift from the origin, a vector $\tilde{c} = (\tilde{a}_1, \tilde{a}_2, \ldots, \tilde{a}_d) \in \mathbb{R}^d$.

Example: Centering a Matrix

Given an initial matrix $A = \begin{bmatrix} 1 & 5 \\ 2 & 3 \\ 3 & 10 \end{bmatrix}$ its center vector is $\tilde{c} = [2 \; 6]$. In Python we can calculate $\tilde{c}$ explicitly, then subtract it

```python
import numpy as np
A= np.array([[1,5],[2,3],[3,10]])

c = A.mean(0)
Ac = A - np.outer(np.ones(3),c)
```

Or use the centering matrix $C_3 = \begin{bmatrix} 0.666 & -0.333 & -0.333 \\ -0.333 & 0.666 & -0.333 \\ -0.333 & -0.333 & 0.666 \end{bmatrix}$

```python
C = np.eye(3) - np.ones(3)/3
Ac = C @ A
```

The resulting centered matrix is $\tilde{A} = \begin{bmatrix} -1 & -1 \\ 0 & -3 \\ 1 & 4 \end{bmatrix}$.

7.6 Multidimensional Scaling

Dimensionality reduction is an abstract problem with input of a high-dimensional data set $P \subset \mathbb{R}^d$ and a goal of finding a corresponding lower dimensional data set $Q \subset \mathbb{R}^k$, where $k \ll d$, and properties of P are preserved in Q. Both low-rank approximations through direct SVD and through PCA are examples of this: $Q = \pi_{V_k}(P)$. However, these techniques require an explicit representation of P to start with. In some cases, we are only presented P through distances. There are two common variants:

- We are provided a set of n objects X, and a bivariate function d : $X \times X \to \mathbb{R}$ that returns a distance between them. For instance, we can put two cities into an airline website, and it may return a dollar amount for the cheapest flight between those two cities. This dollar amount is our "distance."
- We are simply provided a matrix $D \in \mathbb{R}^{n \times n}$, where each entry $D_{i,j}$ is the distance between the ith and jth point. In the first scenario, we can calculate such a matrix D.

Multidimensional scaling (MDS) has the goal of taking such a distance matrix D for n points and giving (typically) low-dimensional Euclidean coordinates to these points so that the resulting points have similar spatial relations to that described in D. If we had some original data set A which resulted in D, we could just apply PCA to find the embedding. It is important to note, in the setting of MDS we are typically just given D, and *not* the original data A. However, as we will show next, we can derive a matrix that will act like AA^T using only D.

Classical MDS

The most common variant of MDS is known as *classical MDS* which uses the centering matrix $C_n = I - \frac{1}{n}\mathbf{1}\mathbf{1}^T$ from PCA, but applies it on both sides of $D^{(2)}$ which squares all entries of D so that $D^{(2)}_{i,j} = D^2_{i,j}$. This *double centering* process creates a matrix $M = -\frac{1}{2}C_n D^{(2)} C_n$. Then, the embedding uses the top k eigenvectors of M and scales them appropriately by the square-root of the eigenvalues. That is let $VLV = M$ be its eigendecomposition. The let V_k and L_k represent the top k eigenvectors and values, respectively. The final point set is $V_k L_k^{1/2} = Q \in \mathbb{R}^{n \times k}$. So the n rows $q_1, \ldots, q_n \in \mathbb{R}^k$ are the embeddings of points to represent distance matrix D. This algorithm is sketched in Algorithm 7.6.1

Algorithm 7.6.1 Classical Multidimensional Scaling: CMDS(D, k)

Set $M = -\frac{1}{2}C_n D^{(2)} C_n$ # *double centering*
Construct $[L, V] = \text{eig}(M)$ # *the eigendocomposition of M*
return $Q = V_k L_k^{1/2}$ # *projection to best k-dimensions*

MDS is often used for k as 2 or 3 so the data can be easily visualized.

There are several other forms of recovering an embedding from distances. This Classical MDS approach is known to minimize a quantity called "stress" which is associated with the inner products rather than the distances. It recovers the distances with no error if we use all n eigenvectors—if they exist. However, as mentioned, there may be less than n eigenvectors, or they may be associated with complex eigenvalues. So if our goal is an embedding into $k = 3$ or $k = 10$, this is not totally assured to work; yet Classical MDS is used a lot nonetheless.

7.6.1 Why does Classical MDS work?

A *similarity matrix* S is an $n \times n$ matrix where entry $S_{i,j}$ is the similarity between the ith and the jth data point. The similarity often associated with Euclidean distance $\|a_i - a_j\|$ is the standard inner product (i.e., dot product) $\langle a_i, a_j \rangle$. In particular, we can expand squared Euclidian distance as

$$\|a_i - a_j\|^2 = \|a_i\|^2 + \|a_j\|^2 - 2\langle a_i, a_j \rangle,$$

and hence

$$\langle a_i, a_j \rangle = \frac{1}{2} \left(\|a_i\|^2 + \|a_j\|^2 - \|a_i - a_j\|^2 \right). \tag{7.1}$$

Next we observe that for the $n \times n$ matrix AA^T the entry $[AA^T]_{i,j} = \langle a_i, a_j \rangle$. So it seems hopeful we can derive AA^T from S using equation (7.1). That is we can set $\|a_i - a_j\|^2 = D_{i,j}^2$. However, we also need values for $\|a_i\|^2$ and $\|a_j\|^2$.

Since the embedding has an arbitrary shift to it (if we add a shift vector s to *all* embedding points, then no distances change), then we can arbitrarily choose a_1 to be at the origin. Then $\|a_1\|^2 = 0$ and $\|a_j\|^2 = \|a_1 - a_j\|^2 = D_{1,j}^2$. Using this assumption and equation (7.1), we can then derive the similarity matrix AA^T. Specifically, each entry is set

$$[AA^T]_{i,j} = \frac{1}{2}(D_{i,1}^2 + D_{j,1}^2 - D_{i,j}^2).$$

Let E_i be an $n \times n$ matrix that is all 0s except the ith column which is all 1s. Then for a matrix $Z \in \mathbb{R}^{n \times n}$, the product $E_i Z$ has each column as the ith column of Z, and the product ZE_i^T has each row as the ith row of Z. Using this notation, we can reformulate the above expression as single matrix operation, where $D^{(2)}$ is define so $D_{i,j}^{(2)} = D_{i,j}^2$, as

$$AA^T = \frac{1}{2}(E_1 D^{(2)} + D^{(2)} E_1^T - D^{(2)}).$$

However, it is now clear there is nothing special about the first row, and we replace each E_1 with any E_i, or indeed the average over them. Using the so-called *mean matrix* $E = \frac{1}{n} \sum_{i=1}^{n} E_i = \frac{1}{n} \mathbf{1}\mathbf{1}^T$, this results in

$$AA^T = \frac{1}{2n} \sum_{i=1}^{n} \left(E_i D^{(2)} + D^{(2)} E_i^T - D^{(2)} \right)$$

$$= -\frac{1}{2} \left(D^{(2)} - \frac{1}{n} \sum_{i=1}^{n} E_i D^{(2)} - D^{(2)} \frac{1}{n} \sum_{i=1}^{n} E_i^T \right)$$

$$= -\frac{1}{2} \left(D^{(2)} - E D^{(2)} - D^{(2)} E \right)$$

$$= -\frac{1}{2} (I - E) D^{(2)} (I - E) + E D^{(2)} E$$

$$= -\frac{1}{2} C_n D^{(2)} C_n + \bar{D}^{(2)}.$$

The last line follows since $(I - E) = C_n$ is the centering matrix and using

$$\bar{D}^{(2)} = E D^{(2)} E = \frac{1}{n^2} \sum_{i=1}^{n} \sum_{j=1}^{n} D_{i,j}^2 \mathbf{1} \mathbf{1}^T$$

which is an $n \times n$ matrix with all entries the same, specifically the average of all squared distances $c = \frac{1}{n^2} \sum_{i=1}^{n} \sum_{j=1}^{n} D_{i,j}^2$.

We have almost derived classical MDS, but need to remove $\bar{D}^{(2)}$. So we consider an alternative $n \times n$ matrix $X = [x_1; x_2; \ldots; x_n]$ as a stacked set of points so that $\langle x_i, x_j \rangle = \langle a_i, a_j \rangle - c$. Thus

$$XX^T = AA^T - \bar{D}^{(2)} = -\frac{1}{2} C_n D^{(2)} C_n = M.$$

Since c is a fixed constant, we observe for any i, j that

$$\|x_i - x_j\|^2 = \|x_i\|^2 + \|x_j\|^2 - 2\langle x_i, x_j \rangle = \|a_i\|^2 - c + \|a_j\|^2 - c - 2(\langle a_i, a_j \rangle - c)$$

$$= \|a_i\|^2 + \|a_j\|^2 - 2\langle a_i, a_j \rangle = \|a_i - a_j\|^2 = D_{i,j}^2.$$

That is X represents the distances in $D^{(2)}$ just as well as A. Why should we use XX^T instead of AA^T? First, it is simpler. But more importantly, it "centers" the data points $x_1, \ldots x_n$. That is if we reapply the centering matrix on X to get $Z = C_n X$, then we recover X again, so the points are already centered:

$$ZZ^T = (C_n X)(X^T C_n^T) = -\frac{1}{2} C_n C_n D^{(2)} C_n C_n = -\frac{1}{2} C_n D^{(2)} C_n = XX^T.$$

This follows since $C_n C_n = (I - E)(I - E) = I^2 + E^2 - 2E = I + E - 2E = C_n$. The key observation is that $E^2 = \frac{1}{n^2} \mathbf{1} \mathbf{1}^T \mathbf{1} \mathbf{1}^T = \frac{1}{n^2} \mathbf{1} n \mathbf{1}^T = E$ since the inner $\mathbf{1}^T \mathbf{1} = n$ is a dot product over two all-ones vectors.

Thus, classical MDS is not only a low-rank Euclidean representation of the distances, but results in data points being centered, as in PCA.

Example: Classical MDS

Consider the example 5×5 distance matrix

$$D = \begin{bmatrix} 0 & 4 & 3 & 7 & 8 \\ 4 & 0 & 1 & 6 & 7 \\ 3 & 1 & 0 & 5 & 7 \\ 7 & 6 & 5 & 0 & 1 \\ 8 & 7 & 7 & 1 & 0 \end{bmatrix}.$$

We can run classical MDS to obtain a $(k = 2)$-dimensional representation point set Q. It is shown in the following plot. Note that it is automatically centered; the mean of Q is the origin.

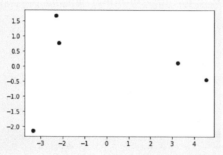

In Python, it is again simple to double center, and generate the eigendecomposition:

```
D = np.array([[0,4,3,7,8],[4,0,1,6,7],[3,1,0,5,7],[7,6,5,0,1],
              [8,7,7,1,0]])
D2 = np.square(D)
C = np.eye(5) - 0.2*np.ones(5)
M = -0.5* C @ D2 @ C

l,V = LA.eig(M)
s = np.real(np.power(l,0.5))
```

Note we converted the eigenvalues l to singular values, and forced them to be real valued, since some numerical error can occur generating small imaginary terms.

Then, we project the data to the most informative two dimensions so we can plot it.

```
V2 = V[:,[0,1]]
s2 = np.diag(s[0:2])
Q = V2 @ s2

import matplotlib.pyplot as plt
plt.plot(Q[:,0],Q[:,1],'ro')
plt.show()
```

7.7 Linear Discriminant Analysis

Another tool that can be used to learn a Euclidian distance for data is *linear discriminant analysis* (or LDA). This term has a few variants, we focus on the multi-class setting. This means we begin with a data set $X \subset \mathbb{R}^d$, and a known partition of X into k classes (or clusters) $S_1, S_2, \ldots, S_k \subset X$, so $\bigcup S_i = X$ and $S_i \cap S_j = \emptyset$ for $i \neq j$.

Let $\mu_i = \frac{1}{|S_i|} \sum_{x \in S_i} x$ be the mean of class i, and let $\Sigma_i = \frac{1}{|S_i|} \sum_{x \in S_i} (x - \mu_i)(x - \mu_i)^T$ by its covariance.

Similarly, we can represent the overall mean as $\mu = \frac{1}{|X|} \sum_{x \in X} x$. Then, we can represent the *between class covariance* as

$$\Sigma_B = \frac{1}{|X|} \sum_{i=1}^{k} |S_i|(\mu_i - \mu)(\mu_i - \mu)^T.$$

In contrast the overall *within class covariance* is

$$\Sigma_W = \frac{1}{|X|} \sum_{i=1}^{k} |S_i| \Sigma_i = \frac{1}{|X|} \sum_{i=1}^{k} \sum_{x \in S_i} (x - \mu_i)(x - \mu_i)^T.$$

Example: Between and Within Class Covariances

The figures shows three color-coded classes with their means: μ_1 for blue, μ_2 for red, and μ_3 for green. The vectors, whose sum of outer-products generates the class covariances Σ_1, Σ_2, and Σ_3 are color-coded as well. The overall mean, and the vectors for the between class covariance Σ_B are shown in black.

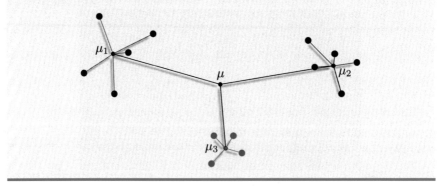

The goal of LDA is a representation of X in a k'-dimensional space that maximizes the between class covariance while minimizing the within class covariance. This is often formalized as finding the set of vectors u which maximize

$$\frac{u^T \Sigma_B u}{u^T \Sigma_W u}.$$

For any $k' \le k - 1$, we can directly find the orthogonal basis $V_{k'} = \{v_1, v_2, \ldots, v_{k'}\}$ that maximizes the above goal with an eigendecomposition. In particular, $V_{k'}$ is the top k' eigenvectors of $\Sigma_W^{-1} \Sigma_B$. Then to obtain the best representation of X, we set the new data set as the projection onto the k'-dimensional space spanned by V_k

$$\tilde{X} = V_{k'}^T X$$

so $\tilde{x} = V_{k'}^T x = (\langle x, v_1 \rangle, \langle x, v_2 \rangle, \ldots, \langle x, v_{k'} \rangle) \in \mathbb{R}^{k'}$.

This retains the dimensions which show difference between the classes, and similarity among the classes. The removed dimensions will tend to show variance within classes without adding much difference between the classes. Conceptually, if the data set can be well-clustered under the k-means clustering formulation, then the $V_{k'}$ (say when rank $k' = k - 1$) describes a subspace with should pass through the k sites $\{s_1 = \mu_1, s_2 = \mu_2, \ldots, s_k = \mu_k\}$; capturing the essential information needed to separate the centers.

7.8 Distance Metric Learning

When using PCA, one should enforce that the input matrix A has the same units in each column (and each row). What should one do if this is not the case?

Let us reexamine the root of the problem: the Euclidean distance. It takes two vectors $p, q \in \mathbb{R}^d$ (perhaps rows of A) and measures:

$$\mathsf{d}_{\mathsf{Euc}}(p, q) = \|p - q\| = \sqrt{\langle p - q, p - q \rangle} = \sqrt{\sum_{i=1}^{d} (p_i - q_i)^2}.$$

If each row of a data set A represents a data point, and each column an attribute, then the operation $(p_i - q_i)^2$ is fine since p_i and q_i have the same units (they quantify the same attribute). However, the $\sum_{i=1}^{d}$ over these terms adds together quantities that may have different units.

The naive solution is to just brush away those units. These are normalization approaches where all values p_i and q_i (in column i) are divided by a constant s_i with the same units as the elements in that column, and maybe adding a constant. In one approach s_i is chosen to *standardize* all values in each column to lie in $[0, 1]$. In the other common approach, the s_i values are chosen to *normalize*: so that the standard deviation in each column is 1. Note that both of these approaches are affected oddly by outliers—a single outlier can significantly change the effect of various data points. Moreover, if new dimensions are added which have virtually the same value for each data point; these values are inflated to be as meaningful as another signal direction. As a result, these normalization approaches can be brittle and affect the meaning of the data in unexpected ways. However, they are also quite common, for better or worse.

If we are to choose a good metric in a principled way, we must know something about which points should be close and which should be far. In the *distance metric learning* problem, we assume that we have two sets of pairs; the close pairs $C \subset X \times X$ and the far pairs $F \subset X \times X$. This process starts with a dataset $X \subset \mathbb{R}^d$, and close and far pairs C and F and tries to find a metric so the close pairs have distance as small as possible, while the far pairs have distance as large as possible.

In particular, we restrict to a Mahalanobis distance defined with respect to a positive semidefinite matrix $M \in \mathbb{R}^{d \times d}$ on points $p, q \in \mathbb{R}^d$ as

$$\mathbf{d}_M(p, q) = \sqrt{(p - q)^T M (p - q)}.$$

So given X and sets of pairs C and F, the goal is to find M to make the close point have small $\mathbf{d}_M$ distance, and far points have large $\mathbf{d}_M$ distance. Specifically, we will consider finding the optimal distance $\mathbf{d}_{M^*}$ as

$$M^* = \max_M \ \min_{\{x_i, x_j\} \in F} \ \mathbf{d}_M(x_i, x_j)^2$$

$$\text{such that} \ \sum_{\{x_i, x_j\} \in C} \mathbf{d}_M(x_i, x_j)^2 \le \kappa.$$

That is we want to maximizes the closest pair in the far set F, while restricting that all pairs in the close set C have their sum of squared distances are at most κ, some constant. We will not explicitly set κ, but rather restrict M in some way so on average it does not cause much stretch. There are other reasonable similar formulations, but this one will allow for simple optimization.

Notational Setup

Let $H = \sum_{\{x_i, x_j\} \in C} (x_i - x_j)(x_i - x_j)^T$; note that this is a sum of *outer* products, so H is in $\mathbb{R}^{d \times d}$. For this to work, we will need to assume that H is full rank; otherwise we do not have enough close pairs to measure. Or we can set $H = H + \delta I$ for a small scalar δ.

Further, we can restrict M to have trace $\text{Tr}(M) = d$ and hence satisfying some constraint on the close points, fixing the scaling of the distance. Recall that the *trace* of a matrix M is the sum of M's eigenvalues. Let $\mathbb{P}$ be the set of all positive semidefinite matrices with trace d; hence the identity matrix I is in $\mathbb{P}$. Also, let

$$\triangle = \{\alpha \in \mathbb{R}^{|F|} \mid \sum \alpha_i = 1 \ \& \ \text{all} \ \alpha_i \ge 0\}.$$

Let $\tau_{i,j} \in F$ (or simply $\tau \in F$ when the indexes are not necessary) to represent a far pair $\{x_i, x_j\}$. And let $X_{\tau_{i,j}} = (x_i - x_j)(x_i - x_j)^T \in \mathbb{R}^{d \times d}$, an outer product. Let $\tilde{X}_\tau = H^{-1/2} X_\tau H^{-1/2}$. It turns out our optimization goal is now equivalent (up to scaling factors, depending on κ) to finding

$$\operatorname*{argmax}_{M \in \mathbb{P}} \min_{\alpha \in \Delta} \sum_{\tau \in F} \alpha_\tau \langle \tilde{X}_\tau, M \rangle.$$

Here $\langle X, M \rangle = \sum_{s,t} X_{s,t} M_{s,t}$, a dot product over matrices, but since X will be related to an outer product between two data points, this makes sense to think of as $\mathbf{d}_M(X)$.

Optimization Procedure

Given the formulation above, we will basically try to find an M which stretches the far points as much as possible while keeping $M \in \mathbb{P}$. We do so using a general procedure referred to as *Frank-Wolfe optimization*, which increases our solution using one data point (in this case a far pair) at a time.

Set $\sigma = d \cdot 10^{-5}$ as a small smoothing parameter. Define a gradient as

$$g_\sigma(M) = \frac{\sum_{\tau \in F} \exp(-\langle \tilde{X}_\tau, M \rangle / \sigma) \tilde{X}_\tau}{\sum_{\tau \in F} \exp(-\langle \tilde{X}_\tau, M \rangle / \sigma)}.$$

Observe this is a weighted average over the $\tilde{X}_\tau$ matrices. Let $v_{\sigma,M}$ be the maximal eigenvector of $g_\sigma(M)$; the direction of maximal gradient.

Then the algorithm is simple. Initialize $M_0 \in \mathbb{P}$ arbitrarily; for instance, as $M_0 = I$. Then repeatedly find for $t = 1, 2, \ldots$ two steps: (1) find $v_t = v_{\mu, M_{t-1}}$, and (2) set $M_t = \frac{t-1}{t} M_{t-1} + \frac{d}{t} v_t v_t^T$, where $v_t v_t^T$ is an outer product. This is summarized in Algorithm 7.8.1.

Algorithm 7.8.1 Optimization for DML

Initialize $M_0 = I$.
for $t = 1, 2, \ldots, T$ **do**
    Set $G = g_\sigma(M_{t-1})$
    Let $v_t = v_{\sigma, M_{t-1}}$; the maximal eigenvector of G.
    Update $M_t = \frac{t-1}{t} M_{t-1} + \frac{d}{t} v_t v_t^T$.
return $M = M_T$.

7.9 Matrix Completion

A common scenario is that a data set P is provided, but is missing some (or many!) of its attributes. Let us focus on the case where $P \in \mathbb{R}^{n \times d}$ that is an $n \times d$ matrix, that we can perhaps think of as n data points in $\mathbb{R}^d$, that is with d attributes each. For instance, Netflix may want to recommend a subset of n movies to d customers. Each movie has only been seen, or even rated, by a small subset of all d customers, so for each (movie, customer) pair (i, j), it could be a rating (saying a score from $[0, 10]$, or some other value based on their measured interaction), denoted $P_{i,j}$. But most pairs (i, j) are empty, there is no rating yet. A recommendation would be based on the predicted score for unseen movies.

The typical notation defines a set $\Omega = \{(i, j) \mid P_{i,j} \neq \emptyset\}$ of the rated (movie, customer) pairs. Then let $\Pi_\Omega(P)$ describe the subset of pairs with scores and $\Pi_\Omega^\perp(P)$ the compliment, the subset of pairs without scores. The goal is to somehow fill in the values $\Pi_\Omega^\perp(P)$.

Example: Missing Data in a Matrix

Consider a matrix where the 5 represent different users, and 4 columns represent movies. The ratings are described numerically, with a large score indicating a better rating. However, some entries there is no rating, and these are denoted as **x**.

$$P = \begin{bmatrix} 3 & 4 & x & 8 \\ 1 & x & 5 & x \\ x & 4 & x & 9 \\ 9 & 7 & 1 & x \\ 2 & 2 & x & x \end{bmatrix} \text{ with mask } \Omega = \begin{bmatrix} 1 & 1 & 0 & 1 \\ 1 & 0 & 1 & 1 \\ 0 & 1 & 0 & 1 \\ 1 & 1 & 1 & 0 \\ 1 & 1 & 0 & 0 \end{bmatrix}.$$

The simplest variants fill in the missing values as the average of all existing values in a row (the average rating in a movie). Or it could be the average of existing scores of a column (average rating of a customer). Or an average of these averages. But these approaches are not particularly helpful for personalizing the rating for a customer (e.g., a customer who likes horror movies but not rom-coms is likely to score things differently than one with the opposite preferences).

A common assumption in this area is that there are some simple "latent factors" which determine a customer's preferences (e.g., they like horror, and thrillers, but not rom-coms). A natural way to capture this is to assume there is some low-rank structure in P. That is we would like to find a low-rank model for P that fits the observed scores $\Pi_\Omega(P)$. The most common formulation looks like ridge regression:

$$P^* = \operatorname*{argmin}_{X \in \mathbb{R}^{n \times d}} \frac{1}{2} \|\Pi_\Omega(P - X)\|_F^2 + \lambda \|X\|_*.$$

Here $\|X\|_*$ is the *nuclear norm* of a matrix, it corresponds to the sum of its singular values (recall squared Frobenius norm is different since it is the sum of *squared* singular values), and it serves as a regularization term which biases the solution toward being low-rank.

A simple, and common way to approach this problem is iterative, and outlined in Algorithm 7.9.1. Start with some guess for X (e.g., average of rows and of columns of P_Ω). Take the svd of X to obtain $USV^T \leftarrow \mathrm{svd}(X)$. Shrink all singular values by λ, but no smaller than 0 (similar to Frequent Directions in Section 11.3, but not on the squared values). This operation ϕ_λ is defined for diagonal matrix S as

$$\phi_\lambda(S) = \text{diag}((S_{11} - \lambda)_+, (S_{22} - \lambda)_+, \ldots, (S_{dd} - \lambda)_+),$$

where $(x - \lambda)_+ = \max\{0, x - \lambda\}$. Then we update $\hat{X} = U\phi_\lambda(S)V^T$; this provides a lower rank estimate since $\phi_\lambda(S)$ will set some of the singular values to 0. Finally, we refill the known values Ω as

$$X \leftarrow \Pi_\Omega(P) + \Pi_\Omega^\perp(\hat{X}),$$

and repeat until things do not change much on an update step (it has "converged").

Algorithm 7.9.1 Matrix Completion(P,Ω,λ)

Initialize X with each $X_{i,j}$ as

$$X_{i,j} = \begin{cases} \Pi_\Omega(P_{i,j}) & \text{if } (i,j) \in \Omega \\ \frac{1}{2}(\text{average}_{i\in[n]}(P_\Omega(P_{i,j})) + \text{average}_{j\in[d]}(P_\Omega(P_{i,j}))) & \text{otherwise} \end{cases}$$

repeat
   $USV^T \leftarrow \text{svd}(X)$
   $\hat{X} \leftarrow U\phi_\lambda(S)V^T$
   $X \leftarrow \Pi_\Omega(P) + \Pi_\Omega^\perp(\hat{X})$
until "converged"
return $\hat{X}$

This typically does not need too many iterations, but if n and d are large, then computing the SVD can be expensive. Various matrix sketching approaches (again see Section 11.3) can be used in its place to estimate a low-rank approximation more efficiently—especially those that pay attention to matrix sparsity. This is appropriate since the rank will be reduced in the ϕ_λ step regardless, and is the point of the modeling.

7.10 Random Projections

We conclude this chapter with a completely different approach to reducing the dimension of a large data set. We consider a set of n points $A \subset \mathbb{R}^d$. The goal is a mapping $v : \mathbb{R}^d \to \mathbb{R}^k$ (with $k \ll d$) so it compresses all of $\mathbb{R}^d$ to $\mathbb{R}^k$. And in particular for any point set $A \subset \mathbb{R}^d$ we want *all* distances preserved so that for all $a, a' \in A$

$$(1 - \varepsilon)\|a - a'\| \le \|v(a) - v(a')\| \le (1 + \varepsilon)\|a - a'\|. \tag{7.2}$$

This is different than the requirements for PCA since we want *all* distances between pairs of points preserved, whereas PCA was asking for an average error to be small—and it was the average cost of the projection to the subspace, and did not directly relate to the distance between points. Thus PCA allows some points to have large error as long as most do not.

The idea to create v is very simple: *choose a linear one at random*! Specifically, to create v, we create k random unit vectors $v_1, v_2, \ldots, v_k$, then project onto the subspace spanned by these vectors. Finally, we need to renormalize by $\sqrt{d/k}$ so the expected norm is preserved.

Algorithm 7.10.1 Random Projection

---
for $i = 1$ **to** k **do**
    Choose $v_i \sim \mathcal{G}_d$ *v_i is drawn from a Gaussian distribution*
    Rescale $v_i = \sqrt{d/k} \cdot v_i / \|v_i\|$
for each points $a \in A$ **do**
    for $i \in [k]$ **do**
        $q_i = \langle a, v_i \rangle$ *a is mapped to $q = (q_1, q_2, \ldots, q_k) = v(a) \in \mathbb{R}^k$ in Q*
return Q

A classic result, named after its authors, called the *Johnson-Lindenstrauss Lemma*, shows that if k is roughly $(1/\varepsilon^2) \log(n/\delta)$ then for all $a, a' \in A$ equation (7.2) is satisfied with probability at least $1 - \delta$. The proof can almost be directly derived via a Chernoff-Hoeffding bound plus Union bound, however, it requires a more careful concentration of measure bound based on Chi-squared random variables. For each distance, each random projection (after appropriate normalization) gives an unbiased estimate; this requires the $1/\varepsilon^2$ term to average these errors so the concentrated difference from the unbiased estimate is small. Then, the union bound over all $\binom{n}{2} < n^2$ distances yields the $\log n$ term.

Interpretation of Bounds

It is pretty amazing that this bound *does not depend on d*. Moreover, it is essentially tight; that is, there are known point sets such that it requires dimensions proportional to $1/\varepsilon^2$ to satisfy equation (7.2).

Although the $\log n$ component can be quite reasonable, the $1/\varepsilon^2$ part can be quite onerous. For instance, if we want error to be within 1% error, we may need k at about $\log n$ times 10,000. It requires *very large d* so that setting $k = 10,000$ is useful. Ultimately, this may be useful when $k > 200$, and not too much precision is needed, and PCA is too slow. Otherwise, SVD or its approximations are often a better choice in practice.

Exercises

We will use one dataset, here:
`https://mathfordata.github.io/data/A.csv`

7.1 Read data set `A.csv` as a matrix $A \in \mathbb{R}^{30 \times 6}$. Compute the SVD of A and report

1. the third right singular vector,
2. the second singular value, and
3. the fourth left singular vector.
4. What is the rank of A?

Compute A_k for $k = 2$.

5. What is $\|A - A_k\|_F^2$?
6. What is $\|A - A_k\|_2^2$?

Center A. Run PCA to find the best two-dimensional subspace B to minimize $\|A - \pi_B(A)\|_F^2$. Report

7. $\|A - \pi_B(A)\|_F^2$ and
8. $\|A - \pi_B(A)\|_2^2$.

7.2 Consider another matrix $A \in \mathbb{R}^{8 \times 4}$ with squared singular values $\sigma_1^2 = 10$, $\sigma_2^2 = 5$, $\sigma_3^2 = 2$, and $\sigma_4^2 = 1$.

1. What is the rank of A?
2. What is $\|A - A_2\|_F^2$, where A_2 is the best rank-2 approximation of A.
3. What is $\|A - A_2\|_2^2$, where A_2 is the best rank-2 approximation of A.
4. What is $\|A\|_2^2$?
5. What is $\|A\|_F^2$?

Let v_1, v_2, v_3, v_4 be the right singular vectors of A.

6. What is $\|Av_2\|^2$?
7. What is $\langle v_1, v_3 \rangle$?
8. What is $\|v_4\|$?
9. What is the second eigenvector of $A^T A$?
10. What is the third eigenvalue of $A^T A$?
11. What is the fourth eigenvalue of AA^T?

Let $a_1 \in \mathbb{R}^4$ be the first row of A.

12. Write a_1 in the basis defined by the right singular vectors of A.

7.3 Consider two matrices A_1 and A_2 both in $\mathbb{R}^{10 \times 3}$. A_1 has singular values $\sigma_1 = 20$, $\sigma_2 = 2$, and $\sigma_3 = 1.5$. A_2 has singular values $\sigma_1 = 8$, $\sigma_2 = 4$, and $\sigma_3 = 0.001$.

1. For which matrix will the power method most likely converge faster to the top eigenvector of $A_1^T A_1$ (or $A_2^T A_2$, respectively), and why?

Given the eigenvectors v_1, v_2, v_3 of $A^T A$. Explain step by step how to recover the following. Specifically, you should write the answers as linear algebraic expressions in terms of v_1, v_2, v_3, and A; it can involve taking norms, matrix multiply, addition, subtraction, but not something more complex like SVD.

2. the second singular value of A
3. the first right singular vector of A
4. the third left singular vector of A

7.4 Describe what will happen if the power method is run after the initialization $u^{(0)}$ is set to be the second eigenvector?

7.5 Consider the 6×6 distance matrix

$$D = \begin{bmatrix} 0 & 1.2 & 3.4 & 5.1 & 6.2 & 7.7 \\ 1.2 & 0 & 2.7 & 4.8 & 6.3 & 8.2 \\ 3.4 & 2.7 & 0 & 4.4 & 5.5 & 3.3 \\ 5.1 & 4.8 & 4.4 & 0 & 2.4 & 2.5 \\ 6.2 & 6.3 & 5.5 & 2.4 & 0 & 0.9 \\ 7.7 & 8.2 & 3.3 & 2.5 & 0.9 & 0 \end{bmatrix}$$

where the entry $D_{i,j}$ measures the distance between the ith and the jth data points.

1. Is this distance a metric?
2. Run classical MDS to create two-dimensional coordinates for each of the 6 data points. Report those coordinates.
3. Plot the two-dimensional coordinates for data points with distances described in D.

7.6 Consider again data set **A** as a matrix $A \in \mathbb{R}^{30 \times 6}$. Treat the first 20 rows as class 1, the next 20 rows (rows 21-40) as class 2, and the last 20 rows as class 3. Use LDA to embed the 60 points into $\mathbb{R}^2$ so as to best separate these classes.

7.7 Consider the DML Optimization Algorithm 7.8.1, after t steps.

1. What is the maximum possible rank of matrix M_t for $t > 0$ and $t \le d$?
2. Show that the trace of M_t is always exactly d; recall that the trace is a linear operator so $\text{Tr}(\alpha A + \beta B) = \alpha \text{Tr}(A) + \beta \text{Tr}(B)$ for scalars α, β and matrices A, B.

7.8 Consider the matrix $X \in \mathbb{R}^{12 \times 7}$ with missing entries denoted as **x**:

$$X = \begin{bmatrix} 9.1 & x & 8.7 & 3.2 & 3.0 & x & 0.8 \\ 8.8 & 8.9 & 8.8 & 3.1 & 2.7 & x & 0.9 \\ 9.0 & 9.1 & x & x & 2.8 & 1.1 & 1.2 \\ 1.7 & x & 1.1 & 1.2 & 1.0 & x & 6.9 \\ 2.0 & 1.9 & x & 0.8 & 1.1 & 7.2 & 6.9 \\ 2.1 & x & 1.9 & 1.0 & x & 6.7 & 7.1 \\ 2.0 & 2.2 & x & 0.8 & 0.8 & x & 7.2 \\ x & 4.2 & 4.0 & x & 7.7 & 1.7 & x \\ 4.2 & 4.1 & x & 7.8 & 8.1 & x & 2.0 \\ 3.8 & x & 3.7 & 8.2 & 8.1 & 2.0 & 1.9 \\ x & 4.2 & 4.0 & x & 8.2 & 2.3 & x \\ 3.9 & 4.3 & 4.2 & x & 7.9 & x & 2.0 \end{bmatrix}.$$

Run the matrix completion algorithm (in Algorithm 7.9.1) to estimate the missing x values.

1. Describe the mask Ω implied by the x symbols.
2. Report the initialized values $X_{i,j}$ in $\Pi_{\Omega}^{\perp}(X)$ by the average of row average, and column column average.
3. Run the **repeat-until** loop for 5 steps, reporting X at the end of each step.

7.9 Consider again the input data set A as a matrix A, representing 30 data points in 6 dimensions. Here, we will compare random projections to best rank-k approximations and PCA (as in Exercise 7.1). *(Typically, a much larger scale is needed to see the benefits of random projection).*

1. Compute the 30×30 matrix D which measures all pairwise distance between points in A (in the original six-dimensional representation).
2. For $k = 2$, find A_k, the best rank-k approximation of A. Compute the distance matrix D_2, and compute both the summed squared error $\|D - D_2\|_F^2$ and the worst case error $\|D - D_2\|_{\infty} = \max_{i,j} |D_{i,j} - (D_2)_{i,j}|$.
3. Find the best two-dimensional subspace B to project A onto with PCA, and its representation $\pi_B(A)$. Use this representation to compute another distance matrix D_B, and again compute $\|D - D_B\|_F^2$ and $\|D - D_B\|_{\infty}$.
4. Randomly project A onto a two-dimensional subspace as a new point set Q (using Algorithm 7.10.1). Compute a new distance matrix D_Q and compute $\|D - D_Q\|_F^2$ and $\|D - D_Q\|_{\infty}$.
5. Since the previous random-projection step was randomized, repeat it 10 times, and calculate the average values for $\|D - D_Q\|_F^2$ and $\|D - D_Q\|_{\infty}$.
6. Increase the dimension used in the random projection until each of $\|D - D_Q\|_F^2$ and $\|D - D_Q\|_{\infty}$ (averaged over 10 trials) match or is less than those errors reported by PCA (it may be larger than the original 6 dimensions).

7.10 Consider a matrix $A \in \mathbb{R}^{n \times d}$ mapped to $Q \in \mathbb{R}^{n \times d}$ by PCA, so itt lies in a k-dimensional subspace. For simplicity assume A is centered, so we can discuss Q in terms of the top k right singular vectors V_k of A.

1. Use the standard notation for the SVD of A as $[U, S, V^T] = \mathsf{svd}(A)$, and its components to describe the set Q.
2. Let $D \in \mathbb{R}^{n \times n}$ be the distance matrix of A, and let $D_Q \in \mathbb{R}^{n \times n}$ be the distance matrix of Q. Use the notation for the representation of Q using the SVD of A to derive an expression for

$$\|D - D_Q\|_F^2 = \sum_{i,j} (D_{i,j}^2 - (D_Q)_{i,j}^2).$$

3. Construct a small example, where $d = 3$ and $k = 2$, and $\|D - D_Q\|_F^2$ is small, but $\|D - D_Q\|_\infty = \max_{i,j} |D_{i,j} - (D_Q)_{i,j}|$ (as is preserved under the Johnson-Lindenstrauss Lemma) is arbitrarily large.
   Hint: For this to work, you need to let n grow large, but only requires a small number of distinct point locations.

Chapter 8
Clustering

Abstract Clustering is the automatic grouping of data points into subsets of similar points. There are numerous ways to define this problem, and most of them are quite messy. And many techniques for clustering actually lack a mathematical formulation. We will initially focus on what is probably the cleanest and most used formulation: assignment-based clustering which includes k-center and the notorious k-means clustering. For background, we will begin with a mathematical detour in Voronoi diagrams. Then we will also describe various other clustering formulations to give insight into the breadth and variety of approaches in this subdomain.

▷ On Clusterability ◁

Given the vast landscape of clustering approaches, consider this overview:

- *When data is easily or naturally clusterable, most clustering algorithms work quickly and well.*
- *When data is not easily or naturally clusterable, then no algorithm will find good clusters.*

This statement is perhaps debatable, and "easily or naturally clusterable" is not defined here. But there are definitely many ways to cluster data, and the most common ones all work on common sense data examples with clusters.

8.1 Voronoi Diagrams

Consider a set $S = \{s_1, s_2, \ldots, s_k\} \subset \mathbb{R}^d$ of sites. We would like to understand how these points can carve up the space $\mathbb{R}^d$.

We now think of this more formally as the *post office problem*. Let these k sites each define the location of a post office. For all points in $\mathbb{R}^d$ (e.g., a point on the map for points in $\mathbb{R}^2$), we would like to assign it to the closest post office. For a fixed

© Springer Nature Switzerland AG 2021

J. M. Phillips, *Mathematical Foundations for Data Analysis*,
Springer Series in the Data Sciences,
https://doi.org/10.1007/978-3-030-62341-8_8

point $x \in \mathbb{R}^d$, we can just check the distance to each post office:

$$\phi_S(x) = \arg \min_{s_i \in S} \|x - s_i\|.$$

However, this may be slow (naively check the distance to all k sites for each point x), and does not provide a general representation or understanding for all points. The "correct" solution to this problem is the Voronoi diagram.

The *Voronoi diagram* decomposes $\mathbb{R}^d$ into k regions (a *Voronoi cell*), one for each site. The region for site s_i is defined.

$$R_i = \{x \in \mathbb{R}^d \mid \phi_S(x) = s_i\}.$$

Example: Voronoi in 1 Dimension

For four points $\{s_1, s_2, s_3, s_4\}$ in $\mathbb{R}^1$, the boundaries between the Voronoi cells are shown. The Voronoi cell for s_3 is highlighted. The points $x \in \mathbb{R}^1$ so that $\phi_S(x) = s_3$ comprises this highlighted region.

If these regions are nicely defined, this solves the post office problem. For any point x, we just need to determine which region it lies in (for instance in $\mathbb{R}^2$, once we have defined these regions, through an extension of binary search, we can locate the region containing any $x \in \mathbb{R}^2$ in only roughly $\log k$ distance computations and comparisons). But what do these regions look like, and what properties do they have.

Voronoi Edges and Vertices

We will start our discussion in $\mathbb{R}^2$. Further, we will assume that the sites S are in *general position*: in this setting, it means that no set of three points lie on a common line, and that no set of four points lie on a common circle.

The boundary between two regions R_i and R_j, called a *Voronoi edge*, is a line or line segment. This edge $e_{i,j}$ is defined as

$$e_{i,j} = \{x \in \mathbb{R}^2 \mid \|x - s_i\| = \|x - s_j\| \leq \|x - s_\ell\| \text{ for all } \ell \neq i, j\},$$

the set of all points equal distance to s_i and s_j, and not closer to any other point s_ℓ.

Geometry of Voronoi Diagram Edge

Why is the set of points $e_{i,j}$ a line segment? If we only have two points in S, then it is the bisector between them. Draw a circle centered at any point x on this bisector, and if it intersects one of s_i or s_j, it will also intersect the other. This is true since we can decompose the squared distance from x to s_i along orthogonal components: along the edge, and perpendicular to the edge from s_i to $\pi_{e_{i,j}}(s_i)$.

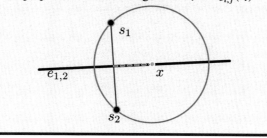

Similarly, a *Voronoi vertex* $v_{i,j,\ell}$ is a point where three sites s_i, s_j, and s_ℓ are all equidistance, and no other points are closer:

$$v_{i,j,\ell} = \{x \in \mathbb{R}^2 \mid \|x - s_i\| = \|x - s_j\| = \|x = s_\ell\| \le \|x - s_k\| \text{ for all } k \ne i, j, \ell\}.$$

This vertex is the intersection (and end point) of three Voronoi edges $e_{i,j}$, $e_{i,\ell}$, and $e_{j,\ell}$. Think of sliding a point x along an edge $e_{i,j}$ and maintaining the circle centered at x and touching s_i and s_j. When this circle grows to where it also touches s_ℓ, then $e_{i,j}$ stops.

Example: Voronoi Diagram

See the following example with $k = 6$ sites in $\mathbb{R}^2$. Notice the following properties: edges may be unbounded, and the same with regions. The circle centered at $v_{1,2,3}$ passes through s_1, s_2, and s_3. Also, Voronoi cell R_3 has $5 = k - 1$ vertices and edges.

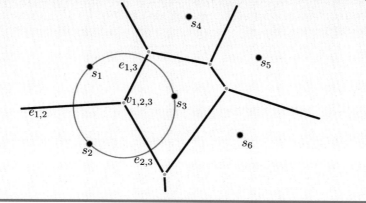

Size Complexity

So how complicated can these Voronoi diagrams get? A single Voronoi cell can have $k-1$ vertices and edges. So can the entire complex be of size that grows quadratically in k? (each of k regions requiring complexity of roughly k)? No. The Voronoi vertices and edges describe a planar graph (i.e., can be drawn in the plane, $\mathbb{R}^2$, with no edges crossing). And planar graphs have asymptotically the same number of edges, faces, and vertices. In particular, *Euler's Formula* for a planar graph with n vertices, m edges, and k faces is that $k + n - m = 2$. To apply this rule to Voronoi diagrams, either one needs to ignore the infinite edges (e.g., $e_{1,2}$ in the example) or connect them all to a single imaginary vertex "at infinite." And *Kuratowski's criteria* says for $n \geq 3$, then $m \leq 3n - 6$. Hence, $k \leq 2n - 4$ for $n \geq 3$. The duality construction to Delauney triangulations (discussed below) will complete the argument. Since there are k faces (the k Voronoi cells, one for each site), then there are also asymptotically roughly $(3/2)k$ edges and $k/2$ vertices.

However, this does not hold in $\mathbb{R}^3$. In particular, for $\mathbb{R}^3$ and $\mathbb{R}^4$, the complexity (number of cells, vertices, edges, faces, etc) grows quadratically in k. This means, there could be roughly as many edges as they are pairs of vertices!

But it can get much worse. In $\mathbb{R}^d$ (for general d) then the complexity is exponential in d, that is asymptotically roughly $k^{\lceil d/2 \rceil}$. This is a lot. Hence, this structure is impractical to construct in high dimensions.

▷ Curse of Dimensionality ◁

The *curse* of dimensionality, refers to when a problem has a nice, simple, and low-complexity structure in low dimensions, but then becomes intractable and unintuitive in high dimensions. For instance, many geometric properties, like the size complexity of Voronoi diagrams, have linear complexity and are easy to draw in low dimensions, but are unintuitive, and have size complexity that grows exponentially as the dimension grows.

Moreover, since this structure is explicitly tied to the post office problem, and the nearest neighbor function ϕ_S, it indicates that in $\mathbb{R}^2$ this function is nicely behaved, but in high dimensions, it is quite complicated.

8.1.1 Delaunay Triangulation

A fascinating aspect of the Voronoi diagram is that it can be converted into a very special graph called the *Delaunay triangulation* where the sites S are vertices. This is the *dual* of the Voronoi diagram.

- Each face R_i of the Voronoi diagram maps to a vertex s_i in the Delaunay triangulation.

- Each vertex $v_{i,j,\ell}$ in the Voronoi diagram maps to a triangular face $f_{i,j,\ell}$ in the Delaunay triangulation.
- Each edge $e_{i,j}$ in the Voronoi diagram maps to an edge $\bar{e}_{i,j}$ in the Delaunay triangulation.

Example: Delaunay Triangulation

See the following example with 6 sites in $\mathbb{R}^2$. Notice that every edge, face, and vertex in the Delaunay triangulation corresponds to a edge, vertex, and face in the Voronoi diagram. Interestingly, the associated edges may not intersect; see $e_{2,6}$ and $\bar{e}_{2,6}$.

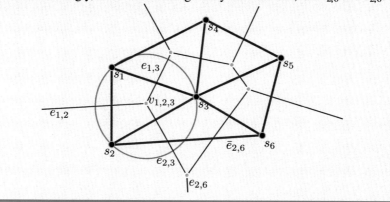

Because of the duality between the Voronoi diagram and the Delaunay triangulation, their complexities are the same. That means the Voronoi diagram is of size roughly k for k sites in $\mathbb{R}^2$, but more generally is of asymptotic size $k^{\lceil d/2 \rceil}$ in $\mathbb{R}^d$.

The existence of the Delaunay triangulation shows that there always exist a triangulation: A graph with vertices of a given set of points $S \subset \mathbb{R}^2$ so that all edges are straightline segments between the vertices, and each face is a triangle. In fact, there are many possible triangulations: one can always simply construct some triangulation greedily, draw any possible edges that does not cross other edges until no more can be drawn.

The Delaunay triangulation, however, is quite special. This is the triangulation that maximizes the smallest angle over all triangles; for meshing applications in graphics and simulation, skinny triangles (with small angles) cause numerical issues, and so these are very useful.

In Circle Property

Another cool way to define the Delaunay triangulation is through the *in circle property*. For any three points, the smallest enclosing ball either has all three points on the boundary, or has two points on the boundary and they are antipodal to each other. Any circle with two points antipodal on the boundary s_i and s_j (i.e., s_i and s_j are on exact opposite spots on the circle), and contains no other points, then the edge $e_{i,j}$ is in the Delaunay triangulation. The set of edges defined by pairs of points defining empty circles is a subset of the Delaunay triangulation called the *Gabriel graph*.

Any circle with three points on its boundary s_i, s_j, and s_ℓ, and no points in its interior, then the face $f_{i,j,\ell}$ is in the Delaunay triangulation, as well as its three edges $e_{i,j}$, $e_{i,\ell}$ and $e_{j,\ell}$. But does not imply those edges are in the Gabriel graph.

For instance, on a quick inspection, (in the example above) it may not be clear if edge $e_{3,5}$ or $e_{4,6}$ should be in the Delaunay triangulation. Clearly, it cannot be both since they cross. But the ball with boundary through s_3, s_4, and s_6 would contain s_5, so the face $f_{3,4,6}$ cannot be in the Delaunay triangulation. On the other hand, the ball with boundary through s_3, s_6, and s_5 does not contain s_4 or any other points in S, so the face $f_{3,5,6}$ is in the Delaunay triangulation.

8.1.2 Connection to Assignment-Based Clustering

So what is the connection to clustering? Given a large set $X \subset \mathbb{R}^d$ of size n, we would like to find a set of k sites S (post office locations) so that each point $x \in X$ is near some post office. This is a proxy problem. So given a set of sites S, determining for each $x \in X$ which site is closest is exactly determined by the Voronoi diagram.

In particular, in *assignment-based clustering*, each cluster is represented by a single site $s \in S$, to which all other points in the cluster are "assigned" by ϕ_S. Given a set X and distance $\mathbf{d} : X \times X \to \mathbb{R}_+$, the output is a set $S = \{s_1, s_2, \ldots, s_k\}$. Using the Voronoi diagram, this implicitly defines a set of clusters where $\phi_S(x) = \arg\min_{s \in S} \mathbf{d}(x, s)$; that is each x is assigned to the closest site $s = \phi_S(x)$. Then, the *k-means clustering problem* is to find the set S of k sites to

$$\text{minimize} \sum_{x \in X} \mathbf{d}(\phi_S(x), x)^2.$$

So we want every point assigned to the closest center, and want to minimize the sum of the squared distances of all such assignments.

There are several other useful variants including:

- the *k-center clustering problem*: minimize $\max_{x \in X} \mathbf{d}(\phi_S(x), x)$
- the *k-median clustering problem*: minimize $\sum_{x \in X} \mathbf{d}(\phi_S(x), x)$
  The *k-mediod* variant is similar to *k*-median, but restricts that the centers S must be a subset of X.

Moreover, the *mixture of Gaussians* approach will allow for more flexibility in the assignment and more modeling power in the shape of each cluster.

8.2 Gonzalez's Algorithm for k-Center Clustering

We begin with what is arguably the simplest and most general clustering algorithm: the *Gonzalez's algorithm*. This algorithm directly maps to the k-center formulation, where again every point is assigned to the closest center, and the goal is to minimize the length of the *longest* distance of any such assignment pairing.

Unfortunately, the k-center clustering problem is NP-hard to solve exactly.[1] In fact, for the general case, it is NP-hard to find a clustering within a factor 2 of the optimal cost!

Luckily, there is a simple, elegant, and efficient algorithm that achieves this factor 2 approximation. That is, for value k and a set X, it finds a set of k sites $\hat{S}$ so

$$\max_{x \in X} \mathbf{d}(\phi_{\hat{S}}(x), x) \le 2 \max_{x \in X} \mathbf{d}(\phi_{S^*}(x), x),$$

where S^* is the set of k sites with optimal cost. That is

$$S^* = \operatorname*{argmin}_{S:|S|=k} \max_{x \in X} \mathbf{d}(\phi_S(x), x).$$

Moreover, in practice, this often works better than the worst case theoretical guarantee. This algorithm, presented as Algorithm 8.2.1, is usually attributed to Teofilo F. Gonzalez (1985), hence the name. It can be described with the following miserly maxim: *Be greedy, and avoid your neighbors!*.

Algorithm 8.2.1 Gonzalez's Algorithm(X, k)

Choose $s_1 \in X$ arbitrarily. Initialize $S = \{s_1\}$.
for $i = 2$ to k **do**
    Set $s_i = \arg\max_{x \in X} \mathbf{d}(x, \phi_S(x))$.
    Include $S = S \cup \{s_i\}$.

The algorithm is iterative, building up the set of sites S over the run of algorithm. It initializes the first site s_1 arbitrarily. And it maintains a set of sites S which have i sites after i steps; that is initially the set only contains the one site s_1 chosen so far. Then it adds to the set the point $x \in X$ which is furthest from any of the current sites; that is the one with largest $\phi_S(x)$ value. This function ϕ_S changes over the course of the algorithm as more sites are added to S.

---

[1] The term *NP-hard* (i.e., non-deterministic polynomial time hard) refers to being as hard as a set of problems in terms of runtime with respect to the size of their input. For these problems, if the correct solution is found, it can be verified quickly (in time polynomial in the size of the input), but to find that correct solution the only known approaches are essentially equivalent to a brute-force search over an exponentially large set of possible solutions, taking time exponential in the input size. It is not known if it *must* take time exponential in the input size, or if there may be a solution which takes time polynomial in the input size (the class of problems P); but in practice, it is often assumed this is not possible. In short, these problems are probably very hard to solve efficiently.

In other words, back to our maxim, it always adds a new site that is furthest from the current set of sites.

≫ Efficient Gonzalez's Algorithm

This algorithm only takes time proportional to kn. There are k rounds, and each round can be done in about n time. We maintain the map $\phi_{S_i}(x)$ for each x; in the algorithm we will just use an array ϕ, where $\phi[j]$ stores the index of the assigned site for the jth point x_j. That is, at any point in the algorithm, point x_j is assigned to $s_{\phi[j]}$. When a new s_i is found, and added to the set of centers, all n assignments $\phi_{S_i}(x)$ can be updated in linear time in n, by checking each distance $\mathbf{d}(x, \phi_{S_{i-1}}(x))$ against $\mathbf{d}(x, s_i)$ and switching the assignment if the latter is smaller. Then, the minimum can be found in the next round on a linear scan (or on the same linear scan).

Algorithm 8.2.2 Efficient Detailed Gonzalez's Algorithm(X, k)

Choose $s_1 \in X$ arbitrarily, and set $\phi[j] = 1$ for all $j \in [n]$
for $i = 2$ to k **do**
    $M = 0,$ $s_i = x_1$
    for $j = 1$ to n **do**
        if $\mathbf{d}(x_j, s_{\phi[j]}) > M$ **then**
            $M = \mathbf{d}(x_j, s_{\phi[j]}),$ $s_i = x_j$
    for $j = 1$ to n **do**
        if $\mathbf{d}(x_j, s_{\phi[j]}) > \mathbf{d}(x_j, s_i)$ **then**
            $\phi[j] = i$

This simple, elegant, and efficient algorithm works for any metric $\mathbf{d}$, and the 2-approximation guarantee will still hold. The resulting set of sites S are a subset of the input points X, this means it does not rely on the input points being part of a nice, easy to visualize, space, e.g., $\mathbb{R}^d$. However, this algorithm biases the choice of centers to be on the "edges" of the dataset, each chosen site is as far away from existing sites as possible, and then is never adjusted. There are heuristics to adjust centers afterwards, including the algorithms for k-means and k-mediod clustering.

Geometry of Gonzalez 2-Approximation

Let $s_{k+1} = \text{argmax}_{x \in X} \mathbf{d}(x, \phi_S(x))$ and let the cost of the solution found be $R = \max_{x \in X} \mathbf{d}(x, \phi_S(x)) = \mathbf{d}(s_j, s_{k+1})$, where $s'_j = \phi_S(s_{k+1})$ (in the picture $s'_j = s_1$). We know that no two sites can be within a distance R; that is $\min_{s, s' \in S} \mathbf{d}(s, s') \geq R$; otherwise, they would not have been selected as a site, since s_{k+1} (inducing the radius R) would have been selected first.

Thus if the optimal set of k sites $\{o_1, o_2, \ldots, o_k\}$ have total cost less than $R/2$, they must each be assigned to a distinct site in S. Assume otherwise that site $s \in S$ (e.g., s_5 in the figure) has none of the optimal sites assigned to it, and the closest optimal site is o_j, assigned to center $s_j \in S$ (e.g., s_2 in the figure). Since we assume $\mathbf{d}(o_j, s_j) < R/2$, and $\mathbf{d}(s, s_j) \geq R$, then by triangle inequality $\mathbf{d}(s_j, o_j) + \mathbf{d}(o_j, s) \geq \mathbf{d}(s_j, s)$ and hence it would imply that $\mathbf{d}(o_j, s) > R/2$, violating the assumption (e.g., o_j must be assigned to s_5).

However, if each distinct site in S has one optimal site assigned to it, then the site around s_j' that has assigned s_{k+1} to it, and s_{k+1} cannot be covered by one center o_j' with radius less than $R/2$. This is again because $\mathbf{d}(s_j', s_{k+1}) = R$ so one of $\mathbf{d}(s_j', o_j')$ and $\mathbf{d}(s_{k+1}, o_j')$ must be at least $R/2$. Since all other sites in S are at distance at least R from s_{k+1}, the same issue occurs in trying to cover them and some optimal site with radius less than $R/2$ as well.

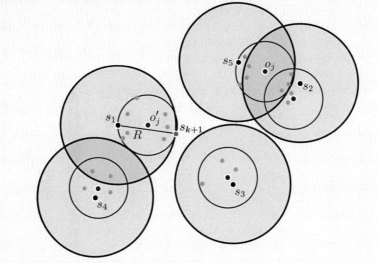

8.3 Lloyd's Algorithm for k-Means Clustering

Probably the most famous clustering formulation is k-means, which we recall is an instance of assignment-based clustering. Specifically, the *k-means clustering problem* is to find the set S of k clusters to minimize

$$\mathsf{cost}(X, S) = \sum_{x \in X} \|\phi_S(x) - x\|^2.$$

So we want every point assigned to the closest site, and want to minimize the sum of the squared distance of all such assignments.

We emphasize the term "k-means" refers to a problem formulation, *not to any one algorithm*. There are many algorithms with aim of solving the k-means problem formulation, exactly or approximately. We will mainly focus on the most common: Lloyd's algorithm. Unfortunately, it is commonly written in data mining literature "the k-means algorithm," which typically should be more clearly stated as Lloyd's algorithm.

8.3.1 Lloyd's Algorithm

When people think of the k-means problem, they usually think of the following algorithm, attributed to Stuart P. Lloyd from a document in 1957, although it was not published until 1982.[2]

The algorithm is again fairly simple and elegant; however, unlike Gonzalez's algorithm for k-center, the sites are all iteratively updated. It initializes with any set of k sites, and then iteratively updates the locations of these sites. Moreover, this assumes that the input data X lies in $\mathbb{R}^d$ and the implicit distance $\mathbf{d}$ is Euclidean; it can be generalized to X in a few other Euclidean-like spaces as well.

As shown in Algorithm 8.3.1, each iteration is split into two steps. In the first step, each point $x \in X$ is explicitly mapped to the closest site $s_i = \phi_S(x)$. In the second step, the set of points $X_i = \{x \in X \mid \phi_S(x) = s_i\}$ which are mapped to the ith site s_i are gathered, and that site is adjusted to the average of all points in X_i.

This second step is why we assume $X \in \mathbb{R}^d$ (that $\mathbf{d}$ is the Euclidean distance) since the average operation can be naturally defined for $\mathbb{R}^d$ (as $\frac{1}{|X_i|} \sum_{x \in X_i} x$). This results in a point in $\mathbb{R}^d$, but not necessarily a point in X. So, differing from Gonzalez's algorithm, this does not in general return a set of sites S which are drawn from the input set X; the sites could be anywhere.

Algorithm 8.3.1 Lloyd's Algorithm(X, k)

Choose k points $S \subset X$	*arbitrarily?*
repeat	
for all $x \in X$: assign x to X_i so $\phi_S(x) = s_i$	*the closest site $s_i \in S$ to x*
for all $s_i \in S$: update $s_i = \frac{1}{|X_i|} \sum_{x \in X_i} x$	*the average of $X_i = \{x \in X \mid \phi_S(x) = s_i\}$*
until (the set S is unchanged, *or other termination condition*)	

Example: Lloyd's Algorithm for k-Means Clustering

After initializing the sites S as $k = 4$ points from the data set, the algorithm assigns data points in X to each site. Then in each round, it updates the site as the average of the assigned points, and reassigned points (represented by the Voronoi cells). After 4 rounds it has converged to the optimal solution.

---

[2] Apparently, the IBM 650 computer Lloyd was using in 1957 did not have enough computational power to run the (very simple) experiments he had planned. This was replaced by the IBM 701, but it did not have quite the same "quantization" functionality as the IBM 650, and the work was forgotten. Lloyd was also worried about some issues regarding the k-means problem not having a unique minimum.

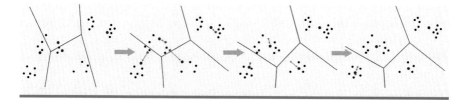

Convergence

This iterative process is continued until some termination condition is obtained. Algorithm 8.3.1 describes this as when the location of S is unchanged between iterates.

In general if the main loop has R iterations (or rounds), then this takes time proportional to Rnk (and can be made closer to $Rn \log k$ with faster nearest neighbor search in some low-dimensional cases). But how large is R; that is, how many iterations do we need?

We argue that the cost function $\mathsf{cost}(X, S)$ always decreases and thus the number of rounds is finite. In fact, each round improves upon the results of the previous round. However, the number of steps until this absolute termination condition is met may be quite large (indeed in certain cases exponentially large!). So in practice such a strict variant is rarely used. Rather, it is common to run for a specified and fixed number of rounds, say $R = 10$ or $R = 20$. Another common approach, as in gradient descent, is to check if the change in $\mathsf{cost}(X, S)$ is less than a sufficiently small threshold.

Proof for Iterative Convergence

The number of rounds is finite. This is true since the cost function $\mathsf{cost}(X, S)$ will always decrease. To see this, write it as a sum over S.

$$\mathsf{cost}(X, S) = \sum_{x \in X} \|\phi_S(x) - x\|^2 \tag{8.1}$$

$$= \sum_{s_i \in S} \sum_{x \in X_i} \|s_i - x\|^2. \tag{8.2}$$

Then in each step of the **repeat-until** loop, this cost must decrease. This holds in the first step since it moves each $x \in X$ to a subset X_i with the corresponding center s_i closer to (or the same distance to) x than before. So in equation (8.1), as $\phi_S(x)$ is updated for each x, the term $\|x - s_i\| = \|x - \phi_S(x)\|$ is reduced (or the same). This holds in the second step since in equation (8.2), for each inner sum $\sum_{x \in X_i} \|s_i - x\|^2$, the single point s_i which minimizes this cost is precisely the average of X_i. This follows by the same argument as in 1-dimension (see Section 1.7.1), since by the Pythagorean formula, the squared distance to the mean can be decomposed along the coordinates. So reassigning s_i as described also decreases the cost (or keeps it the same).

Importantly, if the cost decreases each step, then it cannot have the same set of centers S on two different steps, since that would imply the assignment sets $\{X_i\}$ would also be the same. Thus, in order for this to happen, the cost would need to decrease after the first occurrence, and then increase to obtain the second occurrence, which is not possible.

Since, there are finite ways each set of points can be assigned to different clusters, then, the algorithm terminates in a finite number of steps.

Unfortunately, no matter how the convergence condition is chosen, this algorithm is **not** guaranteed to find the optimal set of sites. It is easy to get stuck in a local minimum, and this may happen somewhat commonly with a large k.

Example: Local Minimum for Lloyd's Algorithm

Two examples are shown for local minimum for Lloyd's algorithm. In these configurations, Lloyd's algorithm is converged, but these are not the optimal k-means cost.

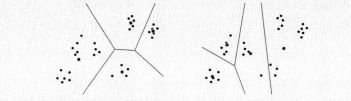

There are two typically ways to help manage the consequences of this property: careful initialization and random restarts.

A random restart, uses the insight that the choice of initialization can be chosen randomly, and this starting point is what determines what minimum is reached. So, then the basic Lloyd's algorithm can be run multiple times (e.g., 3-5 times) with different random initializations. Then, ultimately the set of sites S, across all random restarts, which obtain the lowest $\text{cost}(X, S)$ is returned.

Initialization

The initial paper by Lloyd advocates to choose the initial partition of X into disjoint subsets $X_1, X_2, \ldots, X_k$ *arbitrarily*. However, some choices will not be very good. For instance, if we randomly place each $x \in X$ into some X_i, then (by the central limit theorem) we expect all $s_i = \frac{1}{|X_i|} \sum_{x \in X_i} x$ to all be close to the mean of the full data set $\frac{1}{|X|} \sum_{x \in X} x$.

A bit safer way to initialize the data is to choose a set $S \subset X$ at random. Since each s_i is chosen separately (not as an average of data points), there is no convergence to mean phenomenon. However, even with this initialization, we may run Lloyd's algorithm to completion, and find a sub-optimal solution (*a local minimum!*). However, there are scenarios where such random initialization is unlikely to find the right local minimum, even under several random restarts.

A more principled way to choose an initial set S is to use an algorithm like Gonzalez (see Section 8.2), or k-means++ (see Section 8.3.2). In particular, the initialization by Gonzalez is guaranteed to be within a factor of 2 of the optimal k-center objective. While this objective is different than the k-means objective, it in that sense cannot be too bad; and then Lloyd's algorithm will only improve from that point on. k-means++ is similar, but randomized, and more tuned to the specific k-means objective.

Example: k-Means

To demonstrate k-means we leverage the powerful Python learning and data analysis library sklearn. Unlike several other algorithms in this book, k-means has a number of tricky edge cases. How should one determine convergence? What to do if there is a cluster with no data points? When to do a random restart? The methods in sklearn apply a number of heuristics to these scenarios to handle these issues out of view.

First, we can generate and plot a data set with 4 nicely shaped clusters

```python
import numpy as np
import matplotlib.pyplot as plt
%matplotlib inline

from sklearn.datasets import make_blobs
X, labels = make_blobs(n_samples=100, centers=4,
                       cluster_std=1, random_state=6)
plt.scatter(X[:, 0], X[:, 1], s=20)
plt.show()
```

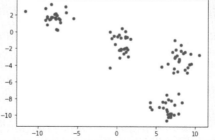

Then in a single command we can heuristically identify the clusters.

```python
from sklearn.cluster import KMeans
kmeans = KMeans(n_clusters=4)
kmeans.fit(X)
labelskm = kmeans.predict(X)
```

And then plot the clusters, by color, as well as the center points in red.

```
plt.scatter(X[:,0], X[:,1], c=labelskm, s=20)
centers = kmeans.cluster_centers_
plt.scatter(centers[:,0], centers[:,1], c='red', s=50)
plt.show()
```

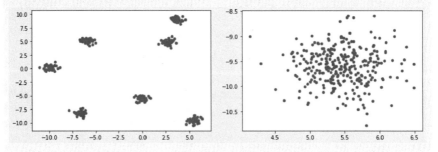

Number of Clusters

So what is the right value of k? Like with PCA, there is no perfect answer toward choosing how many dimensions the subspace should be. When k is not given to you, typically, you would run with many different values of k. Then create a plot of $\text{cost}(S, X)$ as a function of k. This cost will always decrease with larger k; but of course $k = n$ is of no use. At some point, the cost will not decrease much between values (this implies that probably two centers are used in the same grouping of data, so the squared distance to either is similar). Then, this is a good place to choose k.

Example: Elbow Method

We consider two data sets with 300 points in $\mathbb{R}^2$. The first has 7 well-defined clusters, the second has 1 blob of points.

We run Lloyd's for k-means on both sets for values of k ranging from $k = 1$ to $k = 15$, and plot the mean squared error of the result for both. As can be seen, as k increases, the error always decreases, and eventually starts to flatten out. We marked with a red • at the true number of clusters. For the $k = 7$ example (left), the curve decreases rapidly and then levels out, and the true number of clusters is at the "elbow" of the plot. However, when there is only $k = 1$ cluster, the MSE curve does not decrease as rapidly, and there is no real "elbow."

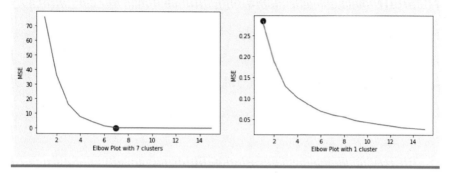

8.3.2 k-Means++

The initialization approach for Lloyd's algorithm, that is most tuned to the k-means objective is known as k-*means++*, or D^2-*sampling*. This algorithm is randomized, unlike Gonzalez, so it is compatible with random restarts. Indeed analyses which argues for approximation guarantees in Lloyd's algorithm require several random restarts with k-means++.

Algorithm 8.3.2 k-Means++ Algorithm(X, k)

Choose $s_1 \in X$ arbitrarily. Initialize $S = \{s_1\}$.
for $i = 2$ to k **do**
  Choose s_i from X with probability proportional to $\mathbf{d}(x, \phi_S(x))^2$.
  Update $S = S \cup \{s_i\}$.

As Algorithm 8.3.2 describes, the structure is like Gonzalez algorithm, but is not completely greedy. It iteratively chooses each next center randomly—the further the squared distances is from an existing center, the more likely it is chosen. For a large set of points (perhaps grouped together) which are far from an existing center, then it is very likely that one (it does not matter so much which one) of them will be chosen as the next center. This makes it likely that any "true" cluster will find some point as a suitable representative.

The critical step in this algorithm, and the difference from Gonzalez, is choosing the new center proportional to some value (the value $\mathbf{d}(x, \phi_S(x))^2$). Recall that this task has an elegant solution called the Partition of Unity, discussed in the context of Importance Sampling (Section 2.4).

8.3.3 k-Mediod Clustering

Like many algorithms in this book, Lloyd's depends on the use of a SSE cost function. Critically, this works because for $X \in \mathbb{R}^d$, then

$$\frac{1}{|X|} \sum_{x \in X} x = \mathsf{average}(X) = \arg \min_{s \in \mathbb{R}^d} \sum_{x \in X} \|s - x\|^2.$$

There are not in general similar properties for other costs functions, or when X is not in $\mathbb{R}^d$. For instance, one may want to solve the k-medians problem where one just minimizes the sum of (non-squared) distances. In particular, this subset has no closed-form solution for $X \in \mathbb{R}^d$ for $d > 1$. Most often gradient descent-type approaches are used for finding an updated center under a k-median objective.

An alternative to the averaging step is to choose

$$s_i = \arg \min_{s \in X_i} \sum_{x \in X_i} \mathbf{d}(x, s)$$

where $\mathbf{d}(x, s)$ is an arbitrary bivariate function (like non-squared Euclidean distance) between x and s. That is, we choose an s from the set X_i. This is particularly useful when X is in a non-Euclidean metric space where averaging may not be well-defined. For the specific case where $\mathbf{d}(x, s) = \|x - s\|$ (for the k-median problem), then this variant of the formulation is called the k-*mediod* clustering problem. Moreover, the most common heuristic is to adapt Lloyd's algorithm by substituting the averaging step with the minimum over all as points in that set, as in Algorithm 8.3.3.

Algorithm 8.3.3 k-Mediod Hueristic(X, k)

Choose k points $S \subset X$ *arbitrarily?*
repeat
    for all $x \in X$: assign x to X_i so $\phi_S(x) = s_i$ *the closest site $s_i \in S$ to x*
    for all $s_i \in S$: update $s_i = \arg\min_{s \in X_i} \sum_{x \in X_i} \mathbf{d}(x, s)$ *the best choice of site from X_i*
until (the set S is unchanged, *or other termination condition*)

This optimization step of choosing each s_i in each iteration is now considerably slower to implement than the averaging step in Lloyd's algorithm. However, that it always chooses s_i so that it is an element of X can be an advantage. This allows the algorithm to generalize to use arbitrary distance functions $\mathbf{d}$ in place of Euclidean distance. Now if the data objects X are customers, documents, or websites an appropriate distance function can be used, and the site at the center of a cluster can be an element of the data set, and thus an interpretable quantity.

Ethical Questions with Clustering and Outliers

The k-mediod clustering formulation is often used because it is general, but also fairly robust to outliers. A single outlier point may be chosen as a separate cluster center for k-center or k-means clustering, with no other points in that cluster. However, for k-mediods or k-median clustering, it is more likely to be added to an existing cluster without as dramatically increasing the cost function. Typically, this property is viewed as a positive for k-mediods in that the resulting cluster subsets are not greatly distorted by a single outlier.

However, if these outliers are then added to a cluster for which it does not fit, it may cause other problems. For instance, if a model is fit to each cluster, then this outlier may distort this model.

Consider the case where each data point is a person applying for a loan. It may be an outlier is included in a cluster, and has had multiple bankruptcies that caused them to default on the loan, and cost previous banks to lose a lot of money. This outlier may distort the model of that cluster so it predicts all of the customers in it are not expected to return on the loan, in expectation. On the other hand, excluding that data point would allow the model to predict more accurately that most of the customers will repay the loan, and be profitable. As a data modeler, what obligations do you have to check for such outliers within a clustering? What is a good way to mitigate the effects of instability in the way data points are clustered? Does the answer to the previous questions remain the same, even if the expected overall profit level for the bank should stay the same?

8.3.4 Soft Clustering

Sometimes it is not desirable to assign each point to exactly one cluster. Instead, we may split a point between one or more clusters, assigning a fractional value to each. This is known as *soft clustering*, whereas the original formulation is known as *hard clustering*.

There are many ways to achieve a soft clustering. For instance, consider the following Voronoi diagram-based approach using natural neighbor interpolation (NNI). Let $V(S)$ be the Voronoi diagram of the sites S (which decomposes $\mathbb{R}^d$). Then construct $V(S \cup x)$ for a particular data point x; the Voronoi diagram of the sites S with the addition of one point x. For the region R_x defined by the point x in $V(S \cup x)$, overlay it on the original Voronoi diagram $V(S)$. This region R_x will overlap with regions R_i in the original Voronoi diagram; compute the volume v_i for the overlap with each such region. Then the fractional weight for x into each site s_i is defined $w_i(x) = v_i / \sum_{i=1}^{k} v_i$.

We can plug any such step into Lloyd's algorithm, and then recalculate s_i as the weighted average of all points partially assigned to the ith cluster.

Example: Soft Clustering

In the following example of $n = 10$ data points X, and $k = 3$ sites S, these sites could either induce a hard or soft cluster assignment.

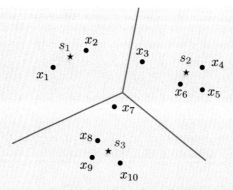

For the hard clustering assignment, it is convenient to represent the assignment as a single array of integers, which indicates the index of the site:

point	1	2	3	4	5	6	7	8	9	10
cluster index	1	1	2	2	2	2	2	3	3	3	3

On the other hand, for a soft clustering assignment, it is better to represent a probability distribution over the sites for each point. That is each point x_i maps to a discrete probability distribution $p_i \in \Delta^k$. For simplicity, the assignment probabilities shown are fairly sparse, with a lot of 0 probabilities; but in other cases (like for mixture of Gaussians shown next) these distributions $p_i \in \Delta_\circ^k$ where no assignment can have 0 probability.

point	1	2	3	4	5	6	7	8	9	10
cluster 1 probability	1	1	0.4	0	0	0	0.3	0	0	0
cluster 2 probability	0	0	0.6	1	1	1	0.3	0	0	0
cluster 3 probability	0	0	0	0	0	0	0.4	1	1	1

8.4 Mixture of Gaussians

The k-means formulation tends to define clusters of roughly equal size. The squared cost discourages points far from any center. It also, does not adapt much to the density of individual centers.

An extension is to fit each cluster X_i with a Gaussian distribution $\mathcal{G}_d(\mu_i, \Sigma_i)$, defined by a mean μ_i and a covariance matrix Σ_i. Recall that the pdf of a d-dimensional *Gaussian distribution* is defined as

$$f_{\mu, \Sigma}(x) = \frac{1}{(2\pi)^{d/2}} \frac{1}{\sqrt{|\Sigma|}} \exp\left(-\frac{1}{2}(x - \mu)^T \Sigma^{-1}(x - \mu)\right)$$

where $|\Sigma|$ is the determinant of Σ. Previously we had mainly considered this distribution where $\Sigma = I$ was the identity matrix, and it was ignored in the notation. For instance, for $d = 2$, and the standard deviation in the x-direction of X is σ_x, and in the y-direction is σ_y, and their correlation is ρ, then

$$\Sigma = \begin{bmatrix} \sigma_x^2 & \rho\sigma_x\sigma_y \\ \rho\sigma_x\sigma_y & \sigma_y^2 \end{bmatrix}.$$

Now the goal is, given a parameter k, find a set of k pdfs $F = \{f_1, f_2, \ldots, f_k\}$ where, $f_i = f_{\mu_i, \Sigma_i}$, to maximize

$$\prod_{x \in X} \max_{f_i \in F} f_i(x),$$

or equivalently to minimize

$$\sum_{x \in X} \min_{f_i \in F} - \log(f_i(x)).$$

For the special case where we restrict that $\Sigma_i = I$ (the identity matrix) for each mixture, then the second formulation (the log-likelihood version) is equivalent to the k-means problem (depending on choice of hard or soft clustering). In particular with $\Sigma = I$,

$$(x - \mu)^T \Sigma^{-1}(x - \mu) = (x - \mu)^T (x - \mu) = \langle x - \mu, x - \mu \rangle = \|x - \mu\|^2;$$

and in the general case $(x - \mu)^T \Sigma^{-1}(x - \mu) = \mathbf{d}_{\Sigma^{-1}}(x, \mu)^2$ is the squared Σ^{-1}-Mahalanobis distance between x and μ.

This hints that we can adapt Lloyd's algorithm toward this problem as well. To replace the first step of the inner loop, we assign each $x \in X$ to the Gaussian which maximizes $f_i(x)$:

$$\text{for all } x \in X: \text{ assign } x \text{ to } X_i \text{ so } i = \underset{i \in 1\ldots k}{\operatorname{argmax}} f_i(x).$$

But for the second step, we need to replace a simple average with an estimation of the best fitting Gaussian to a data set X_i. This is also simple. First, calculate the mean as $\mu_i = \frac{1}{|X_i|} \sum_{x \in X_i} x$. Then calculate the covariance matrix Σ_i of X_i as the sum of outer products

$$\Sigma_i = \sum_{x \in X_i} (x - \mu_i)(x - \mu_i)^T.$$

Indeed the covariance of the Gaussian fit to each X_i is better understood by its principal component analysis. Calculating μ_i, and subtracting from each $x \in X_i$ is the centering step. Letting $\bar{X}_i = \{x - \mu_i \mid x \in X_i\}$, then $\Sigma_i = V S^2 V^T$ where $[U, S, V] = \operatorname{svd}(\bar{X}_i)$. Now, the top principle directions $v_1, v_2, \ldots$ describe the directions of most variance.

8.4.1 Expectation-Maximization

The standard way to fit a mixture of Gaussians actually uses a soft clustering. Each point $x \in X$ is given a weight $w_i(x) = f_i(x)/\sum_i f_i(x)$ for its assignment to each cluster. Then the mean and covariance matrix is estimated using weighted averages. The soft clustering variant of the procedure extending Lloyd's algorithm, described above, is outlined in Algorithm 8.4.1.

Algorithm 8.4.1 EM Algorithm for Mixture of Gaussians

Choose k points $S \subset X$
for all $x \in X$: set $w_i(x) = 1$ for $\phi_S(x) = s_i$, and $w_i(x) = 0$ otherwise
repeat
  for $i \in [1 \ldots k]$ **do**
    Calculate $W_i = \sum_{x \in X} w_i(x)$ *the total weight for cluster i*
    Set $\mu_i = \frac{1}{W_i} \sum_{x \in X} w_i(x)x$ *the weighted average*
    Set $\Sigma_i = \frac{1}{W_i} \sum_{x \in X} w_i(x - \mu_i)(x - \mu_i)^T$ *the weighted covariance*
  for $x \in X$ **do**
    for all $i \in [1 \ldots k]$: set $w_i(x) = f_i(x)/\sum_i f_i(x)$ *partial assignments using $f_i = f_{\mu_i, \Sigma_i}$*
until $(\sum_{x \in X} \sum_{i=1}^{k} -\log(w_i(x) \cdot f_i(x))$ has small change)

This procedure is the classic example of a framework called *expectation-maximization*. This is an alternating optimization procedure, which alternates between maximizing the probability of some model (the partial assignment step) and calculating the most likely model using expectation (the average, covariance estimating step).

Note that expectation-maximization, and even more generally, *alternating optimization*, is a much more general framework. It is particularly useful in situations (like this one) where the true optimization criteria are messy and complex, often non-convex; but it can be broken into two or more steps where each step can be solved with a (near) closed form. Or if there is no closed form, but each part is individually convex, then gradient descent can be invoked to solve each one. Sometimes, as with stochastic gradient descent, it is more efficient and robust to not full solve either step before alternating back to the other one.

8.5 Hierarchical Clustering

Clustering can provide more than a partition, it can also provide a hierarchy of the clusters. That is, at the root of a hierarchy all data points are in the same cluster. And at the leaf nodes, each data point is in its own cluster. Then, the intermediate layers can provide various levels of refinement, depending on how many or how tightly related clusters are desired.

Example: Hierarchy of Clusters

This small example shows the hierarchy of clusters found on $n = 5$ data points. Associated with these clusters and this progression is a tree, showing the local order in which points are merged into clusters.

There are two basic approaches toward this: top-down and bottom-up. The top-down approach starts with all data points in the same cluster, and iteratively partitions the data into 2 (or sometimes more) parts. The main task then, is given a data set, how to split it into two components as best possible. The canonical top-down clustering approach is spectral clustering. This is primarily based on graph ideas, but can be generalized to work with any choice of similarity function **s** (or implicitly a linked distance function **d**), so the base object is an affinity matrix instead of the adjacency matrix of a graph. An in depth discussion is deferred to Section 10.3.

The bottom-up approach often called *hierarchical agglomerative clustering* takes the opposite tact, outlined in Algorithm 8.5.1. Starting with a cluster for each data point, its main task is to determine which pair of clusters to join into a single cluster. As the standard choice is, *the closest two clusters*, then the key element is to study distances between clusters.

Algorithm 8.5.1 Hierarchical Agglomerative Clustering

Each $x_i \in X$ is a separate cluster S_i.
while there exist at least 2 clusters **do**
    Find the *closest* two clusters S_i, S_j
    Merge S_i and S_j into a single cluster

There are many ways to define such a distance $\mathbf{d}_C$ between clusters. Each is defined with respect to another general distance function **d** which can be applied to individual data points, and in some cases also representative objects. The most common are

- *Single Link*: $\mathbf{d}_C(S_i, S_j) = \mathrm{argmin}_{s_i \in S_i, s_j \in S_j} \mathbf{d}(s_i, s_j)$.
  This takes the distance between the closest two points among all points in the clusters. This allows the clusters to adjust to the shape and density of the data, but can lead to oddly shaped clusters as well.

- *Mean Link*: $\mathbf{d}_C(S_i, S_j) = \sum_{s_i \in S_i} \sum_{s_j \in S_j} \mathbf{d}(s_i, s_j)$.
  This takes the average distance between pairs of points in a cluster (written unnormalized). It behaves similar to the k-means objective.
- *Complete Link*: $\mathbf{d}_C(S_i, S_j) = \mathrm{argmax}_{s_i \in S_i, s_j \in S_j} \mathbf{d}(s_i, s_j)$.
  This enforces "round" clusters. It can only join two clusters if all points are similar.
- *Center Link*: $\mathbf{d}_C(S_i, S_j) = \mathbf{d}(c_i, c_j)$ where c_i and c_j are central points representing S_i and S_j.
  The choice of how to define the central point is varied. For instance, it could be the average, median, or representative median as in k-means, k-medians, or k-mediods objectives. Or it could be any other easy-to-determine or robust way to represent the clusters.

These variants of clusters provide additional structural information beyond center-based clusters, in that they also provide which clusters are close to each other, or allow a user to refine to different levels of granularity. However, this comes at the cost of efficiency. For instance, on n data points, these algorithms naively take time proportional to n^3. Some improvements or approximations can reduce this to closer to time proportional to n^2, but may add instability to the results.

Ethical Questions with Forced Hierarchies

Hierarchies can create powerful visuals to help make sense of complex data sets that are compared using a complex distance function over abstract data representations. For instance, phylogentic trees are the dominate way to attempt to present evolutionary connections between species. However, these connections can provide false insights by those over-eager to make new scientific or business connections.

A common way to model and predict the careers of professional athletes is to use their early years to cluster and group athletes with others who had careers before them. Then an up-and-coming athlete can be predicted to have careers similar to those in the same cluster. However, if a new mold of athlete—e.g., basketball players who shoot many 3-pointers, and are very tall—has many junior players, but not senior players to use to predict their career. If they are grouped with other tall players, a coach or general manager may treat them as rebounders minimizing their effectiveness. Or if they are treated as shooting guards, they may be given more value. How might the use of the clustering and its hierarchy be used to mitigate this potential downsides of these possible models?

Such effects are amplified in less scrutinize hiring and management situations. Consider a type of job applicant who has traditionally not be hired in a role, how will their clustering among past applicants be harmful or helpful for their chance at being hired, or placed in a management role? And as a data modeler, what is your role in aiding in these decisions?

8.6 Density-Based Clustering and Outliers

Density-based clustering can be seen as an efficient extension of hierarchical agglom-erative clustering. The basic premise is that data points which are sufficiently closed should be assigned into the same cluster; however, at the behest of efficiency, it omits the building of an explicit hierarchy. It also adds a density requirement.

The most widely used variant is called *DBScan*. The algorithm has two parameters r (a radius, the literature usually uses ε) and τ (a density parameter, the literature usually uses minPts). Now given a data set X and a distance $\mathbf{d} : X \times X \to \mathbb{R}$, a point $x \in X$ is designated a *core point* if at least τ points from X are within a distance r from x. That is, x is a core point if $|\{x' \in X \mid \mathbf{d}(x, x') \le r\}| \ge \tau$. These points are all in clusters.

Next we consider a *graph* (See Chapter 10) $G = (X, E)$, where all data points are edges, and the edge set E has the following property. An edge $e_{i,j} \in E$ between points $x_i, x_j \in X$ if and only if (1) $\mathbf{d}(x_i, x_j) \le \tau$ and (2) at least one of x_i or x_j is a core point.

Now each connected component of G is a cluster. Recall a *connected component* involving a point $x \in X$ includes all points $x' \in X$ that are *reachable* from x. And a point x' is reachable if there is a sequence $x_0, x_1, \ldots, x_k$ where $x_0 = x$, $x_k = x'$ and each consecutive pair x_j, x_{j+1} has an edge $e_{j,j+1} \in E$.

An example data set X of $n = 15$ points. They are clustered using DBScan using a density parameters of $\tau = 3$, and radius r as illustrated in the balls. For instance, x_4 is a core point, since there 3 points in the highlighted ball of radius r around it. All core points are marked, and the induced graph is shown.

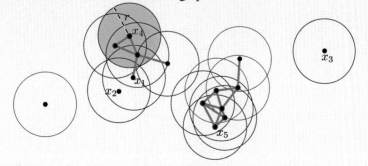

Note that even though x_1 and x_2 are within a distance r, they are not connected since neither is a core point. The points with no edges, like x_2 and x_3, are not in any clusters. And there are two distinct clusters from the two connected components of the graph: point x_5 is in a different cluster than x_1 and x_4, which are in the same cluster.

So all core points (the points which are in dense enough regions) are in clusters, as well as any points close enough to those points. And points are connected in the same clusters if they are connected by a path through dense regions. This allows the clusters to adapt to the shape of the data.

However, there is no automated way to choose the parameters r and τ. This choice will greatly affect the shape, size, and meaning of the clusters. Also, this model also assumes that the notion of density is consistent throughout the dataset.

A major advantage of DBScan is that it can usually be computed efficiently. For each $x \in X$ we need to determine if it is a core points, and find neighbors, but these can usually quickly be accomplished with fast nearest neighbor search algorithms (e.g., LSH in Section 4.6). Then usually the connected components (the clusters) can be recovered without building the entire graph. However, in the worst case, the algorithm will have runtime quadratic in the n, the size of the data set.

8.6.1 Outliers

In general, *outliers are the cause of, and solution to, all of data mining's problems.* That is, for any problem in data mining, one can try to blame it on outliers. And one can claim that if they were to remove the outliers the problem would go away. That is, when data is "nice" (there are no outliers) then the structure we are trying to find should be obvious. There are many solutions, and each has advantages and pitfalls.

Density-Based Approaches

Conceptually regular points have dense neighborhoods, and outlier points have non-dense neighborhoods. One can run a density-based clustering algorithm, and all points which are not in clusters are the outliers. Variants of this idea define density thresholds in other ways, for instance, for each point $x \in X$, counting the number of points within a radius r, evaluating the kernel density estimate $\text{KDE}_X(x)$, or measuring the distance to the kth nearest neighbor. Each has a related version of distance and density threshold.

Reverse Nearest Neighbors

Given a point set X for each point $x \in X$ we can calculate its nearest neighbor p (or more robustly its kth nearest neighbor). Then for p we can calculate its nearest neighbors q in X (or its kth nearest neighbor). Now if $\mathbf{d}(x, p) \approx \mathbf{d}(p, q)$ then x is probably not an outlier.

This approach adapts to the density in different regions of the point set. However, it relies crucially on the choice of distance $\mathbf{d}$. For generic distances this may be quite slow, but for others we can employ LSH-type procedures to make this more efficient.

Furthermore, density (or relative density) may not tell you how far a point is from a model, for instance in the case of regression.

Model and Prune

A more general mechanism to identify outliers requires building a model M as a function of the entire data set X, and then a distance $\mathbf{d}(M(x), x)$ from each data point $x \in X$ to the model. This distance could be $\mathbf{d}(M(x), x) = \phi_S(x)$ from a center-based clustering algorithm, or the residual from a regression model. The algorithm is then:

1. Build a model M of dataset X.
2. For each point $x \in X$ find the residual $r_x = \mathbf{d}(M(x), x)$, where $M(x)$ is the best representation of x in the model M.
3. If r_x is too large, then x is an outlier.

There is no one right way to execute this method. Which model M and distance $\mathbf{d}(M(x), x)$ should we use? What threshold on r_x should we use? For instance, if we remove 3% of the data points, we could iteratively rebuild the model on the remaining points (the "inliers") and then again remove 3% of the points, and rebuild the model until no points are left! Moreover, if we can successfully build a model M we trust, then why do we even need to remove outliers? On the other hand, sometimes running this (model, prune outliers iteration) a small number of steps $(1 - 3$ times) can drastically help refine a model to the core data points.

Ethical Questions with Outlier Removal

Clustering and outlier removal can be performed over population databases (e.g., of a company's customers or users). This may provide a better understanding of those people whom are assigned centrally to a cluster, but those who have a poor fit to a cluster or are marked an outlier, this may not be as favorable. Is it acceptable to not provide any recommendations for some people? Is that better than less accurate ones?

8.7 Mean Shift Clustering

Now for something completely different. Clustering is a very very broad field with no settled upon approach. To demonstrate this, we will quickly review an algorithm called *mean shift clustering*. This algorithm shifts each data point individually to its weighted center of mass. It terminates when all points converge to isolated sets.

First begin with a bivariate kernel function $K : X \times X \to \mathbb{R}$ such as the (unnormalized) Gaussian kernel

$$K(x, p) = \exp(-\|x - p\|^2/\sigma^2)$$

for some given bandwidth parameter σ. The weighted center of mass around each point $p \in X$ is then defined as

$$\mu(p) = \frac{\sum_{x \in X} K(x, p)x}{\sum_{x \in X} K(x, p)}.$$

The algorithm just shifts each point to its center of mass: $p \leftarrow \mu(p)$.

Algorithm 8.7.1 Mean Shift

---

repeat
    for all $p \in X$: calculate $\mu(p) = \frac{\sum_{x \in X} K(x,p)x}{\sum_{x \in X} K(x,p)}$.
    for all $p \in X$: set $p \leftarrow \mu(p)$.
until (the average change $\|p - \mu(p)\|$ is small)

This algorithm does not require a parameter k. However, it has other parameters, most notably the choice of kernel K and its bandwidth σ. With the Gaussian kernel (since it has infinite support, $K(x, p) > 0$ for all x, p), it will only stop when all x are at the same point. Thus, the termination condition is also important. Alternatively, a different kernel with bounded support may terminate automatically (without a specific condition); for this reason (and for speed) truncated Gaussians (i.e., $K_\tau^{\text{trunc}}(x, p) = \{0$ if $\|x - p\| > \tau$; otherwise $K(x, p)\}$) are often used.

This algorithm not only clusters the data, but also is a key technique for *de-noising* data. This is a process that not just removes noise (as often thought of as outliers), but attempts to adjusts points to where they should have been before being perturbed by noise—similar to mapping a point to its cluster center.

Exercises

We will use four datasets, here:
https://mathfordata.github.io/data/P.csv
https://mathfordata.github.io/data/Q.csv
https://mathfordata.github.io/data/C-HAC.csv
https://mathfordata.github.io/data/Ck.csv

8.1 Download data sets P and Q. Both have 120 data points, each in 6 dimensions, can be thought of as data matrices in $\mathbb{R}^{120 \times 6}$. For each, run some algorithm to construct the k-means clustering of them. Diagnose how many clusters you think each data set should have by finding the solution for k equal to 1, 2, 3, ..., 10.

8.2 Draw the Voronoi diagram of the following set of points.

8.3 Consider the drawn data set where the black circles are data points and red stars are the centers. What should you do, if running Lloyd's algorithm for k-means clustering ($k = 2$), and you reach this scenario, where the algorithm terminates?

Can this state ever be reached if initiated with Gonzalez's algorithm?

8.4 Construct a data set X with 5 points in $\mathbb{R}^2$ and a set S of $k = 3$ sites so that Lloyds algorithm will have converged, but there is another set S', of size $k = 3$, so that $\text{cost}(X, S') < \text{cost}(X, S)$. Explain why S' is better than S, but that Lloyds algorithm will not move from S.

8.5 Consider this set of 3 sites: $S = \{s_1 = (0, 0), s_2 = (3, 4), s_3 = (-3, 2)\} \subset \mathbb{R}^2$. We will consider the following 5 data points $X = \{x_1 = (1, 3), x_2 = (-2, 1), x_3 = (10, 6), x_4 = (6, -3), x_5 = (-1, 1)\}$.

For each of the points compute the closest site (under Euclidean distance):

1. $\phi_S(x_1) =$
2. $\phi_S(x_2) =$
3. $\phi_S(x_3) =$
4. $\phi_S(x_4) =$
5. $\phi_S(x_5) =$

Now consider that we have 3 Gaussian distributions defined with each site s_j as a center μ_j. The corresponding standard deviations are $\sigma_1 = 2.0$, $\sigma_2 = 4.0$ and $\sigma_3 = 5.0$, and we assume they are univariate so the covariance matrices are
$$\Sigma_j = \begin{bmatrix} \sigma_j & 0 \\ 0 & \sigma_j \end{bmatrix}.$$

6. Write out the probability density function (its likelihood $f_j(x)$) for each of the Gaussians.

Now we want to assign each x_i to each site in a soft assignment. For each site s_j define the weight of a point as $w_j(x) = f_j(x)/(\sum_{j=1}^3 f_j(x))$. Finally, for each of the points calculate the 3 weights for the 3 sites

7. $w_1(x_1), w_2(x_1), w_3(x_1) =$
8. $w_1(x_2), w_2(x_2), w_3(x_2) =$
9. $w_1(x_3), w_2(x_3), w_3(x_3) =$
10. $w_1(x_4), w_2(x_4), w_3(x_4) =$
11. $w_1(x_5), w_2(x_5), w_3(x_5) =$

8.6 Write down the cost function associated with the Mixture of Gaussians problem and Algorithm 8.4.1 so that at each phase of the algorithm, this cost cannot increase.

8.7 There are many variants of hierarchical clustering; here we explore 3 using dataset C-HAC which we treat as $n = 19$ points in $\mathbb{R}^2$. The key difference in these variants is how to measure the distance $d(S_1, S_2)$ between two clusters S_1 and S_2. We will consider *Single Link*, *Complete Link*, and *Mean Link* with the base distance **d** as Euclidian distance.

1. Run all hierarchical clustering variants on the provided data until there are $k = 4$ clusters, and report the results as sets. Plot the clusters using a different color and marker to denote each clustering assignment.
2. Also run DBScan on this data with radius $r = 5.0$ and density $\tau = 3$.
3. Provide a subjective answer to which variant you believe did the best job, and which was the easiest to compute (think if the data was much larger)? And explain why.

8.8 We will use data set Ck, which we treat as $n = 1040$ points in $\mathbb{R}^2$, to explore assignment-based clustering.

1. Run Gonzalez Algorithm (Algorithm 8.2.1) and k-Means++ (Algorithm 8.3.2) on the data set for $k = 3$. To avoid too much variation in the results, choose the first point in the data file as the first site s_1.

   Report the centers and the subsets (as pictures) for Gonzalez. For both also report:

   - the 3-center cost $\max_{x \in X} \mathbf{d}(x, \phi_S(x))$ and
   - the 3-means cost $\sqrt{\frac{1}{|X|} \sum_{x \in X} (\mathbf{d}(x, \phi_S(x)))^2}$

   (Note this has been normalized so easy to compare to 3-center cost)

2. For k-Means++, the algorithm is randomized, so you will need to report the variation in this algorithm. Run it several trials (at least 20) and plot the *cumulative density function* of the 3-means cost. Also report what fraction of the time the subsets are the same as the result from Gonzalez.

3. Recall that Lloyd's algorithm for k-means clustering starts with a set of k sites S and runs as described in Algorithm 8.3.1.

   a. Run Lloyds Algorithm with S initially with the first 3 points in the data set. Report the final subset and the 3-means cost.

   b. Run Lloyds Algorithm with S initially as the output of the Gonzalez algorithm found above. Report the final subset and the 3-means cost.

   c. Run Lloyds Algorithm with S initially as the output of each run of k-Means++ above. Plot a *cumulative density function* of the 3-means cost. Also report the fraction of the trials that the subsets are the same as the input (where the input is the result of the k-Means++ algorithm).

8.9 k-Means++ (Algorithm 8.3.2) is an inherently iterative algorithm. That is, it recomputes the sampling probabilities after the selection of each site. It would be easier and more efficient to implement if this probability was fixed after the selection of the first site. Identify and describe a scenario for $k = 3$, where this non-iterative variant (the sampling probability is formulated once after s_1 is chosen, but not after s_2 is chosen) will likely give a poor initialization for the k-means formulation. In particular, design a setting where for the non-iterative version as an initialization for Lloyd's algorithm is likely to get stuck in a non-optimal local minimum, but the standard iterative version will likely initialize Lloyd's algorithm so it finds the global minimum.

8.10 Run mean link clustering on data set Ck for 4 iterations. Do so 3 different times for bandwidth parameter σ taking values $\{1, 5, 50\}$.

1. Plot the output at the end of each iteration for each value of σ.
2. Which value of σ does the best job of grouping similar data (finding clusters)? Explain your opinion.

Chapter 9
Classification

Abstract This chapter returns to prediction. Unlike linear regression where we were predicting a numeric value, in this case we are predicting a class: winner or loser, yes or no, positive or negative. Ideas from linear regression can be applied here, but for once least squares is not the dominant error paradigm as it can lead to some structurally inconsistent results. Instead we will organize our discussion around the beautiful geometry of linear classification and let the error model be chosen as more abstract loss functions. This method can be generalized through support vector machines to non-linear classifiers. We follow that with a discussion on learnability, using VC dimension to quantify the sample complexity of classifiers. Again this topic is wide and varied, and the back end of the chapter surveys other common and diverse method in KNN classifiers, decision trees, and neural networks.

This is perhaps *the* central problem in data analysis. For instance, you may want to predict:

- will a sports team win a game?
- will a politician be elected?
- will someone like a movie?
- will someone click on an ad?
- will I get a job? (If you can build a good classifier, then probably yes!)

Each of these is typically solved by building a general purpose classifier (about sports or movies etc), then applying it to the problem in question.

9.1 Linear Classifiers

Our input here is a point set $X \subset \mathbb{R}^d$, where each element $x_i \in X$ also has an associated label y_i. And $y_i \in \{-1, +1\}$.

Like in regression, our goal is prediction and generalization. We assume each $(x_i, y_i) \sim \mu$; that is, each data point pair, is drawn iid from some fixed but unknown

© Springer Nature Switzerland AG 2021
J. M. Phillips, *Mathematical Foundations for Data Analysis*,
Springer Series in the Data Sciences,
https://doi.org/10.1007/978-3-030-62341-8_9

distribution. Then our goal is a function $g : \mathbb{R}^d \to \mathbb{R}$, so that if $y_i = +1$, then $g(x_i) \geq 0$ and if $y_i = -1$, then $g(x_i) \leq 0$.

We will restrict that g is linear. For a data point $x \in \mathbb{R}^d$, written $x = (x^{(1)}, x^{(2)}, \ldots, x^{(d)})$ we enforce that

$$g(x) = \alpha_0 + x^{(1)}\alpha_1 + x^{(2)}\alpha_2 + \ldots + x^{(d)}\alpha_d = \alpha_0 + \sum_{j=1}^{d} x^{(j)}\alpha_j,$$

for some set of scalar parameters $\alpha = (\alpha_0, \alpha_1, \alpha_2, \ldots, \alpha_d)$. Typically, different notation is used: we set $b = \alpha_0$ and $w = (w_1, w_2, \ldots, w_d) = (\alpha_1, \alpha_2, \ldots, \alpha_d) \in \mathbb{R}^d$. Then we write

$$g(x) = b + x^{(1)}w_1 + x^{(2)}w_2 + \ldots + x^{(d)}w_d = \langle w, x \rangle + b.$$

We can now interpret (w, b) as defining a halfspace in $\mathbb{R}^d$. Here, w is the normal of that halfspace boundary (the single direction orthogonal to it) and b is the distance from the origin $\mathbf{0} = (0, 0, \ldots, 0)$ to the halfspace boundary in the direction $w/\|w\|$. Because w is normal to the halfspace boundary, b is also distance from the closest point on the halfspace boundary to the origin (in any direction).

We typically ultimately use w as a unit vector, but it is not important since this can be adjusted by changing b. Let w, b be the desired halfspace with $\|w\| = 1$. Now assume we have another w', b' with $\|w'\| = \beta \neq 1$ and $w = w'/\|w'\|$, so they point in the same direction, and b' set so that they define the same halfspace. This implies $b' = b \cdot \beta$, and $b = b'/\beta$. So the normalization of w can simply be done post-hoc without changing any structure.

Recall, our goal is $g(x) \geq 0$ if $y = +1$ and $g(x) \leq 0$ if $y = -1$. So if x lies directly on the halfspace then $g(x) = 0$.

Example: Linear Separator in $\mathbb{R}^2$

Here, we show a set $X \in \mathbb{R}^2$ of 13 points with 6 labeled $+$ and 7 labeled $-$. A linear classifier perfectly separates them. It is defined with a normal direction w (pointing toward the positive points) and an offset b.

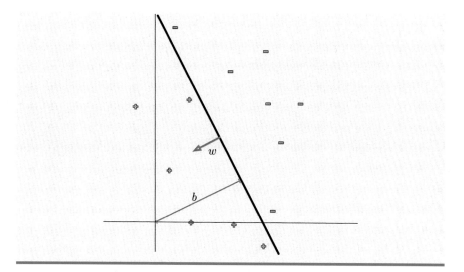

Using techniques we have already learned, we can immediately apply two approaches toward this problem.

Linear Classification via Linear Regression

For each data points $(x_i, y_i) \in \mathbb{R}^d \times \mathbb{R}$, we can immediately represent x_i as the value of d explanatory variables, and y_i as the single dependent variable. Then we can set up a $n \times (d+1)$ matrix $\tilde{X}$, where the ith row is $(1, x_i)$; that is the first coordinate is 1, and the next d coordinates come from the vector x_i. Then with a $y \in \mathbb{R}^n$ vector, we can solve for

$$\alpha = (\tilde{X}^T \tilde{X})^{-1} \tilde{X}^T y$$

we have a set of $d+1$ coefficients $\alpha = (\alpha_0, \alpha_1, \ldots, \alpha_d)$ describing a linear function $g_\alpha : \mathbb{R}^d \to \mathbb{R}$ defined

$$g_\alpha(x) = \langle \alpha, (1, x) \rangle.$$

Hence $b = \alpha_0$ and $w = (\alpha_1, \alpha_2, \ldots, \alpha_d)$. For x such that $g_\alpha(x) > 0$, we predict $y = +1$ and for $g_\alpha(x) < 0$, we predict $y = -1$.

However, this approach is optimizing the wrong problem. It is minimizing how close our predictions $g_\alpha(x)$ is to -1 or $+1$, by minimizing the sum of squared errors. But our goal is to minimize the number of mispredicted values, not the numerical value.

Example: Linear Regression for Classification

We show 6 positive points and 7 negative points in $\mathbb{R}^d$ mapped to $\mathbb{R}^{d+1}$. All of the d-coordinates are mapped to the x-axis. The last coordinate is mapped to the y-axis and is either +1 (a positive point) or −1 (a negative point). Then the best linear regression fit is shown (in purple), and the points where it has y-coordinate 0 defines the boundary of the halfspace (the vertical black line). Note, despite there being a linear separator, this method misclassifies two points because it is optimizing the wrong measure.

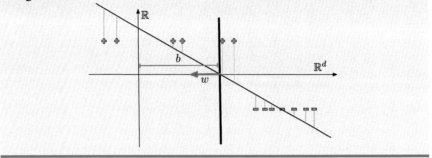

Linear Classification via Gradient Descent

Since, the linear regression SSE cost function is not the correct one, what is the correct one? We might define a cost function Δ

$$\Delta(g_\alpha, (X, y)) = \sum_{i=1}^{n} (1 - \mathbb{1}(\text{sign}(y_i) = \text{sign}(g(x_i))))$$

which uses the *identity function* $\mathbb{1}$ (defined $\mathbb{1}(\text{TRUE}) = 1$ and $\mathbb{1}(\text{FALSE}) = 0$) to represent the *number* of misclassified points. This is what we would like to minimize.

Unfortunately, this function is discrete, so it does not have a useful (or well-defined) gradient. And, it is also not convex. Thus, encoding g_α as a $(d + 1)$-dimensional parameter vector $(b, w) = \alpha$ and running gradient descent is not feasible.

However, most classification algorithms run some variant of gradient descent. To do so we will use a different cost function as a proxy for Δ, called a *loss function*. We explain this next.

9.1.1 Loss Functions

To use gradient descent for classifier learning, we will use a proxy for Δ called a *loss function* $\mathcal{L}$. These are sometimes implied to be convex, and their goal is to approximate Δ. And in most cases, they are *decomposable*, so we can write

$$f(\alpha) = \mathcal{L}(g_\alpha, (X, y)) = \sum_{i=1}^{n} \ell(g_\alpha, (x_i, y_i))$$

$$= \sum_{i=1}^{n} \ell_\alpha(z_i) \text{ where } z_i = y_i g_\alpha(x_i)$$

$$= \sum_{i=1}^{n} f_i(\alpha) \text{ where } f_i(\alpha) = \ell(z_i = y_i g_\alpha(x_i)).$$

Note that the clever expression $z_i = y_i g_\alpha(x_i)$ handles when the function $g_\alpha(x_i)$ correctly predicts the positive or negative example in the same way. If $y_i = +1$, and correctly $g_\alpha(x_i) > 0$, then $z_i > 0$. On the other hand, if $y_i = -1$, and correctly $g_\alpha(x_i) < 0$, then also $z_i > 0$. For instance, the desired cost function, Δ is written as

$$\Delta(z) = \begin{cases} 0 & \text{if } z \geq 0 \\ 1 & \text{if } z < 0. \end{cases}$$

Most *loss functions* $\ell(z)$ which are convex proxies for Δ mainly focus on how to deal with the case $z_i < 0$ (or $z_i < 1$). The most common ones include

- hinge loss: $\ell = \max(0, 1 - z)$
- smoothed hinge loss: $\ell = \begin{cases} 0 & \text{if } z \geq 1 \\ (1 - z)^2/2 & \text{if } 0 < z < 1 \\ \frac{1}{2} - z & \text{if } z \leq 0 \end{cases}$
- squared hinge loss: $\ell = \max(0, 1 - z)^2$
- logistic loss: $\ell = \ln(1 + \exp(-z))$

Geometry of Loss Functions

Loss functions are designed as proxies for the $\{0, 1\}$ cost function Δ, shown in blue. The most common loss functions, shown in red, are convex, mostly differentiable, and in the range of $z \in [-1, 1]$ fairly closely match Δ. Moreover, they do not penalize well-classified data points, those with large z values.

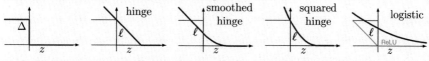

The hinge loss is the closest convex function to Δ; in fact it strictly upper bounds Δ. However, it is non-differentiable at the "hinge-point," (at $z = 1$) so it takes some care to use it in gradient descent. The smoothed hinge loss and squared hinge loss are approximations to this which are differentiable everywhere. The squared hinge loss is quite sensitive to outliers (similar to SSE). The smoothed hinge loss (related to the Huber loss) is a nice combination of these two loss functions.

The logistic loss can be seen as a continuous approximation to the *ReLU (rectified linear unit)* loss function, which is the hinge loss shifted to have the hinge point at $z = 0$. The logistic loss also has easy-to-take derivatives (does not require case analysis) and is smooth everywhere. Minimizing this loss for classification is called *logistic regression*.

The writing of a cost function

$$f(\alpha) = \sum_{i=1}^{n} f_i(\alpha) = \sum_{i=1}^{n} \ell(z_i)$$

where $z_i = y_i g_\alpha(x_i)$ stresses that this is usually used within a gradient descent framework. So the differentiability is of critical importance. And by the chain rule, since each y_i and x_i are constants, then if ∇g_α and $\frac{\mathrm{d}}{\mathrm{d}z}\ell$ are defined, then we can invoke the powerful variants of gradient descent for decomposable functions: batch and stochastic gradient descent.

9.1.2 Cross-Validation and Regularization

Ultimately, in running gradient descent for classification, one typically defines the overall cost function f also using a *regularization* term $r(\alpha)$. For instance, $r(\alpha) = \|\alpha\|^2$ is easy to use (has nice derivatives) and $r(\alpha) = \|\alpha\|_1$ (the L_1 norm) induces sparsity, as discussed in the context of regression. In general, the regularizer typically penalizes larger values of α, resulting in some bias, but less overfitting of the data.

The regularizer $r(\alpha)$ is added to a decomposable loss function $\mathcal{L}(g_\alpha, (X, y)) = \sum_{i=1}^{n} \ell(g_\alpha, (x_i, y_i))$ as

$$f(\alpha) = \mathcal{L}(g_\alpha, (X, y)) + \eta r(\alpha),$$

where $\eta \in \mathbb{R}$ is a *regularization parameter* that controls how drastically to regularize the solution.

Note that this function $f(\alpha)$ is still decomposable, so one can use batch, incremental, or most commonly stochastic gradient descent.

Cross-Validation

Backing up a bit, the *true* goal is not minimizing f or $\mathcal{L}$, but predicting the class for new data points. For this, we again assume all data is drawn iid from some fixed but unknown distribution. To evaluate how well our results generalize, we can use cross-validation (holding out some data from the training, and calculating the expected error of Δ on these held out "testing" data points).

We can also choose the regularization parameter η by choosing the one that results in the best generalization on the test data after training using each on some training data.

Ethical Questions with Data Imbalance and Classification

Generalization goals are typically phrased so a data point has the smallest expected loss. However, this can lead to issues when the data is imbalanced.

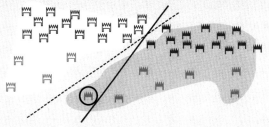

Consider a data set consisting for two types of students applying to college. Let us use hair color as a proxy, and assume each applicant has blue hair or green hair. If each type has different criteria that would make them successful, a classifier can still use the hair color attribute in conjunction with other traits to make useful prediction across both types of applicant. However, if there are significantly more applicants with blue hair, then the classifier will have a smaller expected loss (and even better generalization error) if it increases weights for the traits that correspond with blue-haired applicants succeeding. This will make it more accurate on the blue-haired applicants, but perhaps less accurate on the green-haired applicants, yet more accurate overall on average across all applicants. This provides worse prediction and more uncertainty for the green-haired students due only to the color of their hair; it may even lower their overall acceptance rate.

What can be done as a data analyst to find such discrepancies? And should something be changed in the modeling if these discrepancies are found?

9.2 Perceptron Algorithm

Of the above algorithms, generic linear regression is not solving the correct problem, and gradient descent methods do not really use any structure of the problem. In fact, as we will see we can replace the linear function $g_\alpha(x) = \langle \alpha, (1, x) \rangle$ with any function g (even non-linear ones) as long as we can take the gradient.

Now, we will introduce the *perceptron algorithm* which explicitly uses the linear structure of the problem. (Technically, it only uses the fact that there is an inner product—which we will exploit in generalizations.)

Simplifications

For simplicity, we will make several assumptions about the data. First, we will assume that the best linear classifier (w^*, b^*) defines a halfspace whose boundary passes through the origin. This implies $b^* = 0$, and we can ignore it. This is basically equivalent to for data point $(x'_i, y_i) \in \mathbb{R}^{d'} \times \mathbb{R}$ using $x_i = (1, x'_i) \in \mathbb{R}^d$ where $d' + 1 = d$.

Second, we assume that for all data points (x_i, y_i) that $\|x_i\| \le 1$ (e.g., all data points live in a unit ball). This can be done by choosing the point $x_{\max} \in X$ with largest norm $\|x_{\max}\|$, and dividing all data points by $\|x_{\max}\|$ so that point has norm 1, and all other points have smaller norms.

Finally, we assume that there exists a perfect linear classifier. One that classifies each data point to the correct class. There are variants of the perceptron algorithm to deal with the cases without perfect classifiers, but are beyond the scope of this text.

The Algorithm

Now to run the algorithm, we start with some normal direction w (initialized as any positive point), and then add misclassified points to w one at a time.

Algorithm 9.2.1 Perceptron(X, y)

Initialize $w = y_i x_i$ for any $(x_i, y_i) \in (X, y)$
repeat
     For any (x_i, y_i) such that $y_i \langle x_i, w \rangle < 0$ (is mis-classified) : update $w \leftarrow w + y_i x_i$
until (no mis-classified points   **or** T steps)
return $w \leftarrow w / \|w\|$

Basically, if we find a misclassified point (x_i, y_i) and $y_i = +1$, then we set $w = w + x_i$. This "pushes" w more in the direction of x_i, but also makes it longer. Having w more in the direction of x_i, tends to make it have dot product (with a normalized version of w) closer to 1.

Similar, if we find a misclassified point (x_i, y_i) with $y_i = -1$, the we set $w = w - x_i$; this pushes the negative of w more toward x_i, and thus w more away from x_i, and thus its dot product more likely to be negative.

≫ Detailed Perceptron Algorithm

To implement the Perceptron algorithm, inside the inner loop we need to find some misclassified point (x_i, y_i), if one exists. This can require another implicit loop. A common approach would be to, for some ordering of points $(x_1, y_1), (x_2, y_2), \ldots (x_n, y_n)$ keep an iterator index i that is maintained outside the **repeat-until** loop. It is modularly incremented every step: it loops around to $i = 1$ after $i = n$. That is, the algorithm keeps cycling through the data set, and updating w for each misclassified point if observes.

Algorithm 9.2.2 Detailed Perceptron(X, y)

Initialize $w = y_i x_i$ for any $(x_i, y_i) \in (X, y)$; Set $i = 1; t = 0;$ LAST-UPDATE $= 1$
repeat
    if $(y_i \langle x_i, w \rangle < 0)$ **then**
        $w \leftarrow w + y_i x_i$
        $t = t + 1;$ LAST-UPDATE $= i$
    $i = i + 1 \mod n$
until $(t = T$ or   LAST-UPDATE $= i)$
return $w \leftarrow w / \|w\|$

The Margin

To understand why the perceptron works, we need to introduce the concept of a margin. Given a classifier (w, b), the *margin* is

$$\gamma = \min_{(x_i, y_i) \in X} y_i(\langle w, x_i \rangle + b).$$

It is the minimum distance of any data point x_i to the boundary of the halfspace for a perfect classifier. If the classifier misclassifies a point, under this definition, the margin might be negative. For this framework, the optimal classifier (or the maximum margin classifier) (w^*, b^*) is the one that maximizes the margin

$$(w^*, b^*) = \arg\max_{(w,b)} \min_{(x_i, y_i) \in X} y_i(\langle w, x_i \rangle + b)$$

$$\gamma^* = \min_{(x_i, y_i) \in X} y_i(\langle w^*, x_i \rangle + b^*).$$

A *max-margin classifier* is one that not just classifies all points correctly but does so with the most "margin for error." That is, if we perturbed any data point or the classifier itself, this is the classifier which can account for the most perturbation and still predict all points correctly. It also tends to *generalize* (in the cross-validation sense) to new data better than other perfect classifiers.

Example: Margin of Linear Classifier

For a set X of 13 points in $\mathbb{R}^2$, and a linear classifier defined with (w, b). We illustrate the margin as a pink strip. The margin $\gamma = \min_{(x_i, y_i)} y_i(\langle w, x_i \rangle + b)$ is drawn with an $\leftrightarrow$ for each support point.

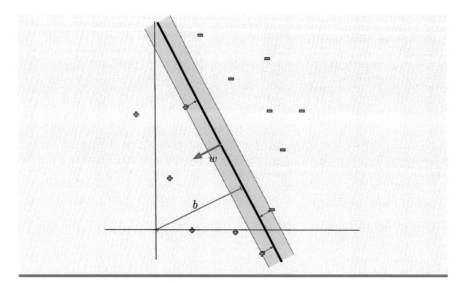

The maximum margin classifier (w^*, b^*) for $X \subset \mathbb{R}^d$ can always be defined uniquely by $d + 1$ points (at least one negative, and at least one positive). These points $S \subset X$ are such that for all $(x_i, y_i) \in S$

$$\gamma^* = y_i (\langle w^*, x_i \rangle + b).$$

These are known as the *support points*, since they "support" the margin strip around the classifier boundary.

Geometry of Why Perceptron Works

Here, we will show that after at most $T = (1/\gamma^*)^2$ steps (where γ^* is the margin of the maximum margin classifier), then there can be no more misclassified points.

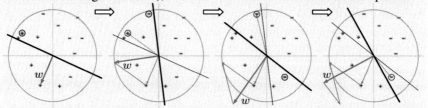

To show this we will bound two terms as a function of t, the number of mistakes found. The terms are $\langle w, w^* \rangle$ and $\|w\|^2 = \langle w, w \rangle$; this is before we ultimately normalize w in the **return** step.

First, we can argue that $\|w\|^2 \leq t$, since each step increases $\|w\|^2$ by at most 1:

$$\langle w + y_i x_i, w + y_i x_i \rangle = \langle w, w \rangle + (y_i)^2 \langle x_i, x_i \rangle + 2 y_i \langle w, x_i \rangle \leq \langle w, w \rangle + 1 + 0.$$

This is true since each $\|x_i\| \leq 1$, and if x_i is misclassified, then $y_i \langle w, x_i \rangle$ is negative.

Second, we can argue that $\langle w, w^* \rangle \geq t\gamma^*$ since each step increases it by at least γ^*. Recall that $\|w^*\| = 1$

$$\langle w + y_i x_i, w^* \rangle = \langle w, w^* \rangle + (y_i)\langle x_i, w^* \rangle \geq \langle w, w^* \rangle + \gamma^*.$$

The inequality follows from the margin of each point being at least γ^* with respect to the max-margin classifier w^*.

Combining these facts ($\langle w, w^* \rangle \geq t\gamma^*$ and $\|w\|^2 \leq t$) together we obtain

$$t\gamma^* \leq \langle w, w^* \rangle \leq \langle w, \frac{w}{\|w\|} \rangle = \|w\| \leq \sqrt{t}.$$

Solving for t yields $t \leq (1/\gamma^*)^2$ as desired.

9.3 Support Vector Machines and Kernels

It turns out, all we need to get any of the above perceptron machinery to work is a well-defined (generalized) inner product. For two vectors $p = (p_1, \ldots, p_d)$, $x = (x_1, \ldots, x_d) \in \mathbb{R}^d$, we have always used as the inner product the standard dot product:

$$\langle p, x \rangle = \sum_{i=1}^{d} p_i \cdot x_i.$$

However, we can define inner products more generally as a similarity function, and typically as a *kernel* $K(p, x)$. For instance, we can use the following non-linear functions

- $K(p, x) = \exp(-\|p - x\|^2/\sigma^2)$ for the Gaussian kernel, with bandwidth σ,
- $K(p, x) = \exp(-\|p - x\|/\sigma)$ for the Laplace kernel, with bandwidth σ, and
- $K(p, x) = (\langle p, x \rangle + c)^r$ for the polynomial kernel of power r, with control parameter $c > 0$.

Are these linear classifiers?

No. In fact, this is the standard way to model several forms of non-linear classifiers. The "decision boundary" is no longer described by the boundary of a halfspace. For the polynomial kernel, the boundary must now be a polynomial surface of degree r. For the Gaussian and Laplace kernel it can be even more complex; the σ parameter essentially controls the curvature of the boundary. As with polynomial regression, these can fit classifiers to more complex types of data, but comes at the expense of mildly more computation cost, and perhaps worse generalization if we allow for too complex of models and overfit to the data.

Geometry of Non-linearity of Polynomial Classifiers

Consider a labeled data set (X, y) with $X \subset \mathbb{R}^1$ that is not linear separable because all of the negatively labeled points are in the middle of two sets of positive points. These are the points drawn along the x-axis.

However, if each x_i is "lifted" to a two-dimensional point $(x_i, x_i^2) \in \mathbb{R}^2$ (with yellow highlight), then it can be linearly separated in this extended space of degree-2 polynomial classifiers. Of course, in the original space $\mathbb{R}^1$, this classifier is not linear.

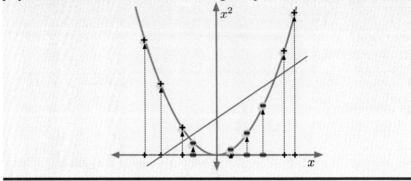

9.3.1 The Dual: Mistake Counter

To use a more general kernel within the perceptron algorithm, we use a different interpretation of how to keep track of the weight vector w. Recall, that each step we increment w by $y_i x_i$ for some misclassified data point (x_i, y_i). Instead we will maintain a length n vector $\alpha = (\alpha_1, \alpha_2, \ldots, \alpha_n)$, where α_i represents the number of times that data point (x_i, y_i) has been misclassified. Then we can rewrite

$$w = \sum_{i=1}^{n} \alpha_i y_i x_i.$$

That is, instead of directly maintaining w, we maintain a length n set of counters α, and keep track of which signed data points $y_i x_i$ were added, so we could reconstruct w as needed. Now in the linear case the function g, applied to a new data point $p \in \mathbb{R}^d$ becomes

$$g(p) = \langle w, p \rangle = \left\langle \sum_{i=1}^{n} \alpha_i y_i x_i, p \right\rangle = \sum_{i=1}^{n} \alpha_i y_i \langle x_i, p \rangle.$$

This seems wasteful to keep such a large vector $\alpha \in \mathbb{R}^n$ around, especially if the number of data points n becomes very large. In contrast, the size of the original version $w \in \mathbb{R}^d$ does not change as the dataset increases. However, we only need to keep track of the non-zero elements of α, and if we run perceptron, there are at most $(1/\gamma^*)^2$ of these. So this is not significantly more space, depending on the relationship between d and $1/\gamma^*$.

The beauty of this form is that now we can easily replace $\langle x_i, p \rangle$ with any other kernel $K(x_i, p)$. That is, the function $g(p)$ now becomes generalized to

$$g(p) = \sum_{i=1}^{n} \alpha_i y_i K(x_i, p).$$

Note this g is precisely a kernel density estimate, with some elements having negative weights (if $y_i = -1$). Then a point p is classified as positive if $g(p) > 0$ and negative otherwise.

Geometry of Kernel SVM

The example shows 6 support points: 3 with a positive label $\{(3.5, 1), (4, 4), (1.5, 2)\}$ as pink +, and 3 with a negative label $\{(0.5, 0.5), (1, 4.5), (1.5, 2.5)\}$ with a black $\ominus$. A Gaussian kernel (with $\sigma = 1$) SVM is used to generate a separator shown as the thick green line.

We also plot the height function (from negative red values to positive blue ones) of the (unnormalized) kernel density estimate of the 6 points. It is actually the difference of the kernel density estimate of the negative labeled points from that of the positive labeled points.

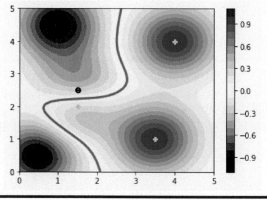

9.3.2 Feature Expansion

If the margin is small, the data is not separable, or we simply do not want to deal with the unknown size of a mistake counter, there is another option to use these non-linear kernels. We can take all data, and apply a non-linear transformation to a higher-dimensional space so the problem is linear again.

For a polynomial kernel, on d-dimensional data points, this is equivalent to the polynomial regression expansion, described in Section 5.3. For a two dimensional data point $p = (p_1, p_2)$, we map this to a five-dimensional space

$$p \mapsto q = (q_1 = p_1, q_2 = p_2, q_3 = p_1^2, q_4 = p_1 p_2, q_5 = p_2^2).$$

Then, we search over a six-dimensional parameter space (b, w) with b a scalar and $w = (w_1, w_2, w_3, w_4, w_5)$ and the kernel defined $K^\uparrow(p, w) = \langle q, w \rangle$. Note we are slightly abusing notation here since in this case, $K^\uparrow : \mathbb{R}^d \times \mathbb{R}^{\dim(w)} \to \mathbb{R}$, where the dimension of $p \in \mathbb{R}^d$ and $w \in \mathbb{R}^{\dim(w)}$ do not match. Now, the z associated with a data point (p, y) as input to a loss function $\ell(z)$ is defined as

$$z = y \cdot (K^\uparrow(p, w) + b) = y \cdot (\langle q, w \rangle + b).$$

Note that the number of free parameters in the feature expansion version of the polynomial kernel is larger than when retaining the kernel in the form $K(x, p) = (\langle x, p \rangle + b)^r$, when it is only $d + 1$. In particular, when $p \in \mathbb{R}^d$ and the polynomial degree r is large, then this dimensionality can get high very quickly; the dimension of q is $\dim(q) = O(d^r)$.

Such expansion is also possible for many other radius basis kernel (e.g., Gaussian, Laplace), but the constructions are only approximate and randomized. Usually it requires the dimensionality of q to be about 100 or more to get a faithful representation.

Generalization

Like in the case of regression, poor generalization on complex models can be an issue. However, this effect can be mostly controlled by a regularization term. In the case of the polynomial kernel limiting the polynomial degree r, prevents too complex a model. With the Gaussian and Laplace, and other similar kernels, the increased dimensionality often does not lead to overfitting, but too small a value of σ may. In both cases these parameters (r and σ) can be appropriately chosen with cross-validation.

Perhaps unintuitively, "overfitting" is probably the wrong way to think about these issues; rather the issue is that the search space of models can be too constrained. When there is enough data to fairly well represent a phenomenon, then generalization can sometimes further improve as the model space becomes larger and "over-complete." Then there may be more than one way to achieve the same minimum of a cost function, and some form of regularization (whether it is a regularization term or stopping gradient descent early or something else) will guide the model found to the simplest near-optimal solution available.

9.3.3 Support Vector Machines

A more general way to work with complex kernels is called a *support vector machine* or SVM. Like with the illustration of the margin for linear classifiers, there are a small number of data points which determine the margin, the *support vectors*. Just these points are enough to determine the optimal margin.

In the case of complex non-linear kernels (e.g., Gaussians), all of the points may be support vectors. Worse, the associated linear expanded space, the result of complete variable expansion, is actually infinite! This means, the true weight vector w would be infinite as well, so there is no feasible way to run gradient descent on its parameters.

However, in most cases the actual number of support vectors is small. Thus, it will be useful to *represent* the weight vector w as a linear combination of these support vectors, without every explicitly constructing them. Consider a linear expansion of a kernel K to an m-dimensional space (think of m as being sufficiently large that it might as well be infinite). However, consider if there are only k support vectors $\{s_1, s_2, \ldots, s_k\} \subset X$ where X is the full data set. Each support vector $s_i \in \mathbb{R}^d$ has a representation $q_i \in \mathbb{R}^m$. But the normal vector $w \in \mathbb{R}^m$ can be written as a linear combination of the q_is; that is, for some parameters $\alpha_1, \alpha_2, \ldots$, we must be able to write

$$w = \sum_{i=1}^{k} \alpha_i q_i.$$

Thus, given the support vectors $S = \{s_1, \ldots, s_k\}$ we can represent w in the span of S (and the origin), reparametrized as a k-dimensional vector $\alpha = (\alpha_1, \alpha_2, \ldots, \alpha_k)$. This α vector is precisely the mistake counter for only the support vectors (the non-zero components), although in this case the coordinates need not be integers.

More concretely, we can apply this machinery without ever constructing the q_i vectors. Each can be implicitly represented as the function $q_i = K(s_i, \cdot)$. Recall, we only ever need to use the q_i in $\langle q_i, w \rangle$. And we can expand w to

$$w = \sum_{i=1}^{k} \alpha_i q_i = \sum_{i=1}^{k} \alpha_i K(s_i, \cdot).$$

Given this expansion, if we consider positive definite kernels, which include Gaussian and Laplace kernels, then they have the so-called *reproducing property*, where each generalized inner product $K(w, p)$ can be written as a weighted sum of inner products with respect to the support vectors:

$$K(w, p) = \sum_{i=1}^{k} \alpha_i K(s_i, p).$$

That is, the reproducing property ensures that for every representation of a classifier w, there is always a sets $\{\alpha_1, \alpha_2, \ldots\}$ and $\{s_1, s_2, \ldots\}$ to satisfy the above equation, and thus can optimize over the choice of α to find the best w. Ultimately, for a data point (p, y), the z in the loss function $\ell(z)$ is defined as

$$z = yK(w, p) = y \sum_{i=1}^{k} \alpha_i K(s_i, p).$$

Solving for SVMs

There are multiple ways to actually optimize SVMs: the task of finding the support vectors $S = \{s_1, \ldots, s_k\}$, and assigning their weights $\alpha = \{\alpha_1, \ldots, \alpha_k\}$. One is to run the kernelized Perceptron algorithm, as outlined above. Alternatively, given a fixed set of support vectors S, one can directly optimize over α using gradient descent, including any loss function and regularizer as before with linear classifiers. Thus, if we do not know S, we can just assume $S = X$, the full data set. Then, we can apply standard gradient descent over $\alpha \in \mathbb{R}^n$. As mentioned, in most cases, most α_i values are 0 (and those close enough to 0 can often be rounded to 0). Only the points with non-zero weights are kept as support vectors.

Alternatively, stochastic gradient descent works like perceptron, and may only use a fraction of the data points. If we use a version of hinge loss, only misclassified points or those near the boundary have a non-zero gradient. The very strongly classified points have zero gradient, and the associated α_i coordinates may remain 0. The proper choice of loss function, and regularizer, can induces sparsity on the α values; and the data points not used are not support vectors.

9.4 Learnability and VC dimension

Models for classifiers can quickly become very complicated, with high degree polynomial models or kernel models. Following this discussion we will explore even more complex models in KNN classifiers, random forests, and deep neural nets. Is there a mathematical model to understand if these are too complicated, other than the data-driven cross-validation? *Yes*, in fact, there are several. The most common one is based on the idea of VC dimension and sample complexity. This provides a way to capture the complexity of the family of classifiers being considered, and relate this to how much data is required to feasibly train a model with such a family.

If you have $n = 10$ data points in 4-dimensions, you should probably not use a degree 3 polynomial. But what about 100 data points, or 1000, or 1 million?

VC dimension

This is best described with notions of combinatorial geometry. Consider a set of data Z and a family of classifiers $\mathcal{H}$ (e.g., linear classifiers in $\mathbb{R}^d$), where each $h \in \mathcal{H}$ maps

$h : Z \to \{-1, +1\}$. If every subset $S \subset Z$ can be induced as $S = \{z \in Z \mid h(s) = +1\}$ for some $h \in \mathcal{H}$, then we say that $\mathcal{H}$ *shatters* Z. Now, given a full data set X and classifier family $\mathcal{H}$, the *VC dimension* is the maximum size subset $Z \subset X$ which $\mathcal{H}$ can shatter.

Geometry of VC dimension for Linear Separators

Linear separators can shatter a set of 3 points in $\mathbb{R}^2$, as shown in the 8 panels below. Each possible subset of these points (including the full and empty sets) is shown in red, as well as the halfspace (in green) which corresponds with that subset. Thus the VC dimension is at least 3.

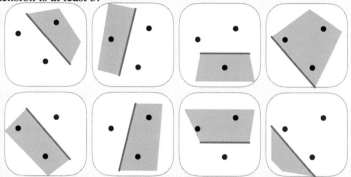

However, linear separators cannot shatter a set of size 4. The two canonical configurations are shown in the panels below, in addition to a labeling which cannot be realized using a linear separator. Thus the VC dimension for points in $\mathbb{R}^2$ and linear separators as $\mathcal{H}$ is exactly 3.

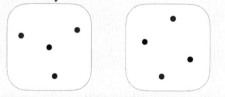

Intuitively, one can think of the VC dimension as (roughly) representing the number of parameters required to describe any one classifier from the family.

Example: VC dimension

The VC dimension ν is well understood for many families of classifiers. Examples on point sets in $\mathbb{R}^d$ include

- Linear separators in $\mathbb{R}^d$; then $\nu = d + 1$.
- Polynomial separators of degree r in $\mathbb{R}^d$; then $\nu = O(d^r)$.
- Kernel SVM (with arbitrarily small bandwidth σ); then $\nu = \infty$, it is unbounded.
- Intervals in $\mathbb{R}^1$; then $\nu = 2$.

- Ball separators in $\mathbb{R}^d$; then $\nu = d + 1$.
- Convex sets in $\mathbb{R}^2$; then $\nu = \infty$, it is unbounded.

Recall that for functions f, g that $g(x) = O(f(x))$ if there exists constants c_1, c_2 so $g(x) \leq c_1 \cdot f(x)$ for all $x > c_2$.

Sample Complexity

When the VC dimension for a family of classifiers is bounded, then we can apply powerful theorems to grapple with the learnability of the problem. These are again modeled under the assumption that data X is drawn iid from an unknown distribution μ. The question is if we learn a classifier $h \in \mathcal{H}$ on X, and apply it to a new point $x \sim \mu$ (not in X, e.g., a test set or yet to be seen data), then how likely are we to misclassify it. We let $\varepsilon \in (0, 1)$ be the acceptable rate of misclassification (e.g., $\varepsilon = 0.01$ means we misclassify it 1% of the time). We also need a probability of failure δ, controlling the probability that X (also a random sample from μ) is a poor representative of μ).

First consider a family of classifiers $\mathcal{H}$ for which there exists a *perfect classifier* $h^* \in \mathcal{H}$, so that for any x, it will always return the correct label. If this is the case and the VC dimension is ν, then $n = O((\nu/\varepsilon)\log(\nu/\delta))$ samples $X \sim \mu$ are sufficient to learn a classifier h which, with probability at least $1 - \delta$, will predict the correct label on a $(1 - \varepsilon)$-fraction of new points drawn from μ.

If there does not exist a perfect classifier, then the situation is more challenging. The best classifier $h^* \in \mathcal{H}$ may still classify α-fraction of new points incorrectly. Then if the VC dimension is ν, drawing $n = O((1/\varepsilon^2)(\nu + \log(1/\delta)))$ samples $X \sim \mu$ is sufficient to learn a classifier h which, with probability at least $1 - \delta$, will misclassify at most $\alpha + \varepsilon$ of new points drawn from μ.

These theorems do not explain *how* to learn these classifiers. In the perfect classifier setting, the perceptron algorithm can be used as long as there is an appropriate inner product, and the margin γ is sufficiently large. Otherwise there may be more than ε error induced in the learning process. This sort of analysis is more useful in warning against certain overly complex classifiers, as an enormous number of training data points may be required to be able to accurately learn the approximately best classifier. Despite that, many modern techniques (like deep neural networks and random forests, introduced below) which have no or enormous VC dimension bounds have gained significant popularity *and* seem to generalize well. While these tend to work best when training on huge data sets, the data set sizes often still do not quite approach that required by the theoretical bounds.

9.5 *k*NN Classifiers

Now for something completely different. There are many ways to define a classifier, and we have just touched on some of them.

The *k*-NN classifier (or *k-nearest neighbors classifier*) works as follows. Choose a scalar parameter k (it will be far simpler to choose k as an odd number, say $k = 5$). Next define a *majority* function maj $: \{-1, +1\}^k \to \{-1, +1\}$. For a set $Y = (y_1, y_2, \ldots, y_k) \in \{-1, +1\}^k$ it is defined as

$$\text{maj}(Y) = \begin{cases} +1 & \text{if more than } k/2 \text{ elements of } Y \text{ are } +1 \\ -1 & \text{if more than } k/2 \text{ elements of } Y \text{ are } -1. \end{cases}$$

Then for a data set X where each element $x_i \in X$ has an associated label $y_i \in \{-1, +1\}$, define a *k*-nearest neighbor function $\phi_{X,k}(q)$ that returns the k distinct points in X which are closest to a query point q. Next let sign report y_i for any input point x_i; for a set of inputs x_i, it returns the set of values y_i.

Finally, the *k*-NN classifier is

$$g(q) = \text{maj}(\text{sign}(\phi_{X,k}(q))).$$

That is, it finds the *k*-nearest neighbors of query point q, and considers all of the class labels of those points, and returns the majority vote of those labels.

A query point q near many other positive points will almost surely return $+1$, and symmetrically for negative points. This classifier works surprisingly well for many problems but relies on a good choice of distance function to define $\phi_{X,k}$.

Unfortunately, the model for the classifier depends on all of X. So it may take a long time to evaluate on a large data set X. In contrast the functions g for non-kernelized methods above take time linear in d to evaluate for points in $\mathbb{R}^d$, and thus are very efficient.

9.6 Decision Trees

A *decision tree* is a general classification tool for multidimensional data (X, y) with $X \in \mathbb{R}^{n \times d}$ and $y \in \{-1, +1\}^n$. The classifier has a binary tree structure, where each non-overlapping set of subtrees, each contain disjoints sets of X. And in particular each node has a rule on $\mathbb{R}^d$ of how to split the subset of X in that subtree into two parts, one for each child. Each leaf node of the tree is given a label -1 or $+1$, based on the majority label from the subset $S \subset X$ in that node. Then to use the tree to classify an unknown value $x \in \mathbb{R}^d$, it starts at the root of tree and follows each rule at the node to choose an appropriate subtree; it follows these subtrees recursively until it reaches a leaf, and then uses the label at the leaf.

So far, this structure should seem basically the same as other trees used for nearest neighbor searching (e.g., kd-trees, an alternative to LSH in Section 4.6) or for clustering (e.g., HAC in Section 8.5 or Spectral Clustering in Section 10.3). Different from most clustering, and similar to kd-trees, decision trees are made tractable by splitting on a single coordinate (out of the d choices) at each node.

We consider a labeled set of $n = 13$ points $X \subset \mathbb{R}^2$, with two classes (-1 as red, and $+1$ as blue). A decision tree with 6 decision boundaries is shown, and each leaf node has exactly one class. The corresponding leaves in the trees are color-coded with the label of the class. The tree on the right corresponds with an axis-aligned space decomposition of $\mathbb{R}^2$ on the left.

In particular, the root node splits at $y = 5$, which is showed on the left with a horizontal line. And the blue leaf node on the far right in the tree contains input points $\{x_6, x_9, x_{12}, x_{13}\}$, all of which are labeled $+1$, and is drawn as the green region with $y \geq 5$ and $x \geq 6$.

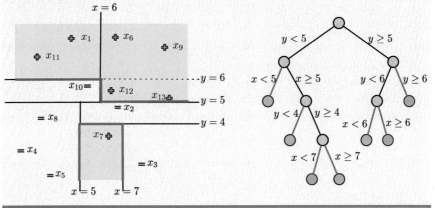

The construction of the decision tree is made top-down. It starts at the root which contains all data X, and then splits recursively until some convergence property is achieved. At each split, each coordinate is considered. For that coordinate, the best split can be found by scanning the possible splits from low to high, and whatever statistic being optimized can be updated easily as a single point is moved from one side to the other in a linear scan. So the remaining considerations are what statistic to use to decide on the coordinate (and the split point), and when to terminate the process.

Splitting Criteria

There are two main criteria used to determine which coordinate to use in a split. Given this criteria, the best split for that criteria is found for each coordinate, and the best one is used. The ultimate goal is to construct the tree so that it is not very deep, but so each leaf node has nearly all of the same labels—or at least is a uniform mix that is not easy to split.

Each criteria is defined for each subset $S \subset X$, which we can categorize in two subsets: the positive labeled $S_+ = \{x_i \in S \mid y_i = +1\}$ and the negative labeled $S_- = \{x_i \in S \mid y_i = -1\}$. Further denote the probability of each class as $p_+ = |S_+|/|S|$ and $p_- = |S_-|/|S|$.

- The *Gini Impurity* should be minimized, and is defined as

$$I_G(S) = p_+(1 - p_+) + p_-(1 - p_-).$$

Each term presents the probability that a member of that class will be chosen, times the probability it will be misclassified. The probability it will be misclassified is based on us choosing a label for that node not just based on the majority weight, but by selecting a random item and using its class.
- The *Information Gain* is based on the entropy of the classes, and maximizes the *entropy* of each subtree set S, defined as

$$H(S) = -p_+ \log p_+ - p_- \log p_-.$$

It is typically defined as a difference, or ratio, between parent's entropy and that of its children, but when the parent is fixed, we can simply maximize the following quantity over both children, weighted by their size. That is for left and right children with sets S_r and S_l the goal is to maximize

$$|S_r| \cdot H(S_r) + |S_l| \cdot H(S_l).$$

And information gain is the value $IG(S) = H(S) + |S_r| \cdot H(S_r) + |S_l| \cdot H(S_l)$.

Regularization, Explainability, and Ensembles

One could continue splitting until each leaf node only has a single label. But this would typically overfit. It is common to regularize the solution by fixing the total number of nodes in a decision tree. At each step, any node can be split, and the one which scores the highest on the splitting criteria (i.e., smallest Gini impurity or largest information gain) is chosen. Alternatively, one can stop splitting when no available split achieves a certain criteria.

Decision trees are quite general, and also extend directly to categorical data, or data that is mixed with some coordinates as real values and some as categorical. It can handle more than two labels. Moreover, it can naturally handle coordinates with different units.

The procedural structure makes them easy to explain, and perhaps audit. The classification decision is made using a trait-by-trait explanation that users even unfamiliar with linear algebra or probability may be able to understand.

On the other hand, more high-powered methods can often provide better prediction results in aggregate. A common way to make decision trees more powerful is to build several decisions trees and then let them vote on the final classification using majority voting like in a k-NN classifier. The two most common ways are *bagging* (which resamples the data, then builds a new tree) and *boosting* (which builds each subsequent tree based on what was poorly classified in previous ones). More complex methods to build and combine trees is the *random forest* technique which often achieves among the best classification results. However, with these more complex combinations, the explainability is sacrificed as a result.

Ethical Questions with Explainable Classifiers

Consider analyzing medical history, generic, dietary, and exposure data related to cancer patients. Your goal is to predict if a patient with a certain medical history will get cancer. After splitting data into a test and training set, on the training data you build a decision tree with a small number of branching points. It achieves 89% accuracy on the test data.

However, you hope you can get a better performance result, and use a random forest approach, which is far more complex. However, it improves the accuracy to 92% on the test data. You need to report back to your medical colleagues to try to provide actionable decisions to improve diagnosis and treatment. Which model should you report? What percentage improvement is worth sacrificing the explainable model of a decision tree to a more opaque model of a random forest or similar approach?

Now consider the application was not medical, such as predicting cancer, but was in predicting which ad to place on a website so people are more likely to click on it. Would your decision on which classifier type to use be the same?

9.7 Neural Networks

A *neural network* is a learning algorithm intuitively based on how a neuron works in the brain. A *neuron* takes in a set of inputs $x = (x_1, x_2, \ldots, x_d) \in \mathbb{R}^d$, weights each input by a corresponding scalar $w = (w_1, w_2, \ldots, w_d)$ and "fires" a signal if the total weight $\sum_{i=1}^{d} w_i x_i$ is greater than some threshold b.

Geometry of Neurons

Our abstraction of a neuron is just a linear classifier. The weights $w = (w_1, \ldots, w_d)$ and decision offset b are just a model w, b which operates on a data point $x \in \mathbb{R}^d$ as $\text{sign}(\langle x, w \rangle - b)$.

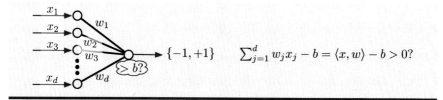

A neural network, is then just a network or directed graph of neurons like these. Typically, these are arranged in layers. In the first layer, there may be d input values $x_1, x_2, \ldots, x_d$. These may provide the input to t neurons (each neuron might use fewer than all inputs). Each layer produces an output $y_1, y_2, \ldots, y_t$. These outputs then serve as the input to the second layer, and so on.

In a neural net, typically each x_i and y_i is restricted to a range $[-1, 1]$ or $[0, 1]$ or $[0, \infty)$, not just the two classes $\{-1, +1\}$. Since a linear function does not guarantee this range of its output (instead of a binary threshold), to achieve this at the output of each node, they typically add an *activation function* $\phi(y)$. Common ones are

- *hyperbolic tangent* : $\phi(y) = \tanh(y) = \frac{e^y - e^{-y}}{e^y + e^{-y}} \in [-1, 1]$
- *sigmoid* : $\phi(y) = \frac{1}{1+e^{-y}} = \frac{e^y}{e^y+1} \in [0, 1]$
- *ReLu* : $\phi(y) = \max(0, y) \in [0, \infty)$.

These functions are not linear, and not binary. They act as a "soft" version of binary. The hyperbolic tangent and sigmoid stretch values near 0 away from 0. Large values stay large (in the context of the range). So it makes most values almost on the boundaries of the range. And importantly, they are differentiable.

The ReLu has become very popular. It is not everywhere differentiable, but is convex. It is basically the most benign version of activation, and is more like the original neuron, in that if the value y is negative, it gets snapped to 0. If it is positive, it keeps its value.

A two-layer neural network is already a powerful way to build classifiers, and with enough neurons in the middle layer, is capable of learning *any function*—its VC dimension is infinite. However, *deep neural nets* with 3 or often many more layers (say 20 or 100 or more) have become very popular due to their effectiveness in many tasks, ranging from image classification to language understanding. To be effective, typically, this requires heavy engineering in how the layers are defined, and how the connections between layers are made.

Example: Deep Neural Network

A deep net is composed of a series of layers. Each layer maps from $h_l \in \mathbb{R}^{k_l}$ to $h_{l+1} \in \mathbb{R}^{k_{l+1}}$, and is composed of k_{l+1} neurons, each followed by an activation function. The input layer is $x = h_0$, the other layers are "hidden" as a user only provides the input x and observes the output in $\{-1, +1\}$. To learn a neural network, one needs to learn each edge weight $W_{i,j}^l$ from $h_l(i)$ to $h_{l+1}(j)$ for each layer l.

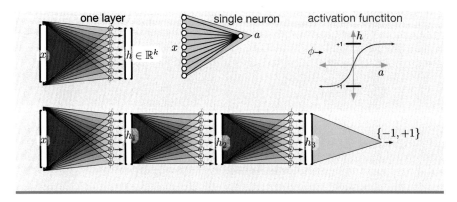

Once the connections are determined, then the goal is to learn the weights on each neuron so that for a given input, a final neuron fires if the input satisfies some pattern (e.g., the input are pixels from a picture, and it fires if the picture contains a car). This is theorized to be "loosely" how the human brain works. Although, neural nets have pretty much diverged in how they learn, compared to attempts to replicate the structure and process of learning in the human brain.

9.7.1 Training with Back-propagation

Given a data set X with labeled data points $(x, y) \in X$ (with $x \in \mathbb{R}^d$ and $y \in \{-1, +1\}$), we already know how to train a single neuron so for input x it tends to fire if $y = 1$ and not fire if $y = -1$. It is just a linear classifier! So, we can use the perceptron algorithm, or gradient descent with a well-chosen loss function.

However, for neural networks to attain more power than simple linear classifiers, they need to be at least two layers, and are often deep (e.g., for "deep learning") with $20, 100$, or more layers. For these networks, the perceptron algorithm no longer works since it does not properly propagate across layers. However, a version of gradient descent called *back-propagation* can be used. In short, it computes the gradient across the edge weights in a network by chaining partial derivatives backwards through the network.

Ultimately the network represents a function $g : \mathbb{R}^d \rightarrow \mathbb{R}$, from which the sign $\text{sign}(g(x))$ predicts the label of x. We can evaluate a given pair (x_i, y_i) using a loss function $\ell(y_i g(x_i))$, and to use within a (stochastic) gradient descent framework, we need to calculate the gradient of $\ell(y_i g(x_i))$. However, there are many model parameters (each edge weight of each neuron in each layer), and they have complicated dependencies. Back-propagation has the insight that the derivative of each model parameter (which compose to form the gradient) can be calculated from the last layer, then the second last layer, and so on. This is an application of the chain rule from calculus. In particular we can write

$$g(x) = h_L = W^L \phi^{L-1}(W^{L-1} \ldots \phi^1(W^1 x) \ldots)$$

as a composition of activation functions ϕ^l and weight matrices W^l, from layer $l = 1$ to layer $l = L$. Each row of W^l contains the weights of one neuron in the lth layer. So each W^l is linear (so convex and differentiable) and each activation function ϕ^l is also chosen to be convex and differentiable.

More specifically, the derivative of parameter $W_{i,j}^l$ is $\nabla_{i,j,l}\ell(yg(x))$ for a data point (x, y) [*Note that we are using i to index into the neural network weight, so we do not use it for the data point*]. We label intermediate results

$$a^l = W^l\phi^{l-1}(W^{l-1}\ldots\phi^1(W^1 x)\ldots) \qquad \text{as the } l\text{th } activation, \text{ and}$$
$$h^l = \phi^l(W^l\phi^{l-1}(W^{l-1}\ldots\phi^1(W^1 x)\ldots)) \qquad \text{as the } l\text{th } hiddenlayer.$$

Now set $z = yg(x) = ya^L$; for notational coherence, we denote the output $g(x)$ as the last activation layer a^L. Then for $l = L$, a^L is a scalar, so we set $i = 1$ for consistent notation, and we can derive

$$\nabla_{1,j,L}\ell(z) = \frac{d\ell(z)}{dW_{1,j}^L} = \frac{d\ell(z)}{da_1^L} \cdot \frac{da_1^L}{dW_{1,j}^L}$$
$$= \ell'(ya_1^L) \cdot h_j^{L-1},$$

where ℓ' is the derivative of the loss function and noting that the derivative $\frac{da_1^L}{dW_{1,j}^L}$ is linear (from a dot product), so is simply h_j^{L-1}. Since each of these terms was chosen to be computable, (since h_j^{L-1} and a_1^L are all computed in the process of computing $g(x)$), then we can derive the gradient for each weight $W_{1,j}^L$ in the last layer. If the neural net was only one layer, then $h_j^{L-1} = x_j$, and this is nothing more than gradient descent for a linear classifier.

Now the derivative of the weights in the second-to-last layer $L - 1$ is computed in a similar way as the last layer L. That is

$$\nabla_{i,j,L-1}\ell(z) = \frac{d\ell(z)}{dW_{i,j}^{L-1}} = \frac{d\ell(z)}{da_1^L} \cdot \frac{da_1^L}{dh_i^{L-1}} \cdot \frac{dh_i^{L-1}}{da_i^{L-1}} \cdot \frac{da_i^{L-1}}{dW_{i,j}^{L-1}}$$
$$= \ell'(ya_1^L) \cdot W_{1,i}^L \cdot (\phi^{L-2})'(a_i^{L-1}) \cdot h_j^{L-2},$$

where $(\phi^{L-2})'$ is the derivative of ϕ^{L-2}, and again each of these terms has been or can be computed. Note that we have computed for each k

$$\frac{d\ell(z)}{da_i^{L-1}} = \frac{d\ell(z)}{da_1^L} \cdot \frac{da_1^L}{dh_i^{L-1}} \cdot \frac{dh_i^{L-1}}{da_i^{L-1}} = \ell'(ya_1^L) \cdot W_{1,i}^L \cdot (\phi^{L-2})'(a_i^{L-1}),$$

which we will reuse.

In the third-to-last layer $L-2$ one more complication arises, where it now depends on each kth activation term (where k was i above) in layer $L-1$, precomputed above, as the sum

$$\nabla_{i,j,L-2}\ell(z) = \frac{d\ell(z)}{dW_{i,j}^{L-1}} = \sum_k \frac{d\ell(z)}{da_k^{L-1}} \cdot \frac{da_k^{L-1}}{dh_i^{L-2}} \cdot \frac{dh_i^{L-2}}{da_i^{L-2}} \cdot \frac{da_i^{L-2}}{dW_{i,j}^{L-2}}$$

$$= \sum_k \frac{d\ell(z)}{da_k^{L-1}} \cdot W_{k,i}^L \cdot (\phi^{L-3})'(a_i^{L-2}) \cdot h_j^{L-3}.$$

And again in this process we have computed each $\frac{d\ell(z)}{da_i^{L-2}}$. The second big insight (other than reusing each $\frac{d\ell(z)}{da_i^l}$) is that in each following layer there is only one sum over the previous activation vector—it is not a sum of sums and so on.

In summary, to compute the derivative for a specific data point (x, y), we "feed forward" x into the network, can compute all activation a^l and hidden layer h^l vectors. Then we use y to evaluate the loss function, and "back propagate" the loss on the network, using the chain rule to compute relevant partial derivatives as we go. Thus, the cost of one step is ultimately proportional to the size of the network, and is reasonably efficient.

Exercises

9.1 Consider the following "loss" function. $\ell_i(z_i) = (1 - z_i)^2/2$, where for a data point (x_i, y_i) and prediction function f, then $z_i = y_i \cdot f(x_i)$. Predict how this might work within a gradient descent algorithm for classification.

9.2 Consider a data set (X, y), where each data point $(x_{1,i}, x_{2,i}, y_i)$ is in $\mathbb{R}^2 \times \{-1, +1\}$. Provide the psuedo-code for the Perceptron Algorithm using a polynomial kernel of degree 2. You can have a generic stopping condition, where the algorithm simply runs for T steps for some parameter T. (*There are several correct ways to do this, but be sure to explain how to use a polynomial kernel clearly.*)

9.3 Consider a set of one-dimensional data points

$$(x_1 = 0, y_1 = +1)\ (x_2 = 1, y_1 = -1)\ (x_3 = 2, y_1 = +1)\ (x_4 = 4, y_1 = +1)$$

$$(x_5 = 6, y_1 = -1)\ (x_6 = 7, y_1 = -1)\ (x_7 = 8, y_1 = +1)\ (x_8 = 9, y_1 = -1)$$

Predict **-1** or **+1** using a kNN (k-nearest neighbor) classifier with $k = 3$ on the following queries.

1. $x = 3$
2. $x = 9$
3. $x = -1$

9.4 Consider the following Algorithm 9.7.1, called the *Double-Perceptron*. We will run this on an input set X consisting of points $X \in \mathbb{R}^{n \times d}$ and corresponding labels $y \in \{-1, +1\}$.

Algorithm 9.7.1 Double-Perceptron(X)

Initialize $w = y_i x_i$ for any $(x_i, y_i) \in (X, y)$
repeat
    For any (x_i, y_i) such that $y_i \langle x_i, w \rangle < 0$ (is mis-classified) : update $w \leftarrow w + 2 \cdot y_i x_i$
until (no mis-classified points **or** T steps)
return $w \leftarrow w/\|w\|$

For each of the following questions, the answer can be **faster, slower, the same,** or **not at all**. And should be accompanied with an explanation.

1. Compared with Algorithm 9.2.1, explain how this algorithm with converge.
2. Next consider transforming the input data set X (not the y labels) so that all coordinates are divided by 2. Now if we run *Double-Perceptron* how will the results compare to regular Perceptron (Algorithm 9.2.1) on the original data set X.
3. Finally, consider taking the original data set (X, y) and multiplying all entries in y by -1, then running the original Perceptron algorithm. How will the convergence compare to running the same Perceptron algorithm, on the original data set.

9.5 Consider a matrix $A \in \mathbb{R}^{n \times 4}$. Each row represents a customer (there are n customers in the database). The first column is the age of the customer in years, the second column is the number of days since the customer entered the database, the third column is the total cost of all purchases ever by the customer in dollars, and the last column is the total profit in dollars generated by the customer. So each column has a different unit.

For each of the following operations, decide if it is **reasonable** or **unreasonable** (think about the units).

1. Run simple linear regression (no regularization) using the first three columns to build a model to predict the fourth column.
2. Use k-means clustering to group the customers into 4 types using Euclidean distance between rows as the distance.
3. Use PCA to find the best two-dimensional subspace, so we can draw the customers in a $\mathbb{R}^2$ in way that has the least projection error.
4. Use the perceptron to build a linear classification model based on the first three columns to predict if the customer will make a profit $+1$ or not -1.
5. Use a decision tree to build a model based on the first three columns to predict if the customer will make a profit $+1$ or not -1.
6. Use a KNN classifier (with Euclidean distance and $k = 3$) to build a model based on the first three columns to predict if the customer will make a profit $+1$ or not -1.

9.6 Consider a "loss" function, called an *double-hinged loss function*

$$\ell_i(z) = \begin{cases} 0 & \text{if } z > 1 \\ 1 - z & \text{if } 0 \le z \le 1 \\ 1 & \text{if } z \le 0 \end{cases}$$

where the overall cost for a dataset (X, y), given a linear function $g(x) = \langle (1, x), \alpha \rangle$ is defined $\mathcal{L}(g, (X, y)) = \sum_{i=1}^{n} \ell_i(y_i \cdot g(x_i))$.

1. What problems might a gradient descent algorithm have when attempting to minimize $\mathcal{L}$ by choosing the best α?
2. Explain if the problem would be better or worse using stochastic gradient descent?

9.7 This question explores linear classifiers for non-linearly separable point sets.

1. Construct and report a set of labeled points (X, y) in $\mathbb{R}^2$ that is not linearly separable (provide a plot).
2. Explain what will happen if you run the perceptron algorithm for a linear classifier on this data set? (Do not set a fixed upper bound on T the number of steps)
3. Describe another algorithm which would provide an acceptable **linear** classifier (but not necessarily a separator) for the set of points.

9.8 Consider a family of classifiers defined by axis-aligned rectangles for data in $\mathbb{R}^2$. What is associated the VC dimension?

9.9 Consider a decision tree algorithm where at each split node a linear SVM is run to construct a linear classifier, but not necessarily aligned with a single coordinate. List at least one advantage and two disadvantages of this algorithm compared to the standard decision tree.

Chapter 10
Graph Structured Data

Abstract A central object in data analysis is a graph $G = (V, E)$ defined by a set of vertices V and edges between those vertices E. The vertices can serve as a proxy for any data type (e.g., social network users, a company's products, and waypoints on a map), and the graph representation then simplifies the structure of these data items down to only their connectivity. This structure can of course be composed with many other structures we have studied (vectors, matrices, etc.), but this chapter will focus specifically on what can be inferred from the graphs, and in particular, focusing on linear algebraic approaches which align well with the rest of this book. In this section, we overview how graphs can be used to model the movement of information, which reveals which vertices are most important. We also show how to cluster or find interesting subgraphs that capture where more interesting or meaningful patterns appear.

Basic Definitions and Models for Graphs

Formally, a *(undirected) graph* $G = (V, E)$ is defined by a set of vertices $V = \{v_1, v_2, \ldots, v_n\}$ and a set of edges $E = \{e_1, e_2, \ldots, e_m\}$ where each edge e_j is an unordered pair of vertices: $e_j = \{v_i, v_{i'}\}$. In a *directed graph*, edges are ordered and $e_j = (v_i, v_{i'})$ indicates that node v_i points to $v_{i'}$, but not the other direction.

Two vertices v_1 and v_k are *connected* if there is a sequence of edges $\langle e_1, \ldots, e_{k-1} \rangle$ such that e_1 contains v_1, e_{k-1} contains v_k, and consecutive edges can be ordered so $e_j = \{v_i, v_{i+1}\}$ and $e_{j+1} = \{v_{i+1}, v_{i+2}\}$ where the second element in e_j is the same as the first in e_{j+1}. The *graph distance* $\mathbf{d}_E(v_i, v_{i'})$ (induced by edges set E) between any two vertices $v_i, v_{i'} \in V$ is the minimum number of edges requires to get from v_i to $v_{i'}$. It is a metric for undirected graphs; for directed graphs it may not be symmetric.

Example: Undirected Graph

Consider graph $G = (V, E)$ with the following vertices: $V = \{a, b, c, d, e, f, g, h\}$ and edges $E = \Big\{\{a, b\}, \{a, c\}, \{a, d\}, \{b, d\}, \{c, d\}, \{c, e\}, \{e, f\}, \{e, g\}, \{f, g\}, \{f, h\}\Big\}$.

© Springer Nature Switzerland AG 2021
J. M. Phillips, *Mathematical Foundations for Data Analysis*,
Springer Series in the Data Sciences,
https://doi.org/10.1007/978-3-030-62341-8_10

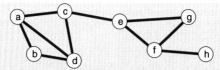

Alternatively, G can be represented as a matrix with 1 if there is an edge, and 0 otherwise.

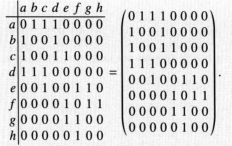

The distance $\mathbf{d}_E(a, g) = 3$, realized by the edges $\{a, c\}$, $\{c, e\}$, and $\{e, g\}$.

We will use this graph as a running example throughout much of this chapter.

Matrix Representations of Graphs

The matrix representation of a graph is essential for understanding many of its structural properties. We initiate this below with the two most basic and essential such representations, but will generalize and extend these in the sections below. It should be noted that in practice, this can be an *extremely* space-inefficient way to store the graph. These matrices have $|V|^2$ cells, however, in many cases the number of actual edges $|E|$ is much closer to the number of vertices (say $|E| \leq 20|V|$ or $|E| \leq 5|V| \log_2 |V|$ or $|E| \leq |V|^{1.2}$ are often not unreasonable). On the small examples that we can present in this text, these distinctions are not meaningful. But many graph data sets in the industry (e.g., where $|V| > 1{,}000{,}000$), this difference can correspond with the data set fitting easily on a laptop, or by requiring a cluster of computers.

The most natural representation of a graph as a matrix is the *adjacency matrix* A. For an n vertex graph, it is an $n \times n$ matrix that records 1 in $A_{i,j}$ for each edge $e_{i,j} = \{v_i, v_j\}$. Entries are 0 otherwise.

The *degree matrix* is a square and diagonal $n \times n$ matrix. This means it only has non-zero entries on the diagonals $D_{i,i}$. In particular, $D_{i,i}$ records the degree of vertex v_i, which is the number of edges containing v_i. In particular, we will sometimes use the matrix $D^{-1/2}$. Recall a square matrix can be multiplied by itself, so in general for a positive integer power p, the matrix operation D^p refers to multiplying D by itself p times as $D \cdot D \cdot \ldots \cdot D$. However, for fractional powers this is less intuitive, and in general is not always defined. For diagonal matrices with positive entries on all diagonal elements, we can simply define D^p as the diagonal matrix replacing each

$D_{i,i}$ with $D_{i,i}^p$. In particular, this means it is easy to invert a diagonal matrix since D^{-1} just replaces each $D_{i,i}$ with $1/D_{i,i}$). One can check that the definition is consistent with the more general definition of a matrix power, that is, $(D^p)^{1/p} = D$.

Example: Adjacency and Degree Matrices

Again using our example graph, we can define the adjacency A and degree matrix D:

$$A = \begin{pmatrix} 0 & 1 & 1 & 1 & 0 & 0 & 0 & 0 \\ 1 & 0 & 0 & 1 & 0 & 0 & 0 & 0 \\ 1 & 0 & 0 & 1 & 1 & 0 & 0 & 0 \\ 1 & 1 & 1 & 0 & 0 & 0 & 0 & 0 \\ 0 & 0 & 1 & 0 & 0 & 1 & 1 & 0 \\ 0 & 0 & 0 & 0 & 1 & 0 & 1 & 1 \\ 0 & 0 & 0 & 0 & 1 & 1 & 0 & 0 \\ 0 & 0 & 0 & 0 & 0 & 1 & 0 & 0 \end{pmatrix} \quad D = \begin{pmatrix} 3 & 0 & 0 & 0 & 0 & 0 & 0 & 0 \\ 0 & 2 & 0 & 0 & 0 & 0 & 0 & 0 \\ 0 & 0 & 3 & 0 & 0 & 0 & 0 & 0 \\ 0 & 0 & 0 & 3 & 0 & 0 & 0 & 0 \\ 0 & 0 & 0 & 0 & 3 & 0 & 0 & 0 \\ 0 & 0 & 0 & 0 & 0 & 3 & 0 & 0 \\ 0 & 0 & 0 & 0 & 0 & 0 & 2 & 0 \\ 0 & 0 & 0 & 0 & 0 & 0 & 0 & 1 \end{pmatrix}.$$

Taking D to the power $-1/2$ is now

$$D^{-1/2} = \begin{pmatrix} 0.577 & 0 & 0 & 0 & 0 & 0 & 0 & 0 \\ 0 & 0.707 & 0 & 0 & 0 & 0 & 0 & 0 \\ 0 & 0 & 0.577 & 0 & 0 & 0 & 0 & 0 \\ 0 & 0 & 0 & 0.577 & 0 & 0 & 0 & 0 \\ 0 & 0 & 0 & 0 & 0.577 & 0 & 0 & 0 \\ 0 & 0 & 0 & 0 & 0 & 0.577 & 0 & 0 \\ 0 & 0 & 0 & 0 & 0 & 0 & 0.707 & 0 \\ 0 & 0 & 0 & 0 & 0 & 0 & 0 & 1 \end{pmatrix}.$$

10.1 Markov Chains

A *Markov chain* (V, P, q) is defined by a set of nodes V, a probability transition matrix P, and an initial state q. We will see soon how P can be induced from the matrix representation of an edge set, so this structure is closely connected to understanding graphs; for now, we describe it in a more general context. Moreover, in many of the most useful and often desirable contexts q is not needed, and the initial part of this discussion will be aimed at showing how to remove the dependence on q.

The point of a Markov chain (V, P, q) is to understand the "movement around" or "flow through" the vertex set V. This movement is governed by the matrix P, but to get a grasp on this, it is important to start with how to think about this, via its initial state q.

The initial state q represents a probability distribution over vertices. For instance, if the mental model of this process is a random walk around the vertices of a graph, then an initial state q (i.e., specifically at a vertex $b \in V$ with probability 1, in our running example) is

$$q^T = [0\ 1\ 0\ 0\ 0\ 0\ 0\ 0].$$

A more general model is when there is not a single vertex which dictates the state, but a distribution over possible vertices. Then if there is a 10% chance of being in state a, a 30% chance of being in state d, and a 60% change of being in state f, this is represented as

$$q^T = [0.1\ 0\ 0\ 0.3\ 0\ 0.6\ 0\ 0].$$

In general we need to enforce that $q \in \Delta^{|V|}$, that is, it represents a probability distribution, so

- each $q[i] \geq 0$
- $\sum_i q[i] = 1$.

Now the *probability transition matrix* P is a column normalized matrix. That is, each column must satisfy a probability distribution; each column $P_j \in \Delta^{|V|}$, so the entrees are non-negative and sum to 1. Each column i represents the probability of where the next vertex would be, conditioned on starting at vertex v_i. That is, entry $P_{i,i'}$ describes that from vertex v_i the probability is $P_{i,i'}$ that the next vertex is $v_{i'}$. Using an adjacency matrix A, we can derive P by normalizing all of the columns, so $P_j = A_j / \|A_j\|_1$.

Example: Probability Transition Matrix

The running example adjacency matrix A can derive a probability transition matrix P as

$$P = \begin{pmatrix}
0 & 1/2 & 1/3 & 1/3 & 0 & 0 & 0 & 0 \\
1/3 & 0 & 0 & 1/3 & 0 & 0 & 0 & 0 \\
1/3 & 0 & 0 & 1/3 & 1/3 & 0 & 0 & 0 \\
1/3 & 1/2 & 1/3 & 0 & 0 & 0 & 0 & 0 \\
0 & 0 & 1/3 & 0 & 0 & 1/3 & 1/2 & 0 \\
0 & 0 & 0 & 0 & 1/3 & 0 & 1/2 & 1 \\
0 & 0 & 0 & 0 & 1/3 & 1/3 & 0 & 0 \\
0 & 0 & 0 & 0 & 0 & 1/3 & 0 & 0
\end{pmatrix}.$$

We can also present this as a graph, where directed edges show the probability transition probability from one node to another. The lack of an edge represents 0 probability of a transition. This sort of representation is already very cluttered, and thus is not scalable to much larger graphs.

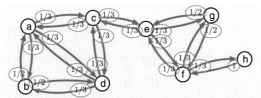

Note that although the initial graph was undirected, this new graph is directed. For instance, edge (a, b) from a to b has probability $1/3$ while edge (b, a) from b to a has probability $1/2$, or more dramatically, edge (f, h) from f to h has probability $1/3$, whereas edge (h, f) from f to h has probability 1; that is, vertex h *always* transitions to f.

Now a given state $q^T = [0\ 1\ 0\ 0\ 0\ 0\ 0\ 0]$ can "transition" to the next state as (using our example P)

$$q_1 = Pq = \begin{bmatrix} \dfrac{1}{2} & 0 & 0 & \dfrac{1}{2} & 0 & 0 & 0 & 0 \end{bmatrix}^T.$$

Then we can get to the next state as

$$q_2 = Pq_1 = PPq = P^2q = \begin{bmatrix} \dfrac{1}{6} & \dfrac{2}{6} & \dfrac{2}{6} & \dfrac{1}{6} & 0 & 0 & 0 & 0 \end{bmatrix}^T$$

and

$$q_3 = Pq_2 = \begin{bmatrix} \dfrac{1}{3} & \dfrac{1}{9} & \dfrac{1}{9} & \dfrac{1}{3} & \dfrac{1}{9} & 0 & 0 & 0 \end{bmatrix}^T.$$

In general we can write $q_n = P^n q$, that is starting with q and "hitting" q on the left n times by P, the transition matrix.

This is called a "Markov" chain after Andrey Markov, because it is a *Markov process*. This term implies that in this sequence of states, each transition decision only depends on its current state, and nothing prior to that (unless it is implicitly encoded in the current state) .

There are *two* ways to think about this Markov chain process.

- It describes a *random walk* of a point starting at one vertex; this corresponds with a single 1 coordinate in q. Then at each step, it decides where to go next randomly based on the appropriate column of P, with the jth column of P encoding the transition probability of the jth vertex. It moves to *exactly one* new state, then repeats.
- It describes the *probability distribution of a random walk*. At each state, we only track the distribution of where it *might* be: this is q_n after n steps. Alternatively, we can consider P^n, then for any initial state q_0, $P^n q_0$ describes the distribution of where q_0 might be after n steps. So entry $P^n_{j,i}$ (jth column, ith row) describes the probability that a point starting in j will be in state i after n steps.

Usually, only one of these two interpretations is considered. They correspond to quite different algorithms and purposes, each with its own advantages. We will discuss both.

10.1.1 Ergodic Markov Chains

A Markov chain is *ergodic* if there exists some t such that for all $n \geq t$, each entry in P^n is positive. This means that from any starting position, after t steps there is *always* a chance that we are in every state. That is, for any q, $q_n = P^n q$ is positive in all entries. It is important to make the distinction in the definition that it is not that we have some positive entry for *some* $n \geq t$, but for *all* $n \geq t$, as we will see.

When is a Markov Chain *NOT* Ergodic?

To characterize when a Markov chain is ergodic, it is simpler to rule out the cases when it is not ergodic, and then if it does not satisfy these properties, it must be ergodic. There are three such non-ergodic properties:

- It is *cyclic*. This means that it alternates between different sets of states every 2 or 3 or in general p steps. This is strict; even a single low probability event that deviates from this makes it not cyclic. The cyclic nature does not need to be on the entire graph; it may only be on a disconnected part of the graph.

Example: Cyclic Probability Transition Matrices

Here are some example cyclic transition matrices:

$$\begin{pmatrix} 0 & 1 \\ 1 & 0 \end{pmatrix} \quad \begin{pmatrix} 0 & 1 & 0 \\ 0 & 0 & 1 \\ 1 & 0 & 0 \end{pmatrix} \quad \begin{pmatrix} 0 & 1/2 & 1/2 & 1/2 & 1/2 & 0 \\ 1/4 & 0 & 0 & 0 & 0 & 1/4 \\ 1/4 & 0 & 0 & 0 & 0 & 1/4 \\ 1/4 & 0 & 0 & 0 & 0 & 1/4 \\ 1/4 & 0 & 0 & 0 & 0 & 1/4 \\ 0 & 1/2 & 1/2 & 1/2 & 1/2 & 0 \end{pmatrix}.$$

- It has *absorbing and transient states*. This corresponds to some Markov chains which can separate V into two classes $A, T \subset V$ so that if a random walk leaves some node in T and lands in a state in A, then it *never* returns to any state in T. In this case, the nodes A are *absorbing*, and the nodes in T are *transient*. Note that this only happens when the initial graph is *directed*, so the walk cannot go backwards on an edge.

Example: Absorbing and Transient Probability Transition Matrices

Here are some examples:

$$
\begin{pmatrix} 1/2 & 0 \\ 1/2 & 1 \end{pmatrix}
\qquad
\begin{pmatrix} 0 & 1 & 0 \\ 1 & 0 & 1 \\ 0 & 0 & 0 \end{pmatrix}
\qquad
\begin{pmatrix}
1/2 & 1/2 & 0 & 0 & 0 & 0 \\
1/2 & 49/100 & 0 & 0 & 0 & 0 \\
0 & 1/100 & 1/4 & 1/4 & 1/4 & 1/4 \\
0 & 0 & 1/4 & 1/4 & 1/4 & 1/4 \\
0 & 0 & 1/4 & 1/4 & 1/4 & 1/4 \\
0 & 0 & 1/4 & 1/4 & 1/4 & 1/4
\end{pmatrix}.
$$

- It is *not connected*. This property indicates that there are two sets of nodes $A, B \subset V$ such that there is no possible way to transition from any node in A to any node in B.

Example: Disconnected Probability Transition Matrices

And some examples are as follows:

$$
\begin{pmatrix} 1 & 0 \\ 0 & 1 \end{pmatrix}
\qquad
\begin{pmatrix} 0 & 1 & 0 \\ 1 & 0 & 0 \\ 0 & 0 & 1 \end{pmatrix}
\qquad
\begin{pmatrix}
1/2 & 1/2 & 0 & 0 & 0 & 0 \\
1/2 & 1/2 & 0 & 0 & 0 & 0 \\
0 & 0 & 1/3 & 1/2 & 1/3 & 0 \\
0 & 0 & 1/3 & 0 & 1/3 & 0 \\
0 & 0 & 1/3 & 1/2 & 1/3 & 0 \\
0 & 0 & 0 & 0 & 0 & 1
\end{pmatrix}.
$$

When it *IS* Ergodic

From now on, we will assume that the Markov chain *is* ergodic. At a couple of critical points, we will show simple modifications to existing chains that ensure this is true.

Now there is an **amazing** property that happens when a Markov chain is ergodic. Let $P^* = P^n$ as $n \to \infty$; this is well-defined, and it will converge to a limiting matrix P^*. Now let $q_* = P^* q$. That is, there is also a *limiting state*, and this *does not* depend on the choice of q.

▷ Ergodic Markov Chains ◁

They have a limiting probability transition matrix ($P^* = P^n$ as $n \to \infty$) and a limiting state $q^* = P^* q$, which is the same for *any* initial state q.

This property has varied and important consequences.

- For all starting states q, the final state is q_* (if we run the chain long enough). So we can analyze such chains, just based on V and P; or if P is derived from an adjacency matrix, this is analyzing properties of the inducing graph.
- As we do a random walk, we will eventually have an expected state precisely described by q_*. So we can analyze the process of such random walks without actually simulating an indefinite number of such processes, as long as we can derive the state q^* and final matrix P^*.
- The final state is stable, specifically $q_* = PP^*q$ thus $q_* = Pq_*$. That is, the probability of being in a state i and leaving to j is the same as being in another state j and arriving at i; this is called the *delicate balance*. Globally, we can generalize this to say the probability of being in any state i and leaving (to any other state) is the same being in any other state and arriving in i. Thus, if a distribution starts in $q_0 = q_*$, it is already in the final distribution. The "further" it starts (e.g., q_0 is more different from q_*), the longer it will take to converge.

Moreover, q_* is the *first* eigenvector of P after normalizing, so the sum of its elements are 1. This *second* eigenvalue λ_2 determines the rate of convergence. The smaller the λ_2, the faster the rate of convergence.

Example: Limiting States

In our example graph,

$$q_* = (0.15, 0.1, 0.15, 0.15, 0.15, 0.15, 0.1, 0.05)^T$$
$$= (\frac{3}{20}, \frac{1}{10}, \frac{3}{20}, \frac{3}{20}, \frac{3}{20}, \frac{3}{20}, \frac{1}{10}, \frac{1}{20})^T.$$

This can be calculated in Python using the first eigenvector of P as

```
import numpy as np
import SciPy.linalg as LA
P = np.array([[0,1/2,1/3,1/3,0,0,0,0],[1/3,0,0,1/3,0,0,0,0],
              [1/3,0,0,1/3,1/3,0,0,0],[1/3,1/2,1/3,0,0,0,0,0],
              [0,0,1/3,0,0,1/3,1/2,0],[0,0,0,0,1/3,0,1/2,1],
              [0,0,0,0,1/3,1/3,0,0],[0,0,0,0,0,1/3,0,0]])

l,V = LA.eig(P)
print(V[:,0]/LA.norm(V[:,0],1))
```

Note that this distribution is *not uniform* and is also not directly proportional to the transition probabilities. It is a global property about the connectivity of the underlying graph. Vertex h which is the most isolated has the smallest probability, and more central nodes have higher probability.

The second eigenvalue of P is 0.875 which is small enough that the convergence is fast. If the part of the graph containing $\{a, b, c, d\}$ was made harder to transition to or from the other part of the graph, this value could be much larger (e.g., 0.99). On the other hand, if another edge was added between say d and f, then this would be much smaller (e.g., 0.5), and the convergence would be even faster. We can retrieve the second eigenvalue in Python as

```
print(l[1].real)
```

10.1.2 Metropolis Algorithm

The *Metropolis algorithm*, sometimes referred to as *Markov Chains Monte Carlo* (MCMC) was developed by Metropolis, Rosenbluth, Rosenbluth, Teller, and Teller in 1953 to help develop the atomic bomb. There is some controversy over who really deserves credit for the invention. But, the lesson is, it pays to have a name that is both cool sounding, and earliest in alphabetical order! This has had enormous influence on statistical physics and in computing Bayesian statistics.

Here each state $v \in V$ has a weight associated with it:

$$w(v) \qquad \text{where} \qquad \sum_{v \in V} w(v) = W.$$

More generally, V may be continuous and then $W = \int_{v \in V} w(v) \, dv$. Then we want to land in a state v with probability $w(v)/W$. But...

- V might be *very* large and W unknown.
- V can be continuous, so there can be *no way* to calculate W. One can think of this as a *probe-only* distribution, since you can measure $\mu(v) = cw(v)$ at any one state at a time where c is some unknown constant (related to W).

Then the goal is to design a special Markov chain so $q_*[v] = w(v)/W$ (without knowing W).

The algorithm sketched in Algorithm 10.1.1 starts with some $v_0 \in V$ so $q = [0\ 0\ 0\ \dots\ 1\ \dots\ 0\ 0]^T$. Then iterate as follows. Choose neighbor u (proportional to $K(v, u)$) where K is some notion of neighborhood/similarity (for instance, a kernel like a Gaussian kernel). And move to u with probability $\min\{1, w(u)/w(v)\}$.

Algorithm 10.1.1 Metropolis on V and w

Initialize $v_0 = [0\ 0\ 0\ \dots\ 1\ \dots\ 0\ 0]^T$.
repeat
    Generate $u \sim K(v, \cdot)$
    if $(w(u) \geq w(v_i))$ **then**
        Set $v_{i+1} = u$
    else
        With probability $w(u)/w(v)$ set $v_{i+1} = u$; otherwise set $v_{i+1} = v_i$.
until "converged"
return $V = \{v_1, v_2, \dots, \}$

This implicitly defines a Markov chain on the state space V. The transition matrix is defined by the algorithm but is not realized as a matrix. Importantly, if the chain is ergodic, then there exists some t such that $i \geq t$, then $\mathbf{Pr}[v_i = v] = w(v)/W$. This value t not only is a limiting notion, but also holds for some finite t (even if V is continuous), through a property called "coupling from the past." However, determining when such a t has occurred analytically placing an upper bound on the value required for t can be challenging.

Often the goal is to create many samples from w, which can then be used as a proxy for the unknown w to estimate various quantities via the concentration of measure bounds. The most formal analysis of such algorithms often dictates that it is run for t steps: take *one* sample, then run for another t steps and take *one* additional sample, and so on repeating tk times to get k samples. This of course seems wasteful but is necessary to strictly ensure samples are independent.

In practice, it is more common to run for $t = 1000$ steps (the "burn in" period), then take the next $k = 5000$ steps as a random sample. This repeats a total of only $t + k$ steps. This second method has "auto-correlation," as samples v_i and v_{i+1} are likely to be "near" each other (either since K is local, or because it did not accept a new state). Officially, we should take only one point every s steps, where s depends on the degree of auto-correlation. But in practice, we take all k samples but treat them (for the purpose of concentration bounds) as k/s samples.

10.2 PageRank

Search engines were revolutionized in the late 1990s when Google was formed, with the *PageRank* algorithm as the basis for ranking webpages within its search engine. Before PageRank, other search engines (e.g., Altavista, Lycos, and Infoseek) and indexes (e.g., Yahoo! and LookSmart) were based almost entirely on a combination of the content of the pages and manually curated lists. These aspects are still used as part of an overall search and ranking method, but PageRank added the perspective of also considering the importance of pages based on the global structure of the webgraph.

Webgraph is a graph where each vertex is a webpage, and directed edges are created when one webpage links to another one. Search engines implicitly had stored these graphs already, since they ran "crawlers." These were algorithms which randomly followed links from one webpage to another, in this case, with the purpose of cataloging the content of each webpage so it could be put into a large nearest neighbor search algorithm (often based on cosine similarity using bag-of-words models or Jaccard similarity using k-gram models).

Intuitively, the PageRank model extended the idea of a crawler to be a "random surfer," someone who randomly browses webpages. The goal was to identify webpages which a random surfer would commonly reach, and to mark them as more important in the search engine. This model can be formalized as a Markov chain, and the importance is given by the limiting state distribution q_*.

However, this only works when the graph is ergodic, and the webgraph is very non-ergodic. It is not connected. And there are webpages which are linked to, but do not link to, anything else; these are absorbing states. Worse, spammers could (and do!) intentionally create such sets of webpages to capture the attention of crawlers and random surfers.

Example: Webgraph

Below is an example directed webgraph. It is disconnected and has absorbing and transient states.

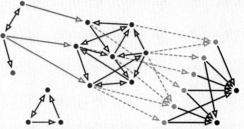

This includes a large part with black and gray edges; this two-layer structure mirrors those used by spammers to attract the traffic of random surfers and automated ranking in search engines. The gray edges are planted links to the gray pages they control. These gray edges might be in comments on blogs, or links on Twitter, or any other click bate. Then the gray pages can be easily updated to direct traffic to the black pages which pay the spammers to get promoted. The effect of this sort of structure can be reduced with PageRank but is not completely eliminated.

So PageRank adds one additional, but crucial, change to the standard Markov chain analysis of the webgraph: teleportation. Roughly for every seven steps (15% of the time), it instructs the random surfer to jump to a completely random node in the webgraph. This makes the graph connected, eliminates absorbing states, increases convergence rates, and jumps out of traps set up by webgraph spammers. That is, it is now ergodic, so the limiting state q_* exists. Moreover, there is an efficient way to implement this without making the webgraph artificially dense, and exploding the memory requirements.

Formalization of the Model

Let P be the $n \times n$ probability transition matrix formed by normalizing the webgraph adjacency matrix. Let Q be another probability transition matrix that is a normalized complete graph. That is, each entry is precisely $1/n$. Now the probability transition matrix which represents the PageRank algorithm is

$$R = (1 - \beta)P + \beta Q,$$

where β is the probability of jumping to a random node. As mentioned a typical setting is $\beta = 0.15$.

Now consider a random surfer whose state is q. Then in one iteration, the state is updated as

$$q_1 = Rq.$$

This operation itself is slow to implement since R is dense, since each entry has value at least β/n. However, we can use the linearity to decompose this as

$$q_1 = Rq = (1 - \beta)Pq + \beta Qq = (1 - \beta)Pq + (\beta/n)\mathbf{1}.$$

Here $\mathbf{1}$ represents the length n, all 1s vector. Since Q has identical columns, it does not matter which column is chosen, and q is eliminated from the second term. More importantly, the second term is now a constant; it just adds β/n to each coordinate created by the sparse matrix multiply $(1 - \beta)Pq$. The state after i rounds can be computed inductively as

$$q_i = Rq_{i-1} = (1 - \beta)Pq_{i-1} + (\beta/n)\mathbf{1}.$$

The limiting state q_* of R is known as the *PageRank vector*. Each coordinate is associated with a webpage and provides an "importance" of that page based on the structure of the webgraph. It is one of the factors that goes into the ranking of pages within Google's search engine.

However, we cannot just yet create q_* based on the first eigenvector of R. Because R is dense, this is too slow to do using standard libraries. But, we can adapt the power method to use the efficient updates in the sparse matrix P. That is, the inductively constructed q_i, for i large enough, is a good approximation of q_*.

Moreover, because the graph has a β-fraction of the complete graph, it is well connected, and the second eigenvalue will be large will converge quickly. Also, after a good estimate of the q_* is found, if the graph is updated, we can use the old estimate to have a good starting point toward the new converged-to distribution. Thus, even on a graph as large as the webgraph, only a few (no more than 50, often much fewer) iterations are required to get a good precision on the estimate of q_* of R.

Ethical Questions with TrustRank

It is generally accepted that some webpages are more trustworthy than others, for instance, Wikipedia pages, top-level pages at reputable universities, established newspapers, and government pages. These sources typically have a formal review process and a reputation to uphold. On the other hand, pages which are parts of blogs, comment sections on news articles, and personal social media pages are typically much less trustworthy.

With this in mind, variants on PageRank have been developed with this trust in mind, as a way to help combat advanced spamming attempts. This works by having trusted pages more frequently teleported to, and less likely to be teleported from.

This is easy to implement by adjusting the $\mathbf{1}$ vector in $(\beta/n)\mathbf{1}$. Denote the limiting vector of this process by the *trustRank* score. It has been suggested that webpages which deviate in their standard PageRank score and their trustRank score are more likely to be spam webpages, and can be down-weighted in the search results.

What are the ethical considerations one should consider when choosing to implement this? And what considerations should be taken in how to build the updated trust weighting?

10.3 Spectral Clustering on Graphs

Spectral clustering is an example of top-down hierarchical clustering. That is, it finds the best way to split the data into two parts, then recurses on both parts until some desired level of resolution has been reached. Within this top-down framework, the key operation of splitting the data is through a mapping to one dimension—a sorted order—and then uses this order to find the best split.

Spectral clustering is most simply defined starting from a graph representation, although we will see how it can use a similarity function $\mathbf{s}$ to construct an equivalent form for any data set. From the graph representation, it defines a special conversion to another matrix called a Laplacian, and uses the eigendecomposition to define the 1-dimensional ordering. For the remainder of this section we will focus on interpreting these graph and matrix representations, and how to use them to define the 1-dimensional ordering.

Typically, on graphs the clustering is a partition of the vertex set. That is, a single hard cluster is a subset $S \subset V$. And to perform top-down clustering, the elements to analyze is a subset $S \subset V$ and its complement $\bar{S} = V \setminus S$.

The edges are then the quantity used to determine how good the clustering is. In general, we want many edges within a cluster, and a few edges between clusters. Define the *volume* of a cluster vol(S) as the number of edges with at least one vertex in S. Also, cut(S, T), the *cut* between two clusters S, T, is defined as the number of edges with one vertex in S and the other in T. Ultimately a good clustering has large vol(S) for each cluster and a small cut(S, T) for each pair of clusters.

Specifically, the *normalized cut* between S and T is

$$\mathsf{ncut}(S, T) = \frac{\mathsf{cut}(S, T)}{\mathsf{vol}(S)} + \frac{\mathsf{cut}(S, T)}{\mathsf{vol}(T)}.$$

And as a result, the key step in top-down graph clustering is to find the cluster S (and complement $T = V \setminus S$) that has the *minimum* ncut(S, T). Dividing by vol(S) and vol(T) prevents this measure from finding either S or T that is too small, and the cut(S, T) in the numerator will ensure a small number of edges are crossing this partition.

Example: Normalized Cut

In the running example graph, clusters $S = \{a, b, c, d\}$ and the singleton cluster with $S' = \{h\}$ both have a small cut value with both $\text{cut}(S', \bar{S}') = 1$ and $\text{cut}(S, \bar{S}) = 1$. The volumes however are very different with $\text{vol}(S) = 6$ and $\text{vol}(\bar{S}) = 5$ while $\text{vol}(S') = 1$ and $\text{vol}(\bar{S}') = 10$.

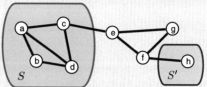

The difference in volumes shows up in the normalized cut score for S and S'. Specifically $\text{ncut}(S', \bar{S}') = 1 + \frac{1}{10} = 1.1$, whereas $\text{ncut}(S, \bar{S}) = \frac{1}{6} + \frac{1}{5} = 0.367$. Overall S results in the smallest normalized cut score; this aligns with how it intuitively is the partition which best divides the vertices in a balanced way without separating along too many edges.

Affinity Matrix

This algorithm will start with the adjacency matrix A of a graph, and then transform it further. However, the adjacency matrix A need not be $0-1$. It can be filled with the *similarity* value defined by a similarity function $\mathbf{s} : X \times X \to [0, 1]$ between elements of a data set X; then A stands for *affinity*. The degree matrix is still diagonal but is now defined as the sum of elements in a row (or column—it must be symmetric). The remainder of the spectral clustering formulation and algorithm will be run the same way; however, we continue the description with the graph representation as it is more clean. However, this generalization allows us to apply spectral clustering to point sets with an arbitrary data set and an appropriate similar measure $\mathbf{s}$.

When the similarity of a pair is very small, it is often a good heuristic to round the values down to 0 in the matrix to allow algorithms to take advantage of fast sparse linear algebraic subroutines.

10.3.1 Laplacians and their EigenStructures

The key step in spectral clustering is found by mapping a graph to its Laplacian, and then using the top eigenvector to guide the normalized cut. The term "spectral" refers to the use of eigenvalues. We start by defining the (unnormalized) *Laplacian matrix* of a graph G with n vertices as $L_0 = D - A$, where as before A is the adjacency (or affinity) matrix and D is the degree matrix. When A is an affinity matrix (each entry is a similarity), D is again a diagonal matrix, where the jth entry is defined as $D_{j,j} = \sum_{i=1; i \neq j}^{n} D_{i,j}$.

Example: Laplacian matrix

The *Laplacian matrix* of our example graph is

$$L_0 = D - A = \begin{pmatrix} 3 & -1 & -1 & -1 & 0 & 0 & 0 & 0 \\ -1 & 2 & 0 & -1 & 0 & 0 & 0 & 0 \\ -1 & 0 & 3 & -1 & -1 & 0 & 0 & 0 \\ -1 & -1 & -1 & 3 & 0 & 0 & 0 & 0 \\ 0 & 0 & -1 & 0 & 3 & -1 & -1 & 0 \\ 0 & 0 & 0 & 0 & -1 & 3 & -1 & -1 \\ 0 & 0 & 0 & 0 & -1 & -1 & 2 & 0 \\ 0 & 0 & 0 & 0 & 0 & -1 & 0 & 1 \end{pmatrix}.$$

Note that the entries in each row and column of L_0 sum up to 0. We can think of D as the *flow* into a vertex, and think of A as the *flow* out of the vertex (related to the Markov chain formulation). This describes a process where the "flow keeps flowing," so it does not get stuck anywhere. That is, as much flows in flows out of each vertex.

It is now useful to consider the eigendecomposition of $L_0 = U_0 \Lambda_0 U_0^T$. The first eigenvalue/vector is not useful: the first eigenvalue is always 0 and the first eigenvector has each element the same. However, the second eigenvalue/vector are important descriptors of the graph. In particular, the second eigenvector of the Laplacian, known as the *Fiedler vector*, can be interpreted as providing a useful 1-dimensional representation of vertices of the graph which preserves connectivity. Taking k eigenvectors provides a k-dimensional representation; however, these are all unit vectors but should not be treated equally. Rather, the vertices can be given a k-dimensional representation using the corresponding part of the eigenvectors u_j scaled by $1/\sqrt{\lambda_j}$, the jth eigenvalue. For eigenvectors $u_j = (u_{j,1}, u_{j,2}, \ldots, u_{j,n})$, the ith vertex can be represented as $(u_{2,i}/\sqrt{\lambda_2}, u_{3,i}/\sqrt{\lambda_3}, \ldots, u_{k+1,i}/\sqrt{\lambda_{k+1}}) \in \mathbb{R}^k$. This scaling correctly implies that the relative importance between two eigenvectors is governed by the ratio between their eigenvalues.

Example: Laplacian-based Embedding

Again following our running example, we compute the eigenvectors and values of the Laplacian L_0. The following table shows the eigenvalues $\lambda_1, \ldots, \lambda_8$ and eigenvectors $u_1, \ldots, u_8$. Note that $\lambda_1 = 0$, and u_1 is a constant vector (each entry is $1/\sqrt{8}$). The second eigenvalue, the Fiedler vector, is $\lambda_2 = 0.278$.

λ	0	**0.278**	1.11	2.31	3.46	4	4.82
U	$1/\sqrt{8}$	$-.36$	0.08	0.10	0.28	0.25	$1/\sqrt{2}$
	$1/\sqrt{8}$	$-.42$	0.18	0.64	$-.38$	0.25	0
	$1/\sqrt{8}$	$-.20$	$-.11$	0.61	0.03	$-.25$	0
	$1/\sqrt{8}$	$-.36$	0.08	0.10	0.28	0.25	$-1/\sqrt{2}$
	$1/\sqrt{8}$	0.17	$-.37$	0.21	$-.54$	$-.25$	0
	$1/\sqrt{8}$	0.36	$-.08$	$-.10$	$-.28$	0.75	0
	$1/\sqrt{8}$	0.31	$-.51$	$-.36$	$-.56$	0.56	0
	$1/\sqrt{8}$	0.50	0.73	0.08	0.11	0.11	0

Next we plot the vertices using the implied 2-dimensional representation from $(u_2/\sqrt{\lambda_2})$ and $(u_3/\sqrt{\lambda_3})$. Note that drawing keeps edge-connected vertices nearby. Moreover, points a and d are directly on top of each other. From the perspective of the graph, they are indistinguishable. Indeed, the eigenstructure does not separate them until u_7.

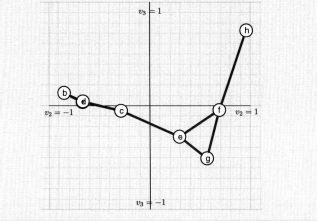

This k-dimensional representation hints at how to perform the cut into two subsets. In fact, the typical strategy is to only use a single eigenvector, u_2. This provides a 1-dimensional representation of the vertices, a sorted order. There are then two common approaches to find the cut.

The first approach is to just select vertices with values less than 0 in S, and those greater or equal to 0 in $\bar{S}$. For large complex graphs, this does not always work as well as the next approach; in particular, there may be two vertices which have values both very close to 0, but one is negative and one is positive. Often, we would like to place these into the same cluster.

The second approach is to consider the cut defined by any threshold in this sorted order. For instance, we can define $S_\tau = \{v_i \in V \mid u_{2,i} \leq \tau\}$ for some threshold τ. In particular, it is easy to find the choice of S_τ which minimizes $\text{ncut}(S_\tau, \bar{S}_\tau)$

by updating this score as τ is increased by incrementally adding points to S_τ (and removing from $\bar{S}_\tau$).

Normalized Laplacian

However, to optimize the cut found using this family of approaches to minimize the normalized cut, it is better to use a different form of the Laplacian known as the normalized Laplacian L of a graph. For a graph G, an identity matrix I, and the graph's diagonal matrix D and adjacency matrix A, its *normalized Laplacian* is defined as

$$L = I - D^{-1/2}AD^{-1/2}.$$

We can also convert L_0 to the normalized Laplacian L using the $D^{-1/2}$ matrix as

$$L = I - D^{-1/2}AD^{-1/2} = D^{-1/2}L_0 D^{-1/2}.$$

The left- and right-multiplication by $D^{-1/2}$ can be thought of as normalizing by the degrees. That is, each entry $P_{i,j}$ of $P = D^{-1/2}AD^{-1/2}$ (and edge (v_i, v_j)) is normalized by the amount of flow in and out of the nodes v_i and v_j of its corresponding edge (v_i, v_j).

Example: Normalized Laplacian

The *normalized Laplacian* matrix of our example graph is $L = I - D^{-1/2}AD^{-1/2} =$

$$\begin{pmatrix}
1 & -0.408 & -0.333 & -0.333 & 0 & 0 & 0 & 0 \\
-0.408 & 1 & 0 & -0.408 & 0 & 0 & 0 & 0 \\
-0.333 & 0 & 1 & -0.333 & -0.333 & 0 & 0 & 0 \\
-0.333 & -0.408 & -0.333 & 1 & 0 & 0 & 0 & 0 \\
0 & 0 & -0.333 & 0 & 1 & -0.333 & -0.408 & 0 \\
0 & 0 & 0 & 0 & -0.333 & 1 & -0.408 & -0.577 \\
0 & 0 & 0 & 0 & -0.408 & -0.408 & 1 & 0 \\
0 & 0 & 0 & 0 & 0 & -0.577 & 0 & 1
\end{pmatrix}.$$

Using that the normalized Laplacian is mechanically the same as using the (unnormalized) Laplacian. The second eigenvector provides a sorted order of the vertices, and the best normalized cut can be found among the subsets according to this sorted order.

Then to complete the spectral clustering, this procedure is recursively repeated on each subset until some designed resolution has been reached.

Example: Normalized Laplacian-based Embedding

The following table shows the eigenvalues $\lambda_1, \ldots, \lambda_8$ and eigenvectors $u_1, \ldots, u_8$ of the normalized Laplacian L for our example graph.

λ	0	**0.125**	0.724	1.00	1.33	1.42	1.66	1.73
V	−.39	0.38	−.09	0.00	0.71	0.26	−.32	0.16
	−.32	0.36	−.27	0.50	0.00	−.51	0.38	−.18
	−.39	0.18	0.36	−.61	0.00	0.03	0.47	−.29
	−.39	0.38	−.09	0.00	−.71	0.26	−.32	0.16
	−.39	−.28	0.48	0.00	0.00	−.57	0.31	0.33
	−.39	−.48	−.29	0.00	0.00	0.05	−.31	−.65
	−.31	−.36	0.27	0.50	0.00	0.51	0.38	−.18
	−.22	−.32	−.61	−.35	0.00	−.07	0.27	0.51

Again we can plot the vertices using the implied 2-dimensional representation from $(u_2/\sqrt{\lambda_2})$ and $(u_3/\sqrt{\lambda_3})$. Again the drawing keeps edge-connected vertices nearby and does not distinguish points a and d.

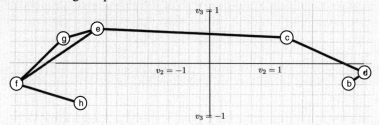

Compared to the plot based on L_0, this one is even more dramatically stretched out along u_2. Also, note that while the suggested cut along u_2 is still at $\tau = 0$, the direction of the orientation is flipped. The vertices $\{a, b, c, d\}$ are now all positive, while the others are negative. The is because the eigendecomposition is not unique in the choice of signs of the eigenvectors, even if eigenvalues are distinct. It is also worth observing that in the second coordinate defined by u_3, vertex f now has a different sign than vertices g and e, because the normalized Laplacian values large cardinality cuts more than the unnormalized Laplacian.

10.4 Communities in Graphs

Finding relationships and communities from a social network has been a holy grail for the Internet data explosion since the late 90s, before Facebook, and even before MySpace. This information is important for targeted advertising, for identifying influential people, and for predicting trends before they happen, or spurring them to make them happen.

At the most general level, a social network is a large directed graph $G = (V, E)$. For decades, psychologists and others studied small-scale networks (100 friends, seniors in a high school). Anything larger was too hard to collect and work with.

Also, mathematicians studied large graphs, famously as properties of random graphs. For instance, the Erdös-Rényi model assumed that each pair of vertices had an edge with probability $p \in (0, 1)$. As p increased as a function $|V|$, they could study properties of the connectedness of the graph: one large connected component forms, then the entire graph is connected, then cycles and more complex structures appear at greater rates.

Ethical Questions with Collecting Large Social Network Graphs

There are many large active and influential social networks (Facebook, Twitter, Instagram, etc.). These are influential because they are an important source of information and news for their users. But in many cases, the full versions of these networks are closely guarded secrets by the companies. These networks are expensive to host and maintain, and most of these hosting companies derive value and profit by applying targeted advertising to the users.

Various example networks are available through inventive and either sanctioned or unsanctioned information retrieval and scraping techniques. Is it ethical to try to scrape public or semi-private posts of users for either academic or entrepreneurial reasons?

Some social media companies attempted to quantify the happiness derived by users based on how they interacted with the social network. This included some approaches by which the study potentially decreased the happiness of the users (e.g., showing only negative posts). Was it useful to perform these experiments, and how and what could go wrong with this ability?

Example: Why do people join groups?

Before large social networks on the Internet, it was a painstaking and noisy process to collect networks. And tracking them overtime was even more difficult. Now it is possible to study the formation of these structures, and collect this data at a large scale on the fly. This allows for more quantitative studies of many questions. An example anthropological question is: *Why do people join groups?*

Consider a group of people C that have tight connections (say in a social network). Consider two people X (Xavier) and Y (Yolonda). Who is more likely to join group C?

- X has three friends in C, all connected.
- Y has three friends in C, none connected.

Before witnessing large social networks, both sides had viable theories with reasonable arguments. Arguments for X were that there was safety and trust in friends who know each other. Arguments for Y were that there was independent support for joining the group; hearing about it from completely different reasons might be more convincing than from a single group of the same size. For static network data, it is probably impossible to distinguish empirically between these cases. But this can be verified by just seeing which scenario is more common as networks form, and users decided to add pre-defined labels indicating a group.

It turns out the answer is: X. Vertices tend to form tightly connected subsets of graphs.

10.4.1 Preferential Attachment

How do edges form in an evolving graph? An important concept is called *preferential attachment*: If edges (a, b) and (a, c) exist in a graph, then it is more likely than random for (b, c) to exist. Reasons include

- b and c somehow already trust each other (through a).
- a may have incentive to bring b and c together.
- All edges may result from common phenomenon (e.g., church group).

This has been widely observed in large networks in many settings. Although the exact model which realizes this, and how the probability is adjusted, is debated, the general property is accepted as a key principle. Notably, this is *not* compatible with the famous mathematical model of Erdös-Rényi graphs which assumes all edges are equally likely. And so the rich mathematics associated with that model needs to be avoided or re-worked to apply to most typical large networks.

A key consequence of this is that the number of triangles in the graph becomes an important parameter in judging the health of the network as a whole or its components.

10.4.2 Betweenness

One way of defining communities of a graph is a procedure through an importance labeling on the graph's vertices or edges. For instance, it could be the PageRank importance assigned to vertices. Once an importance has been assigned, these objects may hold a universal quality or connection, and the parts which are connected despite these objects make up communities. Thus, we can remove all nodes or edges weighted above a threshold, and the remaining connected components may define the true communities of the graph.

A powerful definition of importance on edges which is especially effective in this paradigm is the *betweenness* score. Specifically, the betweenness of an edge (a, b) is defined as

$$\mathrm{betw}(a, b) = \text{fraction of shortest paths that use edge } (a, b).$$

A large score may indicate that an edge does *not* represent a central part of a community. If to get between two communities you need to take this edge, its betweenness score will be high: a "facilitator edge," but not a community edge.

Similarly, the *betweenness of a vertex* is the number of shortest paths that go through this vertex.

Calculating $\mathrm{betw}(a, b)$ for all edges can be time consuming. It typically requires for each vertex, computing the shortest paths to all other vertices (the all pairs shortest path problem). Then for each vertex, its effect on every edge can be calculated by running careful dynamic programming on the directed acyclic graph (DAG) defined by its shortest path.

10.4.3 Modularity

Communities can be defined and formed without providing an entire partition of the vertices, as in a clustering. The common alternative approach is to define a score on each potential community $C \subset V$, and then search for subsets with a large score. The most common score for this approach is called *modularity* and, at a high level it is

$$Q(C) = \text{(fraction of edges in group)} - \text{(expected fraction of edges in group)}.$$

So the higher, the more tightly packed the community is.

More precisely, we use the adjacency matrix A to denote existing edges, where $A_{i,j}$ as 1 if edge (i, j) exists, and 0 otherwise. The expected value of an edge for two nodes i and j with degree d_i and d_j, respectively, is $E_{i,j} = d_i d_j / 2|E|$. Note that this allows self-edges. Now formally,

$$Q(C) = \frac{1}{2|E|} \left[\sum_{i,j \in C} (A_{i,j} - E_{i,j}) \right]$$

and for a set of communities $\mathcal{C}$

$$Q(\mathcal{C}) = \frac{1}{2|E|} \left[\sum_{C \in \mathcal{C}} \sum_{i,j \in C} (A_{i,j} - E_{i,j}) \right].$$

For disjoint communities $\mathcal{C}$, this score can obtain values $Q(\mathcal{C}) \in [-1/2, 1]$. If the number of edges exceeds the expectation defined this way, then it is positive. Typically when $Q \in [0.3, 0.7]$, this is would be deemed a significant grouping.

Finding high modularity subsets can be a challenging search problem. Spectral clustering can efficiently provide good initial guesses—although the goal of this approach was to not rely on such methods. Once an initial guess is formed, or starting from a single vertex community C, then they can be incrementally updated to look for high modularity communities. One can add either one vertex at a time that most increases the score or all vertices which individually increase the score. Alternatively, a random walk, similar to the Metropolis algorithm, can be used to explore the space of communities.

Exercises

10.1 Consider the following probability transition matrix

$$M = \begin{bmatrix} .2 & 0.5 & 0 & 0 & 0 & 0 & 0 & 0 & 0 & 0 \\ .8 & 0.2 & 0 & 0 & 0 & 0 & 0 & 0.1 & 0 & 0 \\ 0 & 0.3 & 0 & 0.3 & 0.2 & 0 & 0 & 0 & 0 & 0 \\ 0 & 0 & 0.5 & 0 & 0.2 & 0.5 & 0 & 0 & 0 & 0 \\ 0 & 0 & 0.5 & 0.3 & 0 & 0.5 & 0 & 0 & 0 & 0 \\ 0 & 0 & 0 & 0.4 & 0.4 & 0 & 0 & 0 & 0 & 0 \\ 0 & 0 & 0 & 0 & 0.2 & 0 & 0 & 0.5 & 0.4 & 0 \\ 0 & 0 & 0 & 0 & 0 & 0 & 0.8 & 0 & 0 & 0.3 \\ 0 & 0 & 0 & 0 & 0 & 0 & 0.2 & 0 & 0 & 0.7 \\ 0 & 0 & 0 & 0 & 0 & 0 & 0 & 0.4 & 0.6 & 0 \end{bmatrix}.$$

We will consider four ways to find $q_* = M^t q_0$ as $t \to \infty$.

Matrix Power: Choose some large enough value t and create M^t. Then apply $q_* = (M^t)q_0$. There are two ways to create M^t; first we can just let $M^{i+1} = M^i * M$, repeating this process $t-1$ times. Alternatively (for simplicity, assume t is a power of 2), in $\log_2 t$ steps create $M^{2i} = M^i * M^i$.

State Propagation: Iterate $q_{i+1} = M * q_i$ for some large enough number t iterations.

Random Walk: Start with a fixed state $q_0 = [0, 0, \ldots, 1, \ldots, 0, 0]^T$ where there is only a 1 at the ith entry, and then transition to a new state with only a 1 in the jth entry by choosing a new location proportional to the values in the ith column of M. Iterate this some large number t_0 of steps to get state q_0'. (This is the *burn in period*.)

Now make t new steps starting at q_0' and record the location after each step. Keep track of how many times you have recorded each location and estimate q_* as the normalized version (recall $\|q_*\|_1 = 1$) of the vector of these counts.

Eigenanalysis: Compute eigendecomposition of M, take the first eigenvector, and L_1-normalize it.

1. Run each method (with $t = 1024$, $q_0 = [1, 0, 0, \ldots, 0]^T$, and $t_0 = 100$ when needed) and report the answers.
2. Rerun the Matrix Power and State Propagation techniques with initialization $q_0 = [0.1, 0.1, \ldots, 0.1]^T$. What value of t is required to get as close to the true answer as the older initial state?
3. Explain at least one **Pro** and one **Con** of each approach. The **Pro** should explain a situation when it is the best option to use. The **Con** should explain why another approach may be better for some situation.
4. Is the Markov chain *ergodic*? Explain why or why not.
5. Each matrix M row and column represents a node of the graph; label these from 1 to 10 starting from the top and from the left. What nodes can be reached from node 4 in one step, and with what probabilities?
6. Repeat the above questions using PageRank with a teleportation rate of $\beta = 0.15$.

10.2 Design a cyclic graph with a cycle length of 4 that also has transient and absorbing states.

10.3 Consider the following 9×9 adjacency matrix

$$
A = \begin{bmatrix}
0 & 1 & 1 & 0 & 0 & 0 & 0 & 0 & 0 \\
1 & 0 & 1 & 1 & 0 & 0 & 0 & 0 & 1 \\
1 & 1 & 0 & 0 & 1 & 1 & 0 & 0 & 0 \\
0 & 1 & 0 & 0 & 1 & 1 & 0 & 0 & 0 \\
0 & 0 & 1 & 1 & 0 & 1 & 0 & 0 & 0 \\
0 & 0 & 1 & 1 & 1 & 0 & 0 & 0 & 0 \\
0 & 0 & 0 & 0 & 0 & 0 & 0 & 1 & 1 \\
0 & 0 & 0 & 0 & 0 & 0 & 1 & 0 & 1 \\
0 & 1 & 0 & 0 & 0 & 0 & 1 & 1 & 0
\end{bmatrix}.
$$

1. Draw a representation of the associated graph with nodes (labeled by index, starting at 1) as circles and edges as line segments.
2. Construct the Laplacian L and Normalized Laplacian L_0 of the associated graph.
3. Report the first eigenvector and associated eigenvalue of L and L_0.
4. Restricting to the orderings induced by these eigenvectors, find the cut S, T which maximizes the normalized cut $\mathrm{ncut}(S, T)$. Report the associated sets and $\mathrm{ncut}(S, T)$ values.

10.4 Consider the following 10×10 adjacency matrix

$$
A_2 = \begin{bmatrix}
0 & 1 & 1 & 0 & 0 & 1 & 0 & 0 & 0 & 0 \\
1 & 0 & 1 & 0 & 1 & 0 & 0 & 0 & 1 & 0 \\
1 & 1 & 1 & 0 & 1 & 1 & 0 & 0 & 0 & 0 \\
0 & 0 & 0 & 0 & 1 & 1 & 0 & 0 & 0 & 0 \\
0 & 1 & 1 & 1 & 0 & 1 & 0 & 0 & 0 & 0 \\
1 & 0 & 1 & 1 & 1 & 0 & 0 & 0 & 0 & 0 \\
0 & 0 & 0 & 0 & 0 & 0 & 0 & 1 & 1 & 1 \\
0 & 0 & 0 & 0 & 0 & 0 & 1 & 1 & 0 & 1 \\
0 & 1 & 0 & 0 & 0 & 0 & 1 & 0 & 1 & 1 \\
0 & 0 & 0 & 0 & 0 & 0 & 1 & 1 & 1 & 0
\end{bmatrix}.
$$

1. Draw a representation of the associated graph with nodes V (labeled by index, starting at 1) as circles and edges as line segments.
2. Find the subset of vertices $C \subset V$ which maximizes the modularity $Q(C)$.

Chapter 11
Big Data and Sketching

Abstract The term *big data* refers to the easy collection and storage of large amounts of data that by virtue of its scale causes rethinking linearly CPU-bound computational approaches and analysis methodologies. These challenges, and the development of approaches to handle them, were in many ways a necessary precursor to the *data science revolution*. For complex phenomena, with small data, the central limit theorem may not reach convergence, and there may not be enough data points to successfully hold out data to faithfully cross-validate. With small data, abstract analysis models are less essential: low-dimensional data does not need its dimension reduced, and data may be clustered or classified by hand (perhaps with the help of plotting).

Most of the data science analysis methods discussed work fine for moderate to large data sets. But when data is humongous, one may need to limit the times that a single data set is read end-to-end. Without specialized strategies to deal with this scale, humongous data may strangely limit analysis to very simplistic approaches. However, there are three main simple and effective strategies to deal with this scale.

1. The first approach is *parallelization*. Instead of using one processor to examine all of the data, this task can be divided among many processors, potentially cutting runtime by that factor. However, in large data settings, it turns out that the dominating cost is data movement (from a hard drive or cloud to the processor, and back). This led to the development of MapReduce-type systems which emphasize the design of algorithms around data movement—and also emphasize the very easy interface to hide many other issues in parallel algorithms.
2. The second approach is *sampling*. If data is enormous, then the central limit theorem, and thus other modeling phenomena, may provably occur at data sizes smaller than what is available. Moreover, most models assume that the data $X \sim \mu$ is drawn iid from an even larger (infinitely defined) unknown distribution μ. Hence, a random sample $S \sim X$ (more or less) satisfies $S \sim \mu$, and hence should retain many of the desired properties necessary for analysis. We covered many sampling properties and mechanisms in Chapter 2.

© Springer Nature Switzerland AG 2021
J. M. Phillips, *Mathematical Foundations for Data Analysis*,
Springer Series in the Data Sciences,
https://doi.org/10.1007/978-3-030-62341-8_11

3. The third approach is *sketching*. This can be thought of as a refined and more flexible version of sampling. Like sampling, this approach also reduces the size footprint of the data, but stores it in some form of data structure S(X). These structures are not lossless; you typically cannot recover the original data, but one can ask queries of the sketch data structure about the data. Moreover, the data structure size typically does not depend on the size of the data, but rather on the accuracy of the answers returned. A random sample is a type of sketch (e.g., retaining an (ε, δ)-PAC style bound with size roughly $(1/\varepsilon^2) \log(1/\delta)$), but as we will see, they can take many other forms.

Sketching is typically associated with the so-called *streaming* setting, where one can make a single pass over the data. This means these sketch data structures can be built in roughly as much time as it takes to read the data once, or can be maintained as data is continually arriving. However, sketches can be considered in other scenarios; for instance, in the distributed data, parallel setting, if we sketch each subset of data, then the algorithms which minimize data movement (e.g., MapReduce) may only need to send the sketches between the parallel processors. These are much smaller and hence become more scalable in this setting as well.

11.1 The Streaming Model

Streaming is a model of computation that emphasizes *space* over all else. The goal is to compute something using as little storage space as possible. So much so that we cannot even store the input. Typically, you get to read the data once, you can update some limited memory based on it, and then let the data go forever! Or sometimes, less dramatically, you can make 2 or more passes on the data.

Formally, there is a stream $A = \langle a_1, a_2, \ldots, a_n \rangle$ of n items. We will mainly consider settings where either each $a_i \in [m]$ (is an element from a large domain of size m) or each $a_i \in \mathbb{R}^d$ is a d-dimensional vector (in some cases $d = 1$, so it is a scalar). This means, the size of each a_i is about $\log m$ bits (to represent an encoding of which element) or d floating-point values (in the vector setting), and just to count exactly how many items you have seen requires space $\log n$ bits. Unless otherwise specified, log is used to represent $\log_2$ that is the base-2 logarithm. In the setting with items from a discrete domain, the goal is usually to compute a summary S_A using space that is only polynomial in $\log n$ and $\log m$ (e.g., $\log^{c_1} n + \log^{c_2} m$ for constants c_1, c_2; note that this is better than an exponential bound where say $2^{\log n} = n$ is basically just storing the entire data set).

Hashing

Many streaming algorithms use separating hash functions to compress data. This is not to be confused with locality-sensitive hash functions discussed in Section 4.6.

A *separating hash function* $h \in \mathcal{H}$ maps from one set $\mathcal{A}$ to another $\mathcal{B}$ (usually $\mathcal{B} = [m]$) so that conditioned on the random choice $h \in \mathcal{H}$, the location $h(x) \in \mathcal{B}$ for any $x \in \mathcal{A}$ is equally likely. And for any two (or more strongly, but harder to achieve, *all*) $x \neq x' \in \mathcal{A}$, the $h(x)$ and $h(x')$ are *independent* (conditioned on the random choice of $h \in \mathcal{H}$). For a fixed $h \in \mathcal{H}$, the map $h(x)$ is *deterministic*, so no matter how many times we call h on argument x, it always returns the same result. The full independence requirement is often hard to achieve in theory, but in practice this can often be relaxed.

In general, we want to construct a hash function $h : \Sigma^k \rightarrow [m]$ where m is a power of two (so we can represent it as $\log_2 m$ bits) and Σ^k is a string of k characters from a set Σ (for example, let us say a set of numbers $\Sigma = \{0, 1, 2, 3, 4, 5, 6, 7, 8, 9\}$, so $x \in \Sigma^k$ can be interpreted as a k-digit number.) This domain Σ^k often represents the sort of elements a_i in the stream A.

1. **SHA-1:** In Python: `hashlib.sha1()`

   This secure hash function takes in a string of bits so $\Sigma = \{0, 1\}$ and k is the number of bits (but many front ends exist that can let Σ be all ASCII characters), and the hash deterministically outputs a set of 160 bits, so $m = 2^{160}$.

   If one desires a smaller value of m, one can simply use any consistent subset of the bits. To create a family of hash functions $\mathcal{H}$ parameterized by some seed a (the *salt*), one can simply concatenate a to all inputs, to obtain a hash $h_a \in \mathcal{H}$. So $h_a(x) = \text{SHA-1}(\text{concat}(a, x))$.

   Functionality for this (or other similar hash functions) should be built in or be available as standard packages for many programming languages. These built-in functions often can take in strings and various other inputs which are then internally converted to bits.

2. **Multiplicative Hashing:** $h_a(x) = \lfloor m \cdot \text{frac}(x * a) \rfloor$

   a is a real number (it should be large with its binary representation, a good mix of 0s and 1s), $\text{frac}(\cdot)$ takes the fractional part of a number, e.g., $\text{frac}(15.234) = 0.234$, and $\lfloor \cdot \rfloor$ takes the integer part of a number, rounding down so $\lfloor 15.234 \rfloor = 15$. It can sometimes be more efficiently implemented as $(xa/2^q) \mod m$ where q is essentially replacing the $\text{frac}(\cdot)$ operation and determining the number of bits precision.

   If a user wants something simple, concrete, and fast to implement themselves, this is a fine choice. In this case, the randomness is the choice of a. Once that is chosen (randomly), then the function $h_a(\cdot)$ is deterministic.

3. **Modular Hashing:** $h(x) = x \mod m$

   *(This is **not recommended**, but is a common first approach. It is listed here to advise against it.)*

   This roughly evenly distributes a number x to $[m]$, however, the problem is that numbers that are both the same   mod m *always* hash to the same location. One can alleviate this problem by using a random large prime $m' < m$. This will leave bin $\{m' + 1, m' + 1, \ldots, m - 1, m\}$ always empty but has less regularity. But still similar numbers, like 71 and 73, will never hash together.

Separating hash functions basically randomly maps data items into a fixed (and often reduced) domain, and some will fall on top of each other. These collisions lead to some error, but if one is careful, the large important items show through. The unimportant items get distributed randomly, and essentially add a layer of noise. Because it is random, this noise can be bounded.

An added benefit of the use of hash functions is the obliviousness. The algorithm is completely independent of the order or content of the data. This means it is easy to change items later if there was a mistake or sketch different parts of data separately, and then combine them together later, or if data is reordered due to race conditions, it has no effect on the output.

11.1.1 Mean and Variance

As a first warm-up task, consider a stream $A = \langle a_1, a_2, \ldots, a_n \rangle$ where each element $a_i \in \mathbb{R}$ is a real value. A common task is to compute the sample mean and sample variance over A.

In the case of mean(A), the mean of A, we cannot simply maintain one value through s elements as $\mathsf{mean}(A_s) = \frac{1}{s}\sum_{i=1}^{s} a_i$. Since when a new element a_{s+1} arrives, averaging it with $\mathsf{mean}(A_s)$ as $\mathsf{mean}(A_{s+1}) = \frac{1}{2}(\mathsf{mean}(A_s) + a_{s+1})$ gives the *wrong* answer. Instead we maintain two values: the sum $M_s = \sum_{i=1}^{s} a_i$ and the total count s. Then we can construct the mean on demand as

$$\mathsf{mean}(A_s) = M_s/s.$$

The sample variance over $A = \mathsf{var}(A)$ is defined as $\frac{1}{n}\sum_{i=1}^{n}(\mathsf{mean}(A) - a_i)^2$. This again appears tricky since we do not know $\mathsf{mean}(A)$ until we have witnessed the entire stream. However, we can maintain three values, the sum of squared values $S_s = \sum_{i=1}^{s} a_i^2$, and the running sum M_s and count s as before. Then the sample variance can be recovered as

$$\mathsf{var}(A_s) = S_s/s + (M_s/s)^2.$$

11.1.2 Reservoir Sampling

Another important task is to draw (or maintain) a random sample from a stream. For uniform random samples (i.e., each element is equally likely), the standard approach is called *reservoir sampling*. It maintains a single sample $x \in A_s = \langle a_1, a_2, \ldots, a_s \rangle$. At the start of the stream (for $s = 1$), it always sets $x = a_1$. Otherwise, on each step it performs the following simple update rule. For the $(s + 1)$th step,

$$x = \begin{cases} a_{s+1} & \text{with probability } 1/(s+1) \\ x & \text{otherwise.} \end{cases}$$

Observe that at this point, element a_{s+1} has been selected as x with probability $1/(s + 1)$ as desired. And by an inductive argument, this holds for every previously witnessed element.

If one desired to maintain k samples $x_1, x_2, \ldots, x_k$, this approach can be extended. If these should be chosen *with replacement* (for iid sampling, some points might be selected more than once), then one can simply maintain k of these samplers in parallel (each with their own randomness).

If these should be chosen *without replacement*, then one can simultaneously maintain the full set $X = \{x_1, x_2, \ldots, x_k\}$, the *reservoir*. Then to initialize the first k elements in the stream, $\{a_1, a_2, \ldots, a_k\}$ are selected as the initial sample. Afterwards, the update step is similar for element a_{s+1}. With probability $k/(s + 1)$, this item a_{s+1} is put in the sample; and if this is the case, one of the current reservoir items is randomly selected to be discarded. Otherwise, the reservoir set X stays the same.

Weighted Sampling

If one desires to maintain a weighted sample (e.g., the elements a_i are selected proportional to assigned weights w_i), then in the single element and *with replacement* setting, the approach is similar. Instead of maintaining a total count s, we maintain a total weight $W_s = \sum_{i=1}^{s} w_i$, and for a single item, it is selected with probability $w_{s+1}/(W_s + w_{s+1}) = w_{s+1}/W_{s+1}$, or, as before, otherwise the selected item stays the same.

For the weighted without-replacement setting, one can use priority sampling, described in Section 2.4. Recall, in this approach, each item a_i is independently assigned a (random) priority ρ_i, and the top k of these are then the selected sample. In the stream, we can also assign each element a_i its priority ρ_i, so then maintaining the largest k of these can be done by comparing each newly observed item's priority ρ_{s+1} to the maintained set. If it is not smaller than *all* of these, then it pushes out the smallest one. In most cases, one only needs to compare to the smallest kept priority value, and so on a discard, this comparison takes constant time.

11.2 Frequent Items

A core data mining problem is to find items that occur more than one would expect. These may be called outliers, anomalies, or other terms. Statistical models can be layered on top of or underneath these notions. In the streaming setting, we consider a very simple version of this problem. There are n elements and they come from a domain $[m]$ (but both n and m might be very large, and we do not want to use them on the order of n or m space). Some items in the domain occur more than once, and we want to find the items which occur the most frequently.

If we can keep a counter for each item in the domain, this is easy. But we will assume m is huge (like all possible IP addresses), and n is also huge, the number of packets passing through a router in a day. So we cannot maintain a counter for each possible item, and we cannot keep track of each individual element that passes through. We must somehow maintain some (approximate) aggregate.

It will be useful to discusses this problem in terms of the *frequency* $f_j = |\{a_i \in A \mid a_i = j\}|$. The term f_j represents the number of elements in the stream that have value j. This notation leads to several useful aggregate values: the *frequency moments*. Let $F_1 = \sum_j f_j = n$ be the total number of elements seen. Let $F_2 = \sqrt{\sum_j f_j^2}$ be the root of the sum of squares of elements counts. Let $F_0 = \sum_j f_j^0$ be the number of distinct elements. Note that typically both F_2 and F_0 are much less than F_1.

Our goal, in a streaming setting, will be to maintain approximate frequency counts $\hat{f}_j$ for *all* items $j \in [m]$. Note, even storing all of these numbers explicitly will require too much space, so we will build a sketch data structure S which stores a few values and implicitly induces values for the remaining indexes j (especially the items with small frequency). Specifically, we desire an *ε-approximate frequency* data structure S so that for any $j \in [n]$, S provides an estimate $S(j) = \hat{f}_j$ which satisfies

$$f_j - \varepsilon n \le \hat{f}_j \le f_j + \varepsilon n.$$

That is, the returned values $\hat{f}_j$ is always with εn (that is ε percent) of the true frequency count.

From another view, a *ϕ-heavy hitter* is an element $j \in [m]$ such that $f_j > \phi n$. We want to build a data structure for ε-approximate ϕ-heavy-hitters so that

- it returns all j such that $f_j > \phi n$;
- it returns no j such that $f_j < \phi n - \varepsilon n$;
- any j such that $\phi n - \varepsilon n \le f_j < \phi n$ can be returned, but might not be.

An ε-approximate frequency data structure can be used to solve the associated heavy hitter problem, which identifies all of the anomalously large elements. For instance, in the case of a router monitoring IP address destinations, such sketch data structures are used to monitor for denial of service attacks, where a huge number of packets are sent to the same IP address overloading its processing capabilities.

Example: Heavy Hitters

We consider a data set with $n = 60$ elements and a domain of size $m = 13$. The frequency of each item $j \in [m]$ is labeled and represented by the height of the blue bars; for instance, $f_5 = 7$, since the 5th blue bar is of height 7.

These frequencies are approximated with red lines; each such approximate frequency $\hat{f}_j$ is within 2 of the true value so $|f_j - \hat{f}_j| \le 2 = \frac{1}{30}n$. Thus this is an $\frac{1}{30}$-approximate frequency.

Finally, we are interested in the 0.2-heavy-hitters. With $\phi = 0.2$, these are the items with $f_j > \phi n = (0.2) \cdot 60 = 12$. These are the bars which are above the green line. Items 3 and 9 are ϕ-heavy hitters, since their frequencies $f_3 = 18$ and $f_9 = 13$ are both greater than 12. The ε-approximate frequencies would report the same items since the same subset of items (3 and 9) are above the frequency threshold of 12. Although, under an ε-approximate ϕ-heavy-hitter it would have been acceptable to report only $j = 3$, since $|\phi n - f_9| \leq \varepsilon n = 2$.

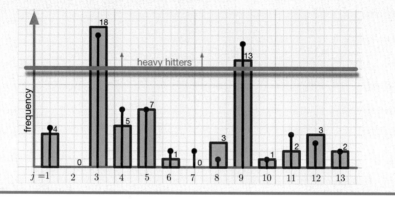

Heavy-Tailed Distributions

For many "Internet-scale" data sets, most of the data objects are associated with a relatively small number of items. However, in other cases, the distribution is more diverse; these are known as *heavy-tailed distributions*.

To illustrate this phenomenon, it is useful to consider a multi set A so each $a_i \in A$ has $a_i = j \in [m]$. For instance, all words in a book. Then the *normalized frequency* of the jth element is defined as $h_j = |\{a_i \in A \mid a_i = j\}|/|A| = f_j/n$. These may follow Zipf's law, where the frequency decays very gradually.

▷ Zipf's Law ◁

Zipf's Law states that: *the frequency of data is inversely proportional to its rank.* This can be quantified so the jth most frequent item will have normalized frequency $h_j \approx c/j$ for a fixed constant c.

Example: Brown Corpus

In the Brown corpus, an enormous English language text corpus, the three most common words are "the" at 7% ($h_{the} = 0.07$); "of" at 3.5% ($h_{of} = 0.035$); and "and" at 2.8% ($h_{and} = 0.028$). So for this application, the constant in Zipf's law is roughly $c = 0.07$.

Heavy tails commonly occur in customer preferences across a large population. For instance, many large Internet companies were built by delivering on the long tail of user preferences (e.g., Amazon for offering more options than local book stores and Netflix for offering more movie options than a rental store.) These effects are often hard to observe at small scales but can become very apparent across large diverse populations.

11.2.1 Warm-Up: Majority

As a warm-up toward ε-approximate frequency estimation, we will start with a simple version of that problem . The *majority* problem outputs an item $j \in [m]$. If there is some $f_j > n/2$, it must output that j. Otherwise it can output anything.

We will accomplish this with $\log m + \log n$ bits of space: one counter c, and one label ℓ. At the first item of the stream a_1, we set the counter as $c = 1$ and the label $\ell = a_1$. Then inductively it adheres to the following simple rules. If it sees a new item a_i with the same label ℓ, it increments the counter. If the label is different, it decrements the counter. If the counter reaches zero, and a new item is observed at a_i (that is $a_i \neq \ell$), then replace the label $\ell = a_i$, and set the counter to 1. The pseudocode is in Algorithm 11.2.1.

Algorithm 11.2.1 Majority(A)

Set $c = 0$ and $\ell = \emptyset$
for $i = 1$ **to** n **do**
   if $(a_i = \ell)$ **then** $c = c + 1$
   else $c = c - 1$
   if $(c < 0)$ **then** $c = 1, \ell = a_i$
return ℓ

Proof for Correctness of the Majority Algorithm

We analyze why this algorithm solves the streaming majority problem. We only need to consider the case where some $j \in [m]$ has $f_j > n/2$ otherwise, anything can be validly returned. In this case, it is useful to think of the counter c as signed: it is negative if the label is not j, and positive otherwise. A j item always increments this signed counter, while a non-j item (say j') could decrement (if the label matches $j' = \ell$, by building up count on the wrong label, or if $j = \ell$ it removes count on the correctly labeled counter) or it could increment (if the label does not match $\ell \neq j$ and $\ell \neq j'$, it makes it less negative). Since there are more than $n/2$ items j (which always increment), then the counter must terminate while positive, and hence the returned label will be j.

11.2.2 Misra-Gries Algorithm

We will next extend the majority algorithm to solve the ε-approximate frequency estimation problem. Basically, it simply uses $k - 1$ counters and labels, instead of one of each, and set $k = 1/\varepsilon$. Thus the total space will be proportional to $(k - 1)(\log n + \log m) \approx (1/\varepsilon)(\log n + \log m)$. We use C, an array of $(k - 1)$ counters $C[1], C[2], \ldots, C[k-1]$. And also use L, an array of $(k-1)$ locations $L[1], L[2], \ldots, L[k - 1]$. There are now three simple rules, with pseudocode in Algorithm 11.2.2:

- If we see a stream element that matches a label, we increment the associated counter.
- If not, and a counter is 0, we reassign the associated label, and increment the counter.
- Finally, if all counters are non-zero, and no labels match, then we decrement *all* counters.

Algorithm 11.2.2 Misra-Gries(A)

Set all $C[i] = 0$ and all $L[i] = \emptyset$
for $i = 1$ **to** n **do**
  if $(a_i = L[j])$ **then**
    $C[j] = C[j] + 1$
  else
    if (some $C[j] = 0$) **then**
      Set $L[j] = a_i$ & $C[j] = 1$
    else
      for $j \in [k - 1]$ **do** $C[j] = C[j] - 1$
return C, L

On a query $q \in [m]$, if $q \in L$ (specifically $L[j] = q$), then return the associated counter $\hat{f}_q = C[j]$. Otherwise return $\hat{f}_q = 0$. This allows the sketch to *implicitly* represent a value for all values $q \in [m]$, with less than m space.

Proof for Bounding the Error in the Misra-Gries Sketch

We can show that this algorithm provides an ε-approximate frequency estimation, and in fact a slightly stronger guarantee for any query $q \in [m]$ so that

$$f_q - \varepsilon n \le \hat{f}_q \le f_q.$$

The upper bound $\hat{f}_q \le f_q$ follows since a counter $C[j]$ representing $L[j] = q$ is only incremented if $a_i = q$. So the estimate $\hat{f}_q$ cannot be an overcount.

The lower bound $f_q - \varepsilon n \le \hat{f}_q$ or equivalently $f_q - n/k \le \hat{f}_q$ can be argued as follows. If a counter $C[j]$ representing $L[j] = q$ is decremented, then $k - 2$ other counters are also decremented, and the current item's count is not recorded. This happens at most n/k times: since each decrement destroys the record of k objects ($k - 1$ stored in counters, and the mismatched current one a_i), and since there are n objects total. Thus a counter $C[j]$ representing $L[j] = q$ is decremented at most n/k times. Thus as desired $f_q - n/k \le \hat{f}_q$.

11.2.3 Count-Min Sketch

In contrast to the Misra-Gries algorithm, we describe a completely different way to solve the ε-approximate frequency estimation problem, called the *count-min sketch*.

Start with t independent (separating) hash functions $\{h_1, \ldots, h_t\}$ where each $h_i : [m] \to [k]$. Now we store a 2-dimensional array of counters for $t = \log(1/\delta)$ and $k = 2/\varepsilon$:

$$
\begin{array}{c||cccc}
h_1 & C_{1,1} & C_{1,2} & \ldots & C_{1,k} \\
h_2 & C_{2,1} & C_{2,2} & \ldots & C_{2,k} \\
\vdots & \vdots & \vdots & \ddots & \vdots \\
h_t & C_{t,1} & C_{t,2} & \ldots & C_{t,k}
\end{array}
$$

To process the stream element $a \in A$, the algorithm plugs the value a into each hash function h_i, maps to a counter $C_{i,h_i(a)}$, and increments that counter. This is specified in Algorithm 11.2.3. That is, each hash function h_i corresponds with a row of counters $C_{i,\cdot}$, and an element a indexes into that row with its hash value $h_i(a)$.

Algorithm 11.2.3 Count-Min(A)

Set all $C_{i,s} = 0$
for $a \in A$ **do**
    for $i = 1$ **to** t **do**
        $C_{i,h_i(a)} = C_{i,h_i(a)} + 1$

After running Algorithm 11.2.3 on a stream A, the *count-min sketch* earns its name, and on a query $q \in [m]$ returns the minimum associated count:

$$\hat{f}_q = \min_{i \in [t]} C_{i,h_i(q)}.$$

That is, for each ith hash function (row in the table), it looks up the appropriate counter $C_{i,h_i(q)}$, and returns the one with minimum value.

Proof for Bounding Error in Count-Min Sketch

Now for a query $q \in [n]$, the value $\hat{f}_q$ is an underestimate ($f_q \leq \hat{f}_q$), since each counter $C_{i,h_i(q)}$ associated with query q has been incremented for each instance of q, and never decremented. Although it may have been incremented for other items, we claim that $\hat{f}_q \leq f_q + W$ for some bounded overcount value W.

Consider just one hash function h_i. The overcount is increased by f_j if $h_i(q) = h_i(j)$. For each $j \in [m]$, this happens with probability $1/k$. We can now bound the overcount using a Markov inequality.

Define a random variable $Y_{i,j}$ that represents the overcount caused on h_i for q because of element $j \in [n]$. That is, for each instance of j, it increments W by 1 with probability $1/k$, and 0 otherwise. Each instance of j has the same value $h_i(j)$, so we need to sum up all these counts, so

$$Y_{i,j} = \begin{cases} f_j & \text{with probability } 1/k \\ 0 & \text{otherwise.} \end{cases}$$

Hence $\mathbf{E}[Y_{i,j}] = f_j/k$. We can then define a random variable $X_i = \sum_{j \in [m]; j \neq q} Y_{i,j}$, which bounds the expected total overcount. By linearity of expectation and $n = \sum_{j=1}^{m} f_j = F_1$, we have

$$\mathbf{E}[X_i] = \mathbf{E}[\sum_{j \neq q} Y_{i,j}] = \sum_{j \neq q} f_j/k \leq F_1/k = \varepsilon F_1/2.$$

Now we recall the Markov inequality. For a random variable $X \geq 0$ and a value $\alpha > 0$, $\mathbf{Pr}[X \geq \alpha] \leq \mathbf{E}[X]/\alpha$. We can apply this with $X_i > 0$, and we set $\alpha = \varepsilon F_1$. Noting that $\mathbf{E}[|X|]/\alpha = (\varepsilon F_1/2)/(\varepsilon F_1) = 1/2$, we have

$$\mathbf{Pr}[X_i \geq \varepsilon F_1] \leq 1/2.$$

But this was for just one hash function h_i. Now we extend this to $t = \log(1/\delta)$ *independent* hash functions to bound the probability that they *all* have too large of an overestimate,

$$\mathbf{Pr}[\hat{f}_q - f_q \geq \varepsilon F_1] = \mathbf{Pr}[\min_i X_i \geq \varepsilon F_1] = \mathbf{Pr}[\forall_{i \in [t]}(X_i \geq \varepsilon F_1)]$$

$$= \prod_{i \in [t]} \mathbf{Pr}[X_i \geq \varepsilon F_1] \leq 1/2^t = \delta.$$

Putting this all together, we have the PAC bound for count-min sketch for any query $q \in [m]$ that

$$f_q \leq \hat{f}_q \leq f_q + \varepsilon F_1$$

where the first inequality always holds, and the second holds with probability at least $1 - \delta$.

Since there are kt counters, and each requires $\log n$ space, the total counter space is $kt \log n$. But we also need to store t hash functions; these can be made to take (roughly) $\log m$ space each. Then since $t = \log(1/\delta)$ and $k = 2/\varepsilon$, it follows the overall total space is roughly $t(k \log n + \log m) = ((2/\varepsilon) \log n + \log m) \log(1/\delta)$.

While the space for the count-min sketch is larger than the Misra-Gries sketch, it has an advantage that its guarantees hold under the so-called *turnstile streaming model*. Under this model, each element $a_i \in A$ can either add one or subtract one from the corpus (like a turnstile at the entrance of a football game), but each count f_j must remain non-negative. Thus count-min has the same guarantees in the turnstile model, but Misra-Gries does not.

11.2.4 Count Sketch

A predecessor of the count-min sketch is the *count sketch*. Its structure is very similar to the count-min sketch; it again maintains a 2-dimensional array of counters, but now with $t = \log(2/\delta)$ and $k = 4/\varepsilon^2$:

$$
\begin{array}{cc||cccc}
h_1 & s_1 & C_{1,1} & C_{1,2} & \cdots & C_{1,k} \\
h_2 & s_2 & C_{2,1} & C_{2,2} & \cdots & C_{2,k} \\
\vdots & \vdots & \vdots & \vdots & \ddots & \vdots \\
h_t & s_t & C_{t,1} & C_{t,2} & \cdots & C_{t,k}
\end{array}
$$

In addition to the t hash functions $h_i : [m] \rightarrow [k]$, it maintains t sign hash functions $s_i : [m] \rightarrow \{-1, +1\}$. Then for a stream element $a \in A$, each hashed-to counter $C_{i,h_i(a)}$ is incremented by $s_i(a)$. So it might add 1 or subtract 1.

Algorithm 11.2.4 Count-Min Sketch(A)

Set all $C_{i,s} = 0$
for $a \in A$ **do**
   for $i = 1$ **to** t **do**
      $C_{i,h_i(a)} = C_{i,h_i(a)} + s_i(a)$

To query this sketch, it takes the median of all values, instead of the minimum.

$$\hat{f}_q = \mathrm{median}_{i \in [t]}\{C_{i,h_i}(q) \cdot s_i(q)\}.$$

Unlike the biased count-min sketch, the other items hashed to the same counter as the query are unbiased. Half the time the values are added, and half the time they are subtracted. So then the median of all rows provides a better estimate. This ensures the following bound with probability at least $1 - \delta$ for all $q \in [m]$:

$$|f_q - \hat{f}_q| \leq \varepsilon F_2.$$

Note that this required k about $1/\varepsilon^2$ instead of $1/\varepsilon$, but usually $F_2 = \sqrt{\sum_j f_j^2}$ is much smaller than $F_1 = \sum_j f_j$, especially for skewed (light-tailed) distributions.

Bloom Filters

There are many variants of these hashing-based sketches, including techniques for locality-sensitive hashing (see Section 4.6) and the extensions for matrices discussed next. Another well-known and related idea is a *Bloom filter*, which approximately maintains a set but not the frequencies of the items. This does not allow false negatives (if an item is in the set, the data structure always reports that it is), but can allow false positives (it sometimes reports an item in the set when it is not).

A Bloom filter processes a stream $A = \langle a_1, a_2, \ldots \rangle$ where each $a_i \in [m]$ is an element of domain of size m. The structure only maintains a single array B of b bits. Each is initialized to 0. It uses k hash functions $\{h_1, h_2, \ldots, h_k\} \in \mathcal{H}$ with each $h_j : [m] \rightarrow [k]$. It then runs streaming Algorithm 11.2.5 on S, where each a_i uses the k hash functions to force k bits as 1.

Algorithm 11.2.5 Bloom-Filter(A)

---
for $a_i \in A$ do
    for $j = 1$ to k do
        Set $B[h_j(a_i)] = 1$

Then on a query to see if $q \in A$, it returns YES only if for *all* $j \in [k]$ that $B[h_j(q)] = 1$. Otherwise it returns No. Since any item which was put in B has all associated hashed-to bits as 1, on a query it will always return YES (no false negatives). However, it may return YES for an item even if it does not appear in the stream. It does not even need to have the exact same hash values from another item in the set; each hash collision could occur because of a different item.

11.3 Matrix Sketching

The singular value decomposition (SVD) can be interpreted as finding the most dominant *directions* in an $n \times d$ matrix A (or n points in $\mathbb{R}^d$), with $n > d$. Recall that the SVD results in

$$[U, S, V] = \mathsf{svd}(A)$$

where $U = [u_1, \ldots, u_n]$, $S = \mathsf{diag}(\sigma_1, \ldots, \sigma_d)$, and $V = [v_1, \ldots, v_d]$, so that $A = USV^T$ and in particular $A = \sum_{j=1}^{d} \sigma_j u_j v_j^T$. To approximate A, we just use the first k components to find $A_k = \sum_{j=1}^{k} \sigma_j u_j v_j^T = U_k S_k V_k^T$ where $U_k = [u_1, \ldots, u_k]$, $S_k = \mathsf{diag}(\sigma_1, \ldots, \sigma_k)$, and $V_k = [v_1, \ldots, v_k]^T$. Then the vectors v_j (starting with smaller indexes) provide the best subspace representation of A.

But, although SVD has been *heavily* optimized on data sets that fit in memory, it can sometimes be improved for very large data sets. The traditional SVD takes proportional to $\min\{nd^2, n^2d\}$ time to compute, which can be prohibitive for large n and/or d.

We next consider computing approximate versions of the SVD in the streaming model. In the setting we focus on, A arrives in the stream, one row a_i at step i. So our input is $\langle a_1, a_2, \ldots, a_i, \ldots, a_n \rangle$, and at any points a_i in the stream, we would like to maintain a sketch of the matrix B which somehow approximates all rows up to that point.

11.3.1 Covariance Matrix Summation

The first regime we focus on is when n is extremely large, but d is moderate. For instance, $n = 100$ million and $d = 100$. The simple approach in a stream is to make one pass using d^2 space, and just maintain the sum of outer products $C = \sum_{i=1} a_i a_i^T$, the $d \times d$ covariance matrix of A exactly.

Algorithm 11.3.1 Summed Covariance

Set C all zeros ($d \times d$) matrix.
for rows (i.e., points) $a_i \in A$ **do**
  $C = C + a_i a_i^T$
return C

We have that at any point i in the stream, where $A_i = [a_1; a_2; \ldots, a_i]$, the maintained matrix C is precisely $C = A_i A_i^T$. Thus the eigenvectors of C are the right singular vectors of A, and the eigenvalues of C are the squared singular values of C. This only requires d^2 space and nd^2 total time, and incurs no error.

We can choose the top k eigenvectors of C as V_k, and on a second pass of the data, project all vectors on a_i onto V_k to obtain the best k-dimensional embedding of the data set.

11.3.2 Frequent Directions

The next regime assumes that n is extremely large (say $n = 100$ million), and that d is also uncomfortably large (say $d = 100$ thousand), and our goal is something like a best rank-k approximation with $k \approx 10$, so $k \ll d \ll n$. In this regime perhaps d is so large that d^2 space is too much, but something close to dk space and ndk time is reasonable. We will not be able to solve things exactly in the streaming setting under these constraints, but we can provide a provable approximation with slightly more space and time.

This approach, called *Frequent Directions*, can be viewed as an extension of the Misra-Gries sketch. We will consider a matrix A one row (one point $a_i \in \mathbb{R}^d$) at a time. We will maintain a matrix B that is $2\ell \times d$, that is it only has 2ℓ rows (directions). We say a row of B is *empty* if it contains all 0s. We maintain that one row of B is always empty at the end of each round (this will always include the last row $B_{2\ell}$).

We initialize with the first $2\ell - 1$ rows a_i of A as B, keeping the last row $B_{2\ell}$ as all zeros. Then on each new row $a_i \in A$, we replace an empty row of B with a_i. If after this step, there are no remaining empty rows, then we need to create more with the following steps. We set $[U, S, V] = \mathsf{svd}(B)$. Now examine $S = \mathrm{diag}(\sigma_1, \ldots, \sigma_{2\ell})$, which is a length 2ℓ diagonal matrix. Then subtract $\delta = \sigma_\ell^2$ from each (squared) entry in S, that is

$$\sigma_j' = \sqrt{\max\{0, \sigma_j^2 - \delta\}} \quad \text{and} \quad S' = \mathrm{diag}(\sqrt{\sigma_1^2 - \delta}, \sqrt{\sigma_2^2 - \delta}, \ldots, \sqrt{\sigma_{\ell-1}^2 - \delta}, 0, \ldots, 0).$$

Now we set $B = S'V^T$. Notice that since S' only has non-zero elements in the first $\ell - 1$ entries on the diagonal, B is at most rank $\ell - 1$ and the last $\ell + 1$ rows are empty.

Algorithm 11.3.2 Frequent Directions

Set B all zeros ($2\ell \times d$) matrix.
for rows (i.e., points) $a_i \in A$ **do**
    Insert a_i into a row of B
    if (B has no empty rows) **then**
      $[U, S, V] = \mathsf{svd}(B)$
      Set $S' = \mathrm{diag}\left(\sqrt{\sigma_1^2 - \sigma_\ell^2}, \sqrt{\sigma_2^2 - \sigma_\ell^2}, \ldots, \sqrt{\sigma_{\ell-1}^2 - \sigma_\ell^2}, 0, \ldots, 0\right)$.
      Set $B = S'V^T$ # the last rows of B will again be empty
return B

The result of Algorithm 11.3.2 is a matrix B such that for any (direction) unit vector $x \in \mathbb{R}^d$

$$0 \le \|Ax\|^2 - \|Bx\|^2 \le \|A - A_k\|_F^2/(\ell - k)$$

and

$$\|A - A\Pi_{B_k}\|_F^2 \le \frac{\ell}{\ell - k}\|A - A_k\|_F^2,$$

for any $k < \ell$, including when $k = 0$, and Π_{B_k} projects onto the span of B_k. So setting $\ell = 1/\varepsilon$, in any direction in $\mathbb{R}^d$, the squared norm in that direction is preserved up to $\varepsilon\|A\|_F^2$ (that is, ε times the total squared norm) using the first bound. Using $\ell = k + 1/\varepsilon$, the squared norm of the direction is preserved up to the squared Frobenius norm of the tail $\varepsilon\|A - A_k\|_F^2$.

In the second bound if we set $\ell = \lceil k/\varepsilon + k \rceil$, then we have $\|A - A\Pi_{B_k}\|_F^2 \leq (1 + \varepsilon)\|A - A_k\|_F^2$. That is, the span B_k, the best rank-k approximation of B, only misses a $(1 + \varepsilon)$-factor more variation (measured by squared Frobenius norm) than the best rank-k subspace A_k.

Geometry of Frequent Directions

Just like with Misra-Gries, when some mass is deleted from one counter it is deleted from all ℓ counters, and none can be negative. So here when one direction has its mass (squared norm) decreased, at least ℓ directions (with non-zero mass) are decreased by the same amount. So no direction can have more than $1/\ell$ fraction of the total squared norm $\|A\|_F^2$ decreased from it.

Finally, since squared norm can be summed independently along any set of **orthogonal** directions (by the Pythagorean theorem), we can subtract each of them without affecting others. Setting $\ell = 1/\varepsilon$ implies that no direction x (e.g., assume $\|x\| = 1$ and measure $\|Ax\|^2$) decreases its squared norm (as $\|Bx\|^2$) by more than $\varepsilon\|A\|_F^2$. By a more careful analysis showing that we only shrink the total norm proportional to the "tail" $\|A - A_k\|_F^2$, we can obtain the bound described above.

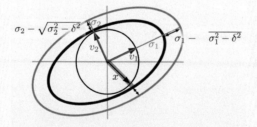

Why do we use SVD? SVD defines the true axis of the ellipse associated with the norm of B (its aggregate shape) at each step. If we shrink along any other basis (or a set of non-orthogonal vectors), we will warp the ellipse and will not be able to ensure that each direction x of B shrinks in squared norm by at most δ_ℓ.

The frequent direction algorithm calls an the SVD operation multiple times, so that it is useful to bound its runtime to ensure it does not take much longer than a single decomposition of A, which would take time proportional to nd^2. In this case, the matrix B which has its SVD computed is only $2\ell \times d$, so this operation can be performed in time proportional to only $d\ell^2$. We do this step every ℓ rows, so it occurs n/ℓ times. In total, this takes time proportional to $n/\ell \cdot d\ell^2 = nd\ell$. So when $\ell \ll d$

(recall $\ell \approx k/\varepsilon$), this is a substantial decrease in running time. Moreover, it only requires $2\ell d$ space, so it can perform these operations without reading data from the disk more than once.

11.3.3 Row Sampling

We next move to a regime where n and d are again both large, and so might be k. But a runtime of ndk may be too large—that is, we can read the data, but maybe a factor of k times reading the data is also large. The next algorithms have runtime slightly more than nd; they are almost as fast as reading the data. In particular, if there are $\mathsf{nnz}(A)$ non-zero entries in a very sparse matrix, and we only need to read these entries, then the runtime is almost proportional to $\mathsf{nnz}(A)$.

The goal is to approximate A up to the accuracy of A_k. But in A_k, the directions v_i are *linear combinations of features*. As a model, this may not make sense; for example, what is a linear combination of genes? What is a linear combination of typical grocery purchases? Instead, our goal is to choose V so that the rows of V are also rows of A.

For each row of $a_i \in A$, set $w_i = \|a_i\|^2$. Then select $\ell = (k/\varepsilon)^2 \cdot \log(1/\delta)$ rows of A, each proportional to w_i. Then "stack" these rows $x_1, x_2, \ldots, x_\ell$ to derive a matrix

$$R = \begin{bmatrix} x_1 \\ x_2 \\ \vdots \\ x_\ell \end{bmatrix}.$$

These ℓ rows will jointly act in place of V_k^T. However since V was orthogonal, the columns $v_i, v_j \in V_k$ were orthogonal. However, the rows of R are not orthogonal, so we need to consider an orthogonalization of them. Let $\Pi_R = R^T(RR^T)^{-1}R$ be the *projection matrix* for R, so that $A_R = A\Pi_R$ describes the *projection* of A onto the subspace of the directions spanned by R. Finally, we claim that

$$\|A - A\Pi_R\|_F \leq \|A - A_k\|_F + \varepsilon\|A\|_F$$

with probability at least $1-\delta$. The analysis roughly follows from importance sampling and concentration of measure bounds. This sketch can be computed in a stream using weighted reservoir sampling or priority sampling.

Leverage Scores

A slightly better bound can be achieved, near the bound from Frequent Directions, if a different sampling probability on the rows is used: the leverage scores. To define these, first consider the SVD: $[U, S, V] = \mathsf{svd}(A)$. Now let U_k be the top k left singular vectors, and let $U_k(i)$ be the ith row of U_k. The *leverage score* of data point

a_i is $s_i = \|U_k(i)\|^2$. Using the leverage scores as sampling weight $w_i = s_i$ achieves

$$\|A - A\Pi_R\|_F \leq (1 + \varepsilon)\|A - A_k\|_F.$$

However, computing the full SVD to derive these weights prevents a streaming algorithm, and part of the motivation for sketching. Yet, there exist ways to approximate these scores, and sample roughly proportional to them in a stream, and moreover, the approximation via selected rows still provides better interpretability.

A significant downside of these row sampling approaches is that the $(1/\varepsilon^2)$ coefficient can be quite large for small error tolerance. If $\varepsilon = 0.01$, meaning 1% error, then this part of the coefficient alone is 10,000. In practice, these approaches are typically most useful until $\ell > 100$ or even larger.

11.3.4 Random Projections and Count Sketch Hashing

Stronger guarantees can be obtained through *random projection*-based sketches (see Section 7.10). The starting point is a random projection matrix $S \in \mathbb{R}^{\ell \times n}$ that maps A to a $\ell \times d$ matrix $B = SA$. As with the random projections approach, each element $S_{i,j}$ of S is drawn iid $S_{i,j} \sim \mathcal{N} \cdot \sqrt{n/\ell}$ from a normal distribution. That is, each row s_i of S is an n-dimensional Gaussian random variable $\sim \mathcal{G}_n$, properly normalized.

Using $\ell \approx k/\varepsilon$ columns in S yields the rank-k approximation result, with a few linear algebra steps. First let $V \in \mathbb{R}^{d \times d}$ be the orthogonal basis defined by the right singular vectors of $B = SA$. Then let $[AV]_k$ be the best rank-k approximation of AV, found using the SVD. Ultimately, we have

$$\|A - [AV]_k V^T\|_F \leq (1 + \varepsilon)\|A - A_k\|_F.$$

A factorized form (like the output of the SVD) can be computed for the product $[AV]_k V^T$ in time which scales linearly in the matrix size nd, and polynomially in $\ell = k/\varepsilon$.

Moreover, this preserves a stronger bound, called an oblivious subspace embedding, using $\ell \approx d/\varepsilon^2$ so, for all $x \in \mathbb{R}^d$

$$(1 - \varepsilon) \leq \frac{\|Ax\|}{\|Bx\|} \leq (1 + \varepsilon).$$

This is a very strong bound that also ensures that given a matrix A of d-dimensional explanatory variables, and a vector b of dependent variables, the result of linear regression on SA and Sb provides a $(1 \pm \varepsilon)$ approximation to the result on the full A and b.

Printed in the United States
by Baker & Taylor Publisher Services

Count Sketch Hashing for Sparse Matrices

By increasing the size of the sketch to $\ell \approx k^2 + k/\varepsilon$ for the rank-k approximation result, or to $\ell \approx d^2/\varepsilon^2$ for the oblivious subspace embedding result, a faster *count sketch*-based approach can be used. In this approach, S has each column S_j as all 0s, except for one randomly chosen entry (this can be viewed as a hash to a row of B) that is either -1 or $+1$ at random. This works just like a count sketch but for matrices.

The runtime of these count-sketch approaches becomes proportional to $\text{nnz}(A)$, the number of non-zeros in A. For very sparse data, such as those generated from bag-of-words approaches, this is as fast as only reading the few relevant entries of the data. Each row is now randomly accumulated onto one row of the output sketch B instead of onto all rows as when using the Gaussian random variables approach. However, this sketch B is not as interpretable as the row selection methods, and in practice for the same space, it often works a bit worse (due to extra factors necessary for the concentration of measure bounds to kick in) than the frequent directions approach.

Exercises

We will use three data sets, here:
`https://mathfordata.github.io/data/S1.txt`
`https://mathfordata.github.io/data/S2.txt`
`https://mathfordata.github.io/data/Z.csv`

11.1 Consider a stream $A = \langle a_1, a_2, \ldots, a_n \rangle$, where each $a_i \in \mathbb{R}$ is a real value. Describe a streaming algorithm that uses a constant number of real or integer values to maintain the standard deviation.

11.2 Consider a stream of characters $A = \langle \text{aaabbccccaaabbb} \rangle$.

1. Run the Misra-Gries algorithm (Algorithm 11.2.2) on this stream with $(k-1) = 2$ counters and labels, and report the state of these two counters and labels after each item appears.
2. Report the guarantees on the estimated counts for characters a, b, c, and d.

11.3 Consider the two data sets S1 and S2, which contain $m = 3{,}000{,}000$ characters and $m = 4{,}000{,}000$ characters, respectively. The order of the file represents the order of the stream.

1. Run the Misra-Gries algorithm (Algorithm 11.2.2) with $(k-1) = 9$ counters on streams S1 and S2. Report the output of the counters at the end of the stream.
   In each stream, use just the counters to report how many objects *might* occur more than 20% of the time, and which must occur more than 20% of the time.
2. Build a count-min sketch (Algorithm 11.2.3) with $k = 10$ counters using $t = 5$ hash functions. Run it on streams S1 and S2.
   For both streams, report the estimated counts for objects a, b, and c. Just from the output of the sketch, which of these objects, with probably $1 - \delta = 31/32$ (that is, assuming the randomness in the algorithm does not do something bad), *might* occur more than 20% of the time?
3. Describe one advantage of the count-min sketch over the Misra-Gries algorithm.

11.4 Consider a data set S1, we will estimate properties of this stream using stream sampling approaches.

1. Run k independent reservoir samplers (each with reservoir size 1) to select k items for a set S. Determine the count of each character selected into S at the end of the stream. Use the counts of each character in S to provide an estimate for the count of those characters in the full sets. Run this for values of $k = \{10, 100, 1000\}$. You may want to repeat with different randomness a few times, and take the average, to get better estimates.
2. Repeat the above experiment with a single reservoir of size k for the same values of k.

3. Now treat the characters as weights based on alphabetical order. So "a" has a weight of 1, "b" has a weight of 2, and ultimately "z" has a weight of 26. Now create k samples (again using $k = \{10, 100, 1000\}$) using k independent weighted reservoir samplers. Compute the estimate for the total weight of the stream using importance sampling (Section 2.4). Repeat a few times, and average, so the estimate concentrates.
4. Repeat this experiment using without-replacement sampling via priority sampling for the same values of k (see Section 2.4.1).

11.5 The exact ϕ-heavy-hitter problem is as follows: return *all* objects that occur more than $\phi\%$ of the time. It cannot return any false positives or any false negatives. In the streaming setting, this requires at least space proportional to m, n if there are n objects that can occur and the stream is of length m.

A 2-pass streaming algorithm is one that is able to read all of the data in order exactly twice, but still only has limited memory. Describe a small space 2-pass streaming algorithm to solve the exact ϕ-heavy hitter problem, i.e., with $\varepsilon = 0$, (say for $\phi = 10\%$ threshold).

11.6 Consider a data set Z, which can be interpreted as matrix $Z \in \mathbb{R}^{n \times d}$ with $n = 10,000$ and $d = 2,000$, where each ith row is an element $z_i \in \mathbb{R}^d$ of a stream $\langle z_1, z_2, \ldots, z_n \rangle$. For each of the following approaches (with $\ell = 10$),

- create a sketch $B \in \mathbb{R}^{\ell \times d}$ or $B^T B \in \mathbb{R}^{d \times d}$,
- report the covariance error $\|Z^T Z - B^T B\|_2$, and
- describe analytically (in terms of n and d, then plug in those values) how much space was required.

The approaches are:

1. Read all of Z and directly calculate $B^T B = Z^T Z$.
2. Summed Covariance (Algorithm 11.3.1) to create $B^T B$.
3. Frequent Directions (Algorithm 11.3.2) with parameter ℓ to create B.
4. Row Sampling (proportional to $\|z_i\|^2$) using ℓ reservoir samplers to create B.
5. Random Projections using ℓ scaled Gaussian random variables to create B.

Index

Symbols
D^2-sampling, 191
L_p norms, 49
$\Delta_\circ^{d-1}$, 65
ε-approximate frequency estimation problem, 269
$\mathbb{S}^d$, 64
ϕ-heavy hitter, 266
ϕ-quantile, 32
ε-approximate frequency, 266
k-center clustering problem, 182
k-means clustering problem, 182, 185
k-means++, 191
k-median clustering problem, 182
k-mediod, 182, 192
k-nearest neighbors classifier, 225
p-stable, 90

A
absorbing, 242
absorbing and transient states, 242
activation, 231
activation function, 229
adjacency matrix, 238
affinity, 250
affinity matrix, 250
aggregate shape, 149, 276
alternating optimization, 196
Andberg similarity, 77
angular distance, 64
approximate frequency, 266
assignment-based clustering, 182

B
back-propagation, 230

backtracking line search, 133
bagging, 228
basis, 51, 55
basis pursuit, 115
batch gradient descent, 138
Bayes' rule, 10
best rank-k approximation, 153
between class covariance, 166
betweenness, 256
betweenness of a vertex, 257
big data, 261
Bloom filter, 273
boosting, 228
Box-Muller transform, 88

C
cardinality, 66
Cauchy distribution, 90
Center Link, 198
centering, 160
centering matrix, 161
central limit theorem, 24
character k-grams, 74
characteristic kernels, 78
Chebyshev inequality, 28
Chernoff-Hoeffding inequality, 29
classical MDS, 162
coefficient of determination, 108
complement of a set, 66
Complete Link, 198
compressed sensing, 119
conditional distribution, 8
conditional probability, 4
connected, 237

© Springer Nature Switzerland AG 2021
J. M. Phillips, *Mathematical Foundations for Data Analysis*,
Springer Series in the Data Sciences,
https://doi.org/10.1007/978-3-030-62341-8